3rd AIChE Regional Process Technology Conference 2011

Galveston, Texas, USA
6-7 October 2011

ISBN: 978-1-5108-1777-7

Printed from e-media with permission by:

Curran Associates, Inc.
57 Morehouse Lane
Red Hook, NY 12571

Printed by Curran Associates, Inc. (2016)

For permission requests, please contact AIChE
at the address below.

AIChE
120 Wall Street, FL 23
New York, NY 10005-4020

Phone: (800) 242-4363
Fax: (203) 775-5177

www.aiche.org

Additional copies of this publication are available from:

Curran Associates, Inc.
57 Morehouse Lane
Red Hook, NY 12571 USA
Phone: 845-758-0400
Fax: 845-758-2634
Email: curran@proceedings.com
Web: www.proceedings.com

TABLE OF CONTENTS

OPENING KEYNOTE ADDRESS

The Arc of Invention: Slipping Free of Our Mind-made Manacles 1
John Lienhard

(1) ENERGY EFFICIENCY

BRCP Energy Initiatives & Critical Exchanger Cleaning Program 14
Stacy Wilson
The World Turns to Elliott 21
Doug Fisher
System-Wide Optimization and Control 32
Quin Dennis
Reducing Emissions in Plant Flaring Operations 49
Brian Duck

(2) CATALYSIS AND REACTION ENGINEERING

Modular MicroReactor Technology - A Novel Approach for Process Intensification from Lab to Production 73
Bharat Marwaha, Maria Krusche
The State of the Working Catalyst on Alkali-Promoted Pt/Al2O3 and Pt/TiO2 for the Water Gas Shift Reaction 89
Jorge Pazmino, Jun Wang, Mayank Shekhar, W. Damion Williams, Jeffrey T. Miller, M. Cem Akatay, Eric A. Stach, W. Nicholas Delgass, Fabio H. Ribeiro
Accurate Estimation of Si Accumulation in Coker Naphtha Hydrotreaters Using an Improved Direct Injection ICP (ICPDIN) Analysis and Hot Loop Feed Sampling 90
Thienan D. Tran, Patrick J. Gripka, Lawrence S. Kraus

(3) CARBON CAPTURE AND STORAGE WORKSHOP

CCS Overview, Carbon Capture and Pipeline Transportation 99
Gerald Hill
Commercial CO_2 Sequestration 113
Dwight Peters
Monitoring, Verification and Accounting (MVA) in Retention and EOR Settings 126
Michael H. Young, Susan D. Horovka

(4) DEVELOPMENTS IN PROCESS SAFETY

The Need for Constructive and Rational Dialogue on Risk 143
M. Sam Mannan
Lessons Learned at the U.S. Chemical Safety Board 144
John Bresland
Assuring Protection of Health, Safety, and Environment in Engineering Projects 161
Ralph Ellis, Alan English, Jack Chosnek, Victor H. Edwards, Myra Ray
How to Ensure HAZOP Quality? 167
Sanjeev Saraf, Subodh Medhekar

(5) DISTILLATION AND SEPARATION

Troubleshooting and Improving Operation of a Main Fractionating Tower 174
Lowell Pless, Shane Lacasse
Dividing Wall Column - Reliable Solution for Efficient Refinery Separations 175
Bhari Bhujan Singh, Bill Townsend
Extending the Operating Range of Cross-Flow Trays 188
Izak Nieuwoudt
Thoughts on Troubleshooting: Tips from Those Who Do It Best 196
Ruth Sands
Divided Wall Columns - Tray Technology 214
Daniel R. Summers

(6) ENGINEERING ETHICS AND PROFESSIONALISM

TBPE Board Update, Professional Practice, and Engineering Ethics 226
C. W. Clark

SOUTH TEXAS SECTION SOCIAL HOUR AND DINNER PROGRAM (TICKETED EVENT)

The First Professor of Chemical Engineering in Space 237
Albert Sacco Jr.

(7) CHEMICAL PROCESS TECHNOLOGY

Algenol's Direct to Ethanol Process 238
Ben McCool, Pradeep Sharma
Ultra-low NO_x Burner Arrangements in Furnace Revamps: Utilization of CFD Modeling to Prevent and Mitigate Adverse Furnace Flow Patterns 239
Jaime Erazo, Shirley Chen, Bailey Hendrix, Kirk Wendel
Exceptional Events Management Applied to Continuous Pharmaceutical Manufacturing: A Strategy to Detect, Diagnose and Mitigate Multiple Exceptional Events 251
Intan Munirah Hamdan, Gintaras V. Reklaitis, Venkat Venkatasubramanian
Improving Ethylene Charge Gas Compressor Performance 253
Wendy Young
An Introduction to Intellectual Property and Patent Law 254
Aron Griffith

(8) PETROLEUM REFINING TECHNOLOGY 1

Latest Advances in Hydrocracking Catalysts: Exceed Your Hydrocracker Potential 275
Raul Adarme, Ward Koester
Processing Heavy Crudes in Existing Refineries 286
Kevin Turini, James Turner, Arthur Chu, Shankar Vaidyanathan
Gasoline Blend Property Control and Optimization 296
Clint Dolby, Robert B. Hawkins
Step-Out Improvements in Heavy Oil Conversion 305
Irv Wiehe
When Things Don't Go Right 314
James Turner

(9) PETROLEUM REFINING TECHNOLOGY 2

Advances in Digital Automation: Lowering Refinery Operating Costs 327
Tim Olsen
Examining the Value Chain in Production of Renewable Diesel 347
Bob Riley

UOP FCC Innovations Developed Using Sophisticated Engineering Tools 348
Lisa Wolschlag
Maintain Peak HEN Performance Under Fouling Conditions 349
Jimmy D. Kumana, Simon J. Pugh, Edward M. Ishiyama
The Outlook for Energy: A View to 2030 366
Ram Seetharam

(10) THERMODYNAMICS AND PROCESS SIMULATION

Insight Into Phase Behavior of Complex Fluids 378
Walter G. Chapman
Infinite Dilution Revisited: Filling Predictive Gaps 397
Kenneth R. Cox
Prediction of Mineral Scaling Using a Comprehensive Thermodynamic Model 418
Andre Anderko, P. Wang, R. D. Springer, M. M. Lencka, J. J. Kosinski
Use of API Standards for Phase Equilibrium and Thermophysical Properties in the Simulation of Refinery Processes 432
Calvin F. Spencer, Arjun Gujar, Todd J. Willman
Thermodynamic Modeling of Composition Dependence of Relative Volatility 448
Kai Fischer, Eliane Kuhne, Eric Hendriks

(11) TECHNICAL SESSIONS FOR YOUNG PROFESSIONALS

Introduction to Process Safety Principles 462
Julie D. White, Angela E. Summers
Delayed Coking Basics 470
Matthew Johnson
Process Simulator Basics…That Sometimes You Need to Learn on Your Own 484
David Hill
Selective Catalytic Reduction Reactors and Their Uses 496
Bryan Pourchot
Author Index

THE ARC OF INVENTION:
SLIPPING FREE OF OUR MIND-MADE MANACLES

by John H. Lienhard

for the South Texas Section of AIChE Regional Process Technology Conference, Moody Gardens Hotel, Galveston.
8:30 AM Thursday, October 6, 2011

It's a pleasure to be here with people who create chemical processes. Well, it's a pleasure to talk with people who turn their inventive abilities on *any* problem. But for you, invention takes a curious turn. You invent means for producing the product. But what you invent stays behind. That puts you in the same boat as the defense squad on a football team. All eyes focus on the quarterback, while the game is yours to control. And that's a blessing and a curse:

The bad news: While you need the same smarts as the inventor of the product itself, you're relatively invisible. The good news: You own the success of all the many processes you've created. And that gives you an odd sort of freedom.

It's that mental freedom I want to talk about this morning. Freedom and invention are really two sides of one coin. They're both about getting outside of the box. And the boxes, of course, are ones we all encase ourselves in. That's why I take my title from William Blake's poem, *London*. It begins,

I wander thro' each charter'd street,
Near where the charter'd Thames does flow,
And mark in every face I meet
Marks of weakness, marks of woe.

In every cry of every man,
In every Infant's cry of fear,
In every voice, in every ban,
The mind-forg'd manacles I hear.

It's a terrible thing to be locked in those mind-made chains. But then -- not everyone really *wants* out. Twenty-two years after Blake wrote that, Lord Byron wrote his *The Prisoner of Chillon.* And listen to what that prisoner has to say:

My very chains and I grew friends,
So much a long communion tends
To make us what we are:—even I
Regain'd my freedom with a sigh.

Byron knew that Blake's mind-forged manacles are a security blanket that most of us will cling to.

Small wonder: Mental freedom and invention are about seeing things that'd normally give us no reason for turning our heads -- or at least seeing things that often seem, at first, to make no sense. Even things that offend our senses. That's why I'm as edgy as anyone about leaving my various boxes. I might see what I don't want to see, *out there.*

Years ago a car hit me, then fled the scene. A passing driver, a lawyer, saw it happen and called for help. Later he told me that he would never again regard eye-witness testimony the same way. What he'd seen was *that* unclear in his own mind.

Exactly! We have to encode reality to fit it into our memory. Our minds have nothing like the storage capacity for a full 64-mm Technicolor record of our lives. Instead, we compress experience into algorithms. When we see something foreign to our experience, we absolutely have to begin by trying to fit it into one of our familiar boxes.

Writer William Langewiesche plays that idea in his book, *Inside the Sky*. Langewiesche has piloted every sort of light and ultra-light airplane. So he's acutely aware of how hard it is for anyone to interpret Earth when they first see it from above. Most of his passengers are people who've spent years looking at clouds from 35,000 feet. But they have a hard time interpreting the view from only a few thousand feet. Mountains look wrong; plains and oceans can be hard to tell apart.

Years ago, a friend flew my wife and me, very low, over Eastern Washington's wheat fields in his Piper Cub. When we abruptly passed the rim of the Snake River Canyon, Earth fell and away we seemed to hurtle a thousand feet upward on a magic elevator. And I have to ask: Was that distortion of reality *exhilaration* or was it *terror?*

Artists have always meddled with reality in that way. Many people find Dali or Picasso very disturbing. Good artists know they only find reality after they've seen how to bend theirs, and our, vision.

I recently saw a lovely aerial photo of a desert caravan -- dark camels standing out sharply against the brown sand. Only something about the perspective was out of whack. I Finally realized I wasn't seeing camels at all. I was seeing shadows cast by the slanting light -- the tops of the camels themselves were barely visible. That was just one more minor upsetting, and exciting, journey outside the box: I'd been played for a fool, *and* I'd been given a delicious sunburst of surprise.

Again: A while back I was stroking a horse. As she nuzzled me I noticed I could photograph myself in the reflection from her eyes. The unreality of those pictures drew a very mixed reaction. A landscape framed by a horse's eyelashes hardly fits anyone's memory categories. Even I found the pictures unnerving.

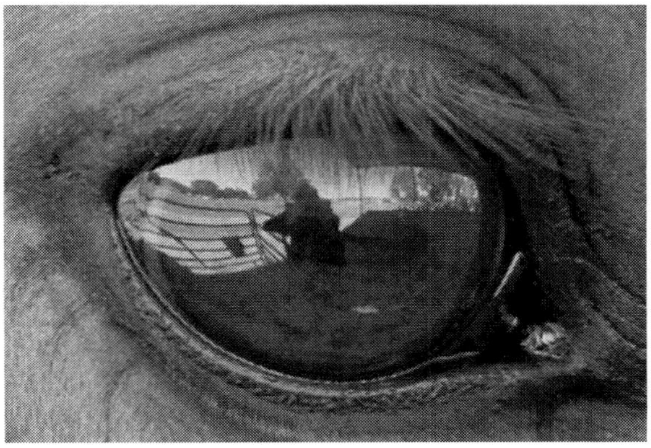

My point, then, is that we really do have a deep need for being able to trust what we see. Do we really want reality to lurch off into regions between the hard earth and the blue sky? We're in trouble if we can't stay anchored -- if we cannot color inside the lines. Yet invention has to involve cutting anchor chains. It demands that we see things which appear not to fit. How to do that? Well, I'll suggest my favorite exercise: Walk your neighborhood with a camera. Look at everything comfortable and familiar. Then set yourself the task of bringing back one photo of something which, properly framed and isolated, makes no sense to the viewer.

If we can tune our eye to see what others do not see, if we can bring back a piece of unreal reality, then we've remade the world and we've remade ourselves. As I say, it's exhilarating, but messing with reality like that really does give us cause to be wary.

So we build our boxes (all of us do) and we pull them around us like blankets. It's fine to tell each another to think outside the box. But we first need to understand, very well, why we want to stay inside the box. We need to acknowledge that our mind-forg'd manacles serve a purpose.

Now, that said, invention will happen only when we find our way out of our self-made constraints. That's hard to do, so let me tell you about a very useful study of this problem. It's the report of a conference that took place in Sigtuna, Sweden, back in 1978.

Thirty-one creative scholars were called there to have a try at finding those common elements of the creative process. They talked and wrote -- personally and self-revealingly. They free-associated as they looked for commonality in the way they worked. Finally, they listed five stages in the creative process:

First, the arrival of a new idea.

Second, sharing the idea. *Exposing it to criticism by trusted friends.*

Third, the hard work of analysis. *Fleshing out and testing the idea.*

Fourth, creating a language of exposition.
Learning to teach the idea to others.

Fifth, going public with the developed idea.

Now, you all know very well the importance of staying anchored in the real world of pressure and problems. So I ask you to notice here, how these stages move back and forth from public to private -- from solitary to social.

All the participants wrote, sometimes poignantly, about the tension between needs for intellectual companionship and for isolation. Most had a lot to say about *places.* They all spoke of walks and neighborhoods -- cafes, libraries, and bookstores.

All of them craved creative solitude -- some special place of quiet. Often that place was a personal office -- often a place occupied by other people. One geographer said,

> *Although I am a loner, I do not like to be alone. [I further my creative dream-like thinking] by being alone in big peaceful crowds. I like to mix with people moving on the main streets ... looking at things ... seeing faces.*

The place changed with the creative stage. In **Stage One**, these people had all found means for claiming a kind of dream state or reverie. Ideas arrived in a peculiar loneliness fed by beauty of one sort or another. It could be the beauty of remote nature or it might be the beauty of the presence of other people. In either case, it had to be *solitary.*

But their new ideas immediately needed community nurture. In **Stage Two** we call on those we trust to help us breathe life into our idea. Without human contact, it'll probably die.

(This is where we might wonder about the so-called *brainstorming* process. Brainstorming is dangerous, unless it's done in a very small group of people who trust each other deeply. It can get competitive. It can be bent by personal agendas. One Alpha in the group can skew the process.)

Stage Three: This is the analytical stage and, once again, it's lonely. Now the place should be a familiar and comfortable workplace where thought is uninterrupted. Here's where we write the equations, erect the logic, do the cost/benefit analysis. Here's where we summon up every constraint and counter-argument we possibly can.

But the analyzed and tested idea that comes out of our office is still far from complete.

Stage Four is terribly important. Here's where we complete the idea by developing a language for telling it to others. (And to explain it to ourselves! How often have all of us seen our own idea -- an idea that looked so good in broad brush -- crumble to dust when we try to explain it fully. Boy! I have.)

Stage Five is where we lay our idea before a wider public -- people who don't give a rap about us or our feelings. Here's where we meet the *show-me* public of skeptics and doubters.

Notice what all this does to the myth of the lonely intellectual. Creativity, with all its need for retreat and isolation, needs the nourishment of a community. But the Sigtuna participants identified two kinds of community. First, the close comfortable kind and, make no mistake, that's absolutely needed for this process to work. The Sigtuna group treasured that sort of community.

The other kind of community, equally important, is the community of harsher critics. And taking it on is the *last* stage. I constantly encounter inventors who've challenged that community without properly finishing **Stage Four.**

Those are the inventors I mentioned a moment ago -- the ones who don't know what hit them, the ones frustrated and unhappy with the unwelcoming world. You all know people who've been chewed up and spat out when they hit **Stage Five**. To them, I find myself trying to explain that any inventor is the establishment's enemy. If they've failed to do all the needed diplomacy, they'll be received as enemies. That's the way the world is wired together. And maybe it's the way things *should be.*

In any case, these conferees understood, at a very deep level, how cooperation and invention lie up against one another. And I think we can as well. Just watch how they moved back and forth from solitary -- then to one person or a small group, on to a large public, and solitary again.

But none of this matters if we cannot reach that first stage. So I'll tell you about a visit, years ago, to our Museum of Fine Arts. I wandered through the place, letting my mind wind down -- gradually putting the work-a-day churn behind me.

When I got to the exhibit of modern Texas art, I asked a security guard, "With all the hours you spend here, which is your favorite work?" "Sir," he said, "as a matter of policy, we don't express opinions on the art." Then he smiled and said. "I'm going to break that rule and tell you. Actually, I like all the art."

I laughed at his self-deprecatory, fool-no-one, diplomacy. It put me in the right frame to enjoy the good humor of modern art. When we passed again, I said, "You know, you have a future in politics." He pulled a long face: "Now why should you have such a low opinion of me!" Then he said,

> *I need to give you the real answer to your question. The finest art in this room isn't hanging on the walls. It's the people who come through here. People aren't the same in here as they are*

walking through a supermarket, you know. They're different. They're open. They come here to be changed.

I was astonished. Of course! That's what it was all about. He'd gone right to the center of it. Long ago, Louis Pasteur talked about creativity. He said, "Chance favors only the prepared mind." But while Pasteur offered scant guidance as to *how* we prepare our minds, that security guard *did* offer guidance.

People *do not* move through museums the way they move through supermarkets. Art is a lot more than just a product served up for our pleasure. It's a trigger for our imaginations. We all live in need of ideas. We all have problems to solve. At some point, most of us realize that, a problem in need of a creative solution cannot be attacked with purely methodical tools. Method takes us down familiar roads. We need to see the shrubbery-shrouded side roads that we ignore by habit.

The hardest thing we do is leave the highway and float above the land. Music, theater, sculpture -- the movies -- they all cut us loose from the road of method and common sense. The so-called creative leap is not a leap in the dark -- without antecedents or stimulus. Rather, it happens when we find a liminal state, on the outer edge of awareness, a state where ideas arrive without order or hierarchy.

In that mental world, cow paths are just as important as freeways. That security guard realized how, inside the museum, we lay aside our shopping lists of needs to be met. Here was a place where it was possible to leave our supermarket lives and go wandering the woods. He'd seen how, with their lists left outside, people were ready to be taken into alien spaces -- onto byroads of thought -- into places close at hand that they weren't aware of.

Mary Shelley gave us remarkable first-hand account of this process in the introduction to her 1819 edition of her book *Frankenstein*. In it she recalls how, as a girl of only 19, she'd invented the story. And her words make a fine tutorial on creative thinking.

Many and long (she says) were the conversations between Lord Byron and Shelley. During one of these, the principle of life [was discussed -- whether it could ever be] discovered. They talked of

> *the experiments of Dr. Darwin who preserved a piece of vermicelli till by some extraordinary means it began to move ...*
>
> *Perhaps a corpse [could] be re-animated; galvanism [hints of] such things: perhaps the component parts of a creature might be manufactured, brought together, and endued with vital warmth.*

Then she goes on to say,

> *Night waned upon this [talk. She says, Even] the witching hour had gone by. When I placed my head on my pillow, I did not sleep, nor could I be said to think. My imagination, unbidden, possessed and guided me, gifting the successive images that arose in my mind with a vividness far beyond the usual bounds of reverie. I saw -- with shut eyes, but acute mental vision, -- [a pale] student of the unhallowed arts kneeling beside the thing he had put together. I saw the hideous phantasm of a man stretched out, and then, on the working of some powerful engine, show signs of life, and stir with an uneasy, half vital motion.*

People like Shelley were giving us the early 19th-century concept of invention that really took hold. And it served us well for roughly a century and a half. They pointed out that we all experience that moment of conscious detachment just before sleep.

As Mary Shelley reached that moment, she was still replaying the evening's conversation in her mind. Suddenly, bits of talk fused into the most perfect tale of horror ever written. The Romantic poets that she ran with used the word *reverie* to describe that state. And I'm pretty sure all of you recognize it. Haven't every one of us in this room sometime jumped out of bed to go write down the revelation we've just had? Or kicked ourselves for failing to do so.

How do we carry that process over into your high-pressure world? That's the hard part. Jean Paul Sartre described that need with great accuracy when he said,

> *Let us not look for the door, and the way out, anywhere but in the wall against which we are living. Instead, let us seek the respite where it is -- in the very thick of the battle.*

We have to be able to find mental repose, all right, but we need to find it while we're standing right in the middle of the marketplace.

So, let me remind you of what might be a familiar story to many of you. It's the one about the benzene molecule. Physicist Hans von Baeyer credits Friederich Kekule with discovering benzene's structure, although several others had at least part of the story before he did. But that's beside the point because, whether or not Kekule was first, our interest is in *how* he came upon the idea for himself. It begins in 1850 with young architecture student Kekule testifying before a grand jury in Giessen, Germany.

The charred body of a neighbor lady had been found in her burned out room. People thought she was the victim of spontaneous combustion, brought on by drinking too much liquor.

The great chemist, Justus von Liebig, testified at the trial. He made it quite clear that the lady would've died long before she'd drunk enough alcohol to make her flammable.

Then Kekule's testimony incriminated a servant who'd been stealing from the lady. He identified her distinctive ring, which turned up in the servant's possession. Together, Liebig's and Kekule's testimony convicted the scoundrel of murdering the lady.

The trial left its mark on the young Kekule. He gave up architecture and took up the study of chemistry with Liebig. The lady's odd ring also lingered in Kekule's mind. That ring had carried the old alchemy seal showing two intertwined serpents, biting each other's tails.

Fifteen years later, we find Kekule working with Liebig on the new chemical called *benzene*. The same kind of intense world of research and development that you inhabit.

The rules of chemical combination were just taking their place in that world. Benzene had to be an arrangement of six carbon, and six hydrogen. No one could figure out any arrangement of the 12 atoms that would satisfy the valence rules. It looked impossible.

One night, Kekule dozed in his chair by the fire, trying to solve the riddle. As he nodded, half-waking, half-sleeping, he dreamt of the twining serpents on that old ring, whirling in the flames.

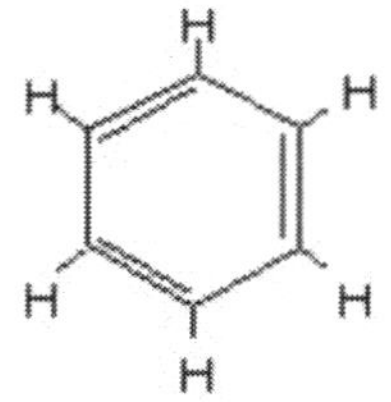

Suddenly, in his thoughts, the serpents caught each other's tails and formed a circle and Kekule saw the answer. The carbon atoms formed a hexagonal *ring* with alternating single and double bonds. Each carbon held its own hydrogen atom – "like charms on a bracelet," says von Baeyer. It was a structure utterly alien to anything else in known chemistry.

We're seldom given such a clear account of the moment an idea reveals itself. And, in this case, *two* things come out of Kekule's story. One is the importance of bringing new tools to the table. Kekule had an architect's spatial and structural sense. That gave him a place to stand outside of chemistry. He had another way of looking at things.

And he had that terribly important ability to admit the problem he was working on into his subconscious mind. He was able to turn his waking dreams loose upon it. The problem is; that's *not* a handbook procedure. Every fiber of our being wants to move in a straight line toward the answers we need. We want, we expect, to deduce answers. But invention is the fruit of recognition, not deduction.

So I repeat Pasteur's wonderful remark that "Chance favors only the prepared mind." What's astonishing is just how much stuff floats by if we only have the wits to catch it.

Finally, one more wrinkle in all this: It is that late 20th-century mathematics finally confirmed something we'd always known in our bones -- that no human future can ever be predicted. Insofar as we can cast the prediction of even modestly complex futures in mathematical terms, the outcome is hopelessly dependent upon specifying the present with perfect accuracy. And that *always* turns out to be impossible.

I like to think about the example of pouring cream into tea -- how the cream quickly spreads and diffuses in increasingly complicated patterns. Suppose we wanted to reproduce one such pattern. We fill a large glass tank with tea and wait until it seems absolutely still.

Then we clamp a pipette of cream above the tank. We release a tenth of a cubic centimeter from a height of three centimeters and photograph the resulting pattern even so little as one second after impact. When that's done, we replace the tea and repeat the experiment.

Finally, we compare the two photos and, sure enough, the two patterns of cream diffusing into tea are quite different.

Even if we could write the terribly complex time-dependent equations for the fluid motion, minuscule changes in the way the two drops hit the surface would cause the two calculations of the diffusion to start diverging almost immediately.

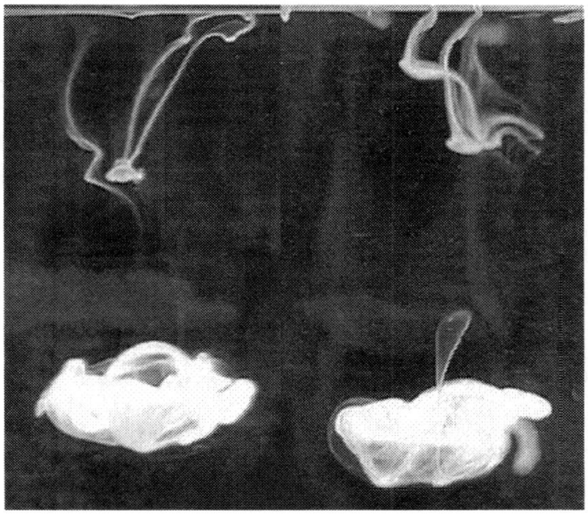

The droplet size will always differ slightly (if only on a microscopic level). Their speed of impact with the water can never be made exactly the same. Each droplet distorts differently as it falls. All those differences can be made tiny, but the resulting effect on the two patterns will not be tiny at all. If we wait five or ten seconds instead of one, we'll see almost no relation between the two results.

We have a popular shorthand for that fact: We call it the *Butterfly Effect.* And that awesome susceptibility to minuscule changes in initial conditions applies in all of human affairs. If I'd chosen to eat an egg instead of cereal at breakfast this morning, I would've literally altered the course of history.

So, even if we accept the fact that we live in a time of rapid change, we can never know where change is taking us or how to use change well. Planning is too often a matter of extrapolating the present in ways that cannot *conceivably* take account of inventive intervention.

Invention that does *not* send us off in new directions is no invention at all. If we *could* predict the future, it'd have to be in a world where we lived without the wonderful wildcard of human creativity. And that'd be a terrible thing.

That means that the only kind of planning that can make any sense must ultimately be contingency planning -- deciding how to react to an unfolding present. It's a matter of knowing how to react intelligently to any of the infinity of complex futures that keep opening up.

Not only does planning founder because it's a way we try predict the future by stipulating it; the same is true of *engineering design*. Any design process that does not have contingency thinking built into it is doomed. If it doesn't fail, it'll at least be repeatedly derailed before it's finished.

The upshot of all this is that we, as engineers, are forever condemned to create a future that we cannot know. It means that our designs will never work the way we intend them to. It means that we have to spend our lives starting down countless pathways that go nowhere near our real goal.

> Edison intended his *phonograph* to serve only the function of stenography.
>
> The *post-it note* was invented before anyone figured out its uses.
>
> The *telephone* was intended to be no more than a simplification of the telegraph. An instrument of business, not a tool of social cement.
>
> The purpose Orville Wright saw for his *airplane* was sport.
>
> James Watt was appalled that anyone would ever use a *steam engine* in a moving vehicle.
>
> Lisa Meitner had no thought of weaponry when she wrote her paper showing that *atomic fission* could be used to create huge amounts of useful energy.

Perhaps the most exciting thing of all is exactly this -- that we only learn after the fact just what it is we've planned -- just what it is that we've created.

I began with two Romantic poets. I shall finish with a third, since the Romantics understood the minatory terrors of the trip better than any group I've ever encountered.

The same year Byron wrote *Prisoner of Chillon*, Coleridge gave us *Kubla Kahn*. He ended that poem abruptly -- breaking off from his vision of Xanadu to tell of the frightening arrival of the creative hero -- or daemon. He describes that terrifyingly beautiful creature -- just loosed from own box. Coleridge says,

Weave a circle round him thrice,
And close your eyes with holy dread,

Well, that has a ring of Romantic excess, of course. But invention and inventors really do bring disruption with them. Breaking out of the box is a messy business.

That's the reason we like to talk about *Innovation*. It literally means adding just a bit to what's already been invented. It means not going too far outside our box. So I won't use it. If we want a better process, then let's break all the way out of our box and *invent* one.

Some Sources

For the Mary Shelley quotes, see, e.g., Jennings, H., *Pandaemonium: 1660-1886: The Coming of the Machine as Seen by Contemporary Observers*. New York: The Free Press, 1985, "Frankenstein," pp. 141-142. (I haven't shown my fairly extensive cuts in Mary Shelley's text. The full text of *Frankenstein: A Modern Prometheus,* with this preface, may be read online at: http://etext.virginia.edu/etcbin/toccer-new2?id=SheFran.sgm&images=images/modeng&data=/texts/english/modeng/parsed&tag=public&part=front

W. Langewiesche, *Inside the Sky: A Meditation on Flight.* (New York: Pantheon Books, 1998).

For the Sigtuna conference report, see: Buttimer, A., *Creativity and Context*. CWK Gleerup: The Royal University of Lund, Dept. of Geography, 1983.

Von Baeyer, H.C., A Dream Come True. *The Sciences*, Jan./Feb., 1989, pp. 6-8.

J. Gleick, *CHAOS: Making a New Science* (New York: Penguin Books, 1987.)

The complete poems of the Romantic poets that I've quoted may be read online as follows:

London, by William Blake: http://www.readbookonline.net/readOnLine/2370/

The Prisoner of Chillon, by Lord Byron: http://classiclit.about.com/library/bl-etexts/lbyron/bl-lbyron-prisoner.htm

Xanadu - Kubla Khan, by Samuel Taylor Coleridge: http://www.poetry-online.org/coleridge_kubla_khan.htm

The photos above are all by John H. Lienhard.

ExxonMobil

Taking on the world's toughest energy challenges.™

BRCP Energy Initiatives & Critical Exchanger Cleaning Program

Stacey B Wilson

2011 AIChE Regional Process Technology Conference

October 6, 2011

Overview

- BRCP & Energy Management
- ACC Energy Efficiency Recognition
- Energy Critical Exchanger (ECE) Selection
- Ongoing Monitoring/ Maintenance/ Responsibilities
- Conclusion

2

ExxonMobil

Taking on the world's toughest energy challenges.™

Baton Rouge Chemical Plant

- Modern petrochemical processing and manufacturing facility located on a 150-acre tract
- Nearly 6.1 billion pounds of petrochemical products are manufactured annually, including:
  - Methyl Ethyl Ketone
  - Isopropyl Alcohol
  - Ethylene/Propylene
  - Isoprene
  - Vistalon Rubber
- Feedstock: Refinery Gas Oil, Refinery Cat Gas, Ethane & Propane, Higher Olefins, Powerformate
- Materials produced at the plant are shipped by pipeline, railroad cars, tank trucks, tankers, barges and trucks

3

ExxonMobil
Taking on the world's toughest energy challenges.

BRCP Energy Management

- ExxonMobil energy program
  - Foundation = "Global Energy Management Systems" (G-EMS)
  - Company and site management set energy efficiency as a high priority
  - Set and steward continual improvement targets
  - Resource and support improve projects by business
- Importance of structured program
  - Inherent system to sustain progress
  - Ownership by business teams to demonstrate progress
  - Plan developed regardless of business environment

4

ExxonMobil
Taking on the world's toughest energy challenges.

2009 BRCP ACC Awards

11 Awards, including 2 exceptional merit

Category	Project	MBtu/yr savings	T CO2/year
Exceptional Merit	Elastomers Block Operation	850,000	51,000
	Butadiene DMC	112,000	6,500
Foundational G-EMS Activities	Steam system efficiency improvements	1,340,000	78,000
	Aromatics SWAT / Energy Initiatives	150,000	8,000
	Intermediates KEV Implementation	260,000	15,000
	Energy Critical Heat Exchanger Program	100,000	26,000
	TLE Chemical Cleaning	140,000	8,000
Business Improve Projects	Refinery Gas Recovery G-EMS projects	860,000	100,000
	Phthalic Anhydride Incinerators	90,000	5,000
	Plasticizer Batch Reactor Delay	20,000	1,000
	Isopropyl Alcohol Vat Reboiler	30,000	2,000
		Total:	*Total:*
	TOTAL SITE REDUCTIONS	**3,950,000 Mbtu/yr**	**300 kTons CO_2/year**

5

ExxonMobil
Taking on the world's toughest energy challenges.

2010 BRCP ACC Awards

6 Awards

Category	Project	MBtu/yr savings	T CO2/year
Foundational G-EMS Activities	Steam system optimization	420,000	25,000
	Furnace Energy Improvements	200,000	12,000
Business Improve Projects	Co-product Unit Optimization	18,000	1,000
	Aromatics DMC Application	64,000	5,000
	Vistalon™ Slurry Optimization	45,000	5,000
	Building Lighting Upgrades	12,000	1,000
		Total:	*Total:*
	TOTAL SITE REDUCTIONS	**759,000 Mbtu/yr**	**49 kTons CO_2/year**

6

ExxonMobil
Taking on the world's toughest energy challenges.

ECE Program Development

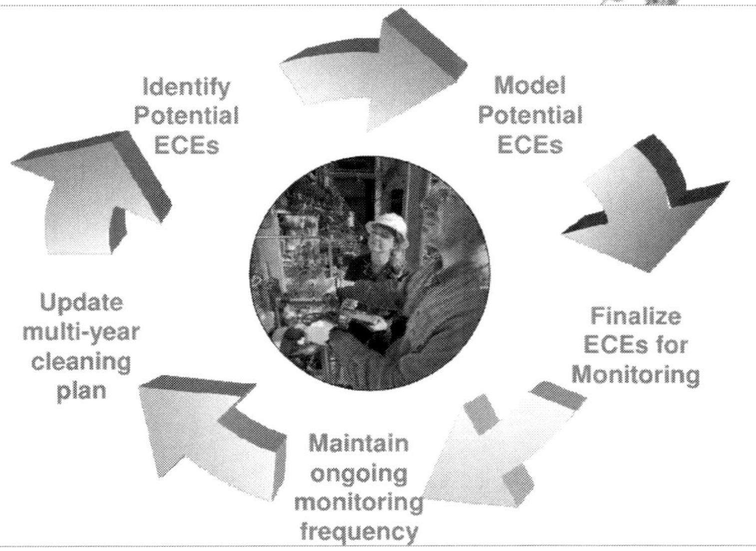

ECE Selection Process

- Focus on energy intense areas
  - Large duty exchangers
  - Extreme temperatures
  - Known heavy fouling services
- Evaluate exchanger service
  - Needs to impact utility consumption/ generation
  - Enables process critical equipment functionality
- Consider multiple exchanger integrated circuits by function
  - Economizer efficiencies reduce additional utility consumption
  - Higher level monitoring can indicate non-optimal deviations
- Do not ignore maintenance costs and production credits/ debits

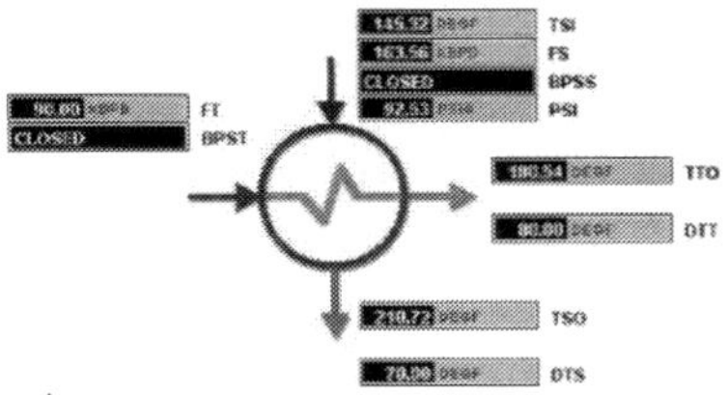

Exchanger Monitoring Tool

- Single input and output formats
- Standardized for consistency
  - Input: plant data
  - Output: status summary, historical graphs
  - Calculation methods, utility prices
- One time initial setup per ECE
  - Single/Multi Phase
  - Cleaning data (cost, online vs. offline, previous cleaning dates)
- Ongoing responsibility of technical resource to update at set frequency
  - Ensure data reliability
  - Assess all financial benefits

Exchanger Monitoring Tool - Output

Unit: EXAMPLE
Heat Exchanger: EX-01

Shellside: Utility
Tubeside: Process

Can this heat exchanger be cleaned on the run? Y

Monitoring Period: Start Date 11/12/09
End Date 2/16/11

Dates Last Cleaned: 11/12/09 3/1/08

Requirements met for SEP funded Cleaning?
As of 2/16/11 Funding Requirements Met

Data Quality: Satisfactory
Current LOC: $ 101,243.68
Current B:C ratio 3.37 : 1
Current LPS Price $ 5.00

Notes: Example only

Data Source

Shell		
	T in	Tag
	T out	Tag
	Flow	Tag
Tube		
	T in	Tag
	T out	Tag
	Flow	Tag

Monitoring Tool – Automated Trending

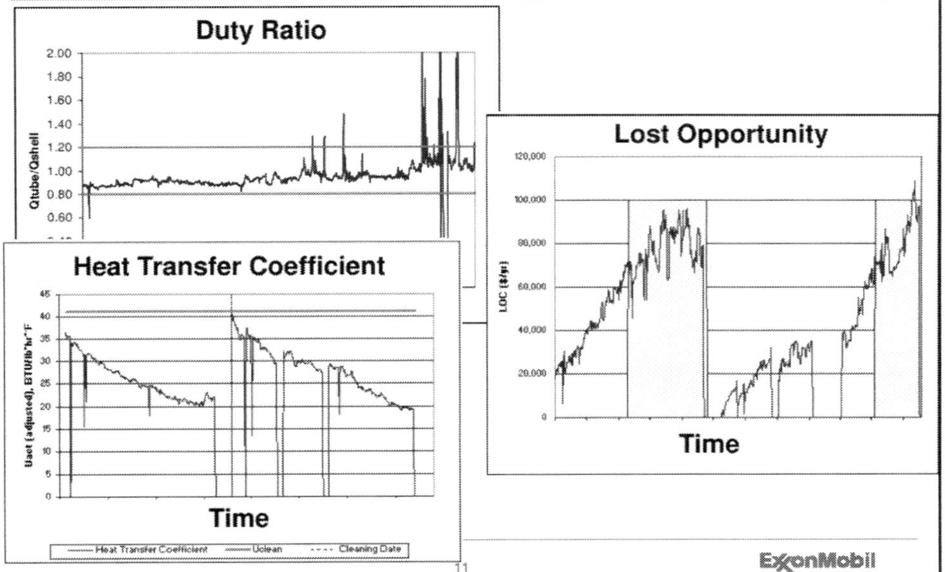

Roles & Responsibilities

- Technical:
  - Develop list of exchangers that meet criteria
  - Use Heat Exchanger Monitoring Tool
  - Monitor tool routinely
  - Identify ready to be cleaned HEx
  - Provide updates to multi-year outlook
- Coordination:
  - Own current monitoring tool template
  - Steward HEx to be cleaned for funding
  - Develop/update multi-year cleaning schedule
- Management:
  - Provide Benefit to Cost Ratio target for cleaning
  - Help ensure optimum budget / annual cleaning

Conclusions

- Success factors
  - Centralized development with technical support
  - Standardized work effort across the site
  - Stewardship of performance and results
  - Management support of energy projects and programs

- Path forward
  - Continual funding and support
  - Enabling further program automation, standardization
  - Sharing best practices

13

ExxonMobil
Taking on the world's toughest energy challenges.

ELLIOTT GROUP
EBARA CORPORATION

CONTACT INFORMATION

- Doug Fisher, Marketing Engineer
- Elliott Group Ebara Corporation
- Office: 724-600-8483
- Cell: 724-600-9521
- E-mail: dfisher@elliott-turbo.com

ELLIOTT GROUP
EBARA CORPORATION

Elliott Company

- **Established in 1895**
- **A Major Player in The Global Turbomachinery Market**
- **Design, Manufacture, Test and Service Process and Power Turbomachinery**
- **Equipment in service**
  - **3,455 Multistage Centrifugal Compressors**
  - **620 Single Stage Centrifugal Compressors**
  - **2,420 Multistage Steam Turbines**
  - **52 Hot Gas Expanders**
  - **30,000 YR Single Stage Steam Turbines**

ELLIOTT GROUP
EBARA CORPORATION

The Elliott Service Story

- Elliott services all kinds of equipment…**Regardless of OEM!**
  - Centrifugal compressors
  - Axial compressors
  - Steam turbines
  - Gas turbines
  - Gear boxes
  - Expanders
  - Blowers
  - Pumps
  - Fans

ELLIOTT GROUP
EBARA CORPORATION

COMPRESSOR RERATES

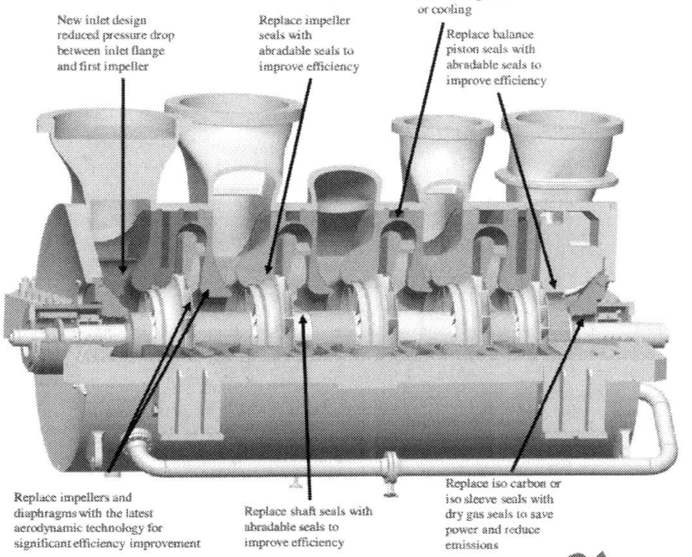

ELLIOTT GROUP
EBARA CORPORATION

Energy $avings for Centrifugal Compressors

1. Impellers and stage redesign
2. Compressor inlet redesign
3. Abradable interstage seals
4. Abradable balance piston seal
5. Anti-foulant coating
6. Gas seals
7. Variable Frequency Drive

ELLIOTT GROUP
EBARA CORPORATION

The first impellers were not very efficient...

1. Impeller & Stage Redesign

- Earliest impellers (1950's & 1960's)
  - Three part design
  - Two dimensional blade
  - Blades were riveted to hub & cover

- Later improvements (1970's & 1980's)
  - Inducers
  - Three dimensional blades
  - Vaned diffusers

- Latest improvements (early 1990's)
  - Redesigned inlet
  - CFD to further improve blade design

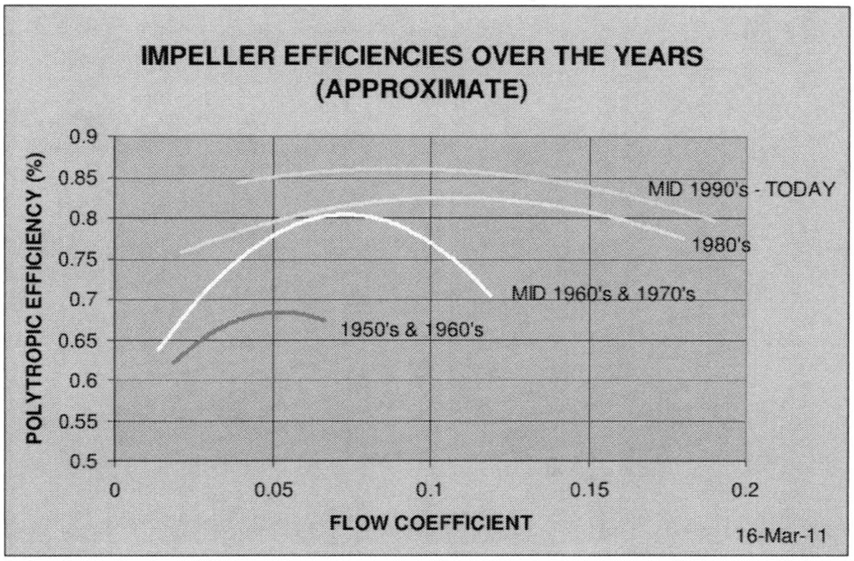

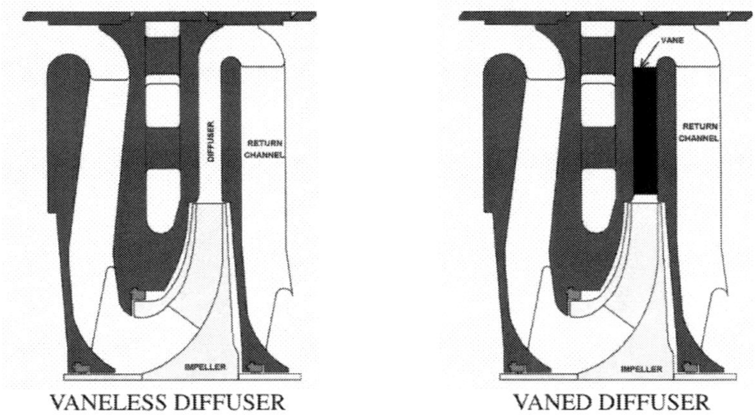

VANELESS DIFFUSER VANED DIFFUSER

Vaned Diffuser	
Pros	Cons
+ Boosts efficiency	- Reduced flow range
+ Increased head	- More surface area for foulant

ELLIOTT GROUP
EBARA CORPORATION

2. COMPRESSOR INLET REDESIGN

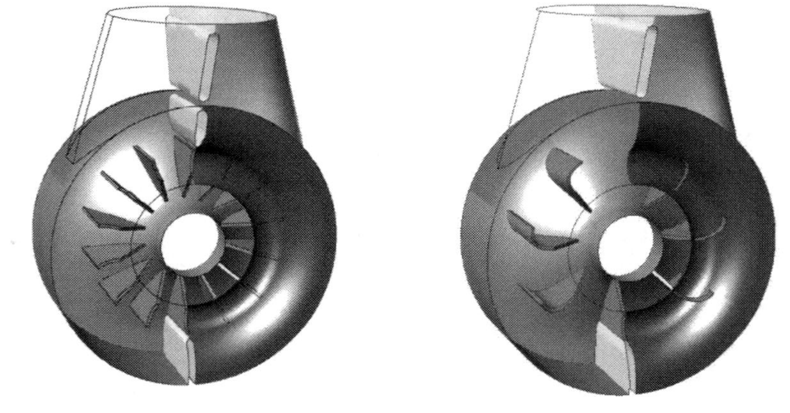

STRAIGHT RADIAL INLET VANES　　CURVED RADIAL INLET VANES

ELLIOTT GROUP
EBARA CORPORATION

3. ABRADABLE INTERSTAGE SEALS

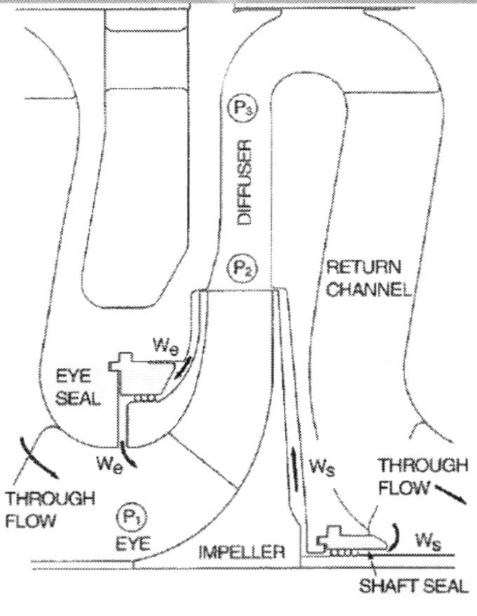

ELLIOTT GROUP
EBARA CORPORATION

4. ABRADABLE BALANCE PISTON SEAL

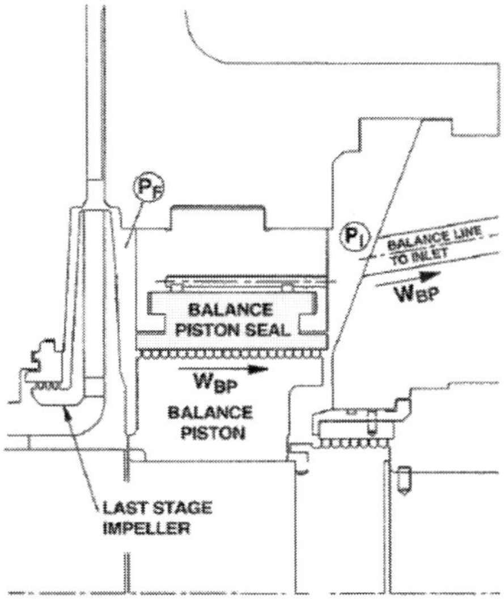

ELLIOTT GROUP
EBARA CORPORATION

5. ANTI-FOULANT COATING

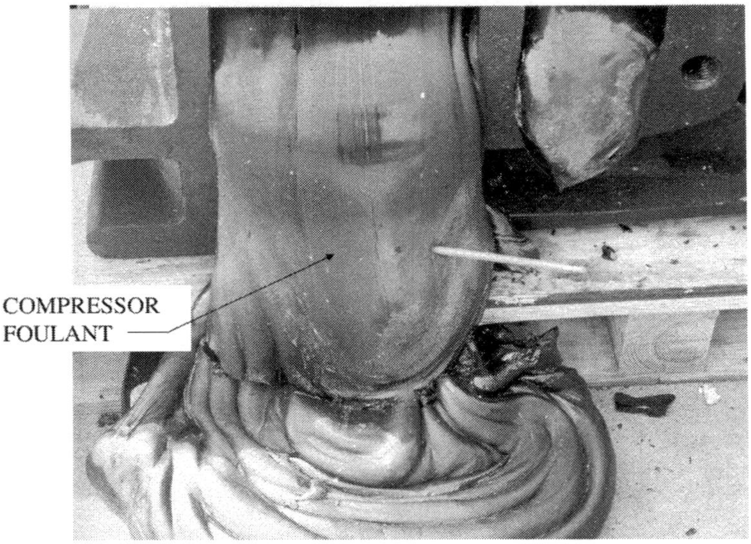

ELLIOTT GROUP
EBARA CORPORATION

Pos-E-Coat on a compressor rotor

ELLIOTT GROUP
EBARA CORPORATION

Pos-E-Coat on the stationary components

ELLIOTT GROUP
EBARA CORPORATION

6. GAS SEALS

GAS SEAL CONVERSION SAVES POWER!

- Gas seals have no oil shear, so the power consumed by a gas seal is a fraction of that of an oil seal. Savings could be 80 HP on a large compressor!
- Gas seals allow for the elimination or downsizing of oil pumps, reservoir heaters, degassing tank heaters, etc.
- Gas seals can reduce emissions. In one case, a methane compressor operator saved over $300,000 in one year by converting from oil seals to gas seals!

ELLIOTT GROUP
EBARA CORPORATION

7. VARIABLE FREQUENCY DRIVE

- Constant speed compressor with multiple operating points will usually either be suction throttled or have recycled flow to meet the desired operating point.
- A variable frequency drive (VFD) can save power by changing speed and allowing the compressor to operate at a more efficient point.

ELLIOTT GROUP
EBARA CORPORATION

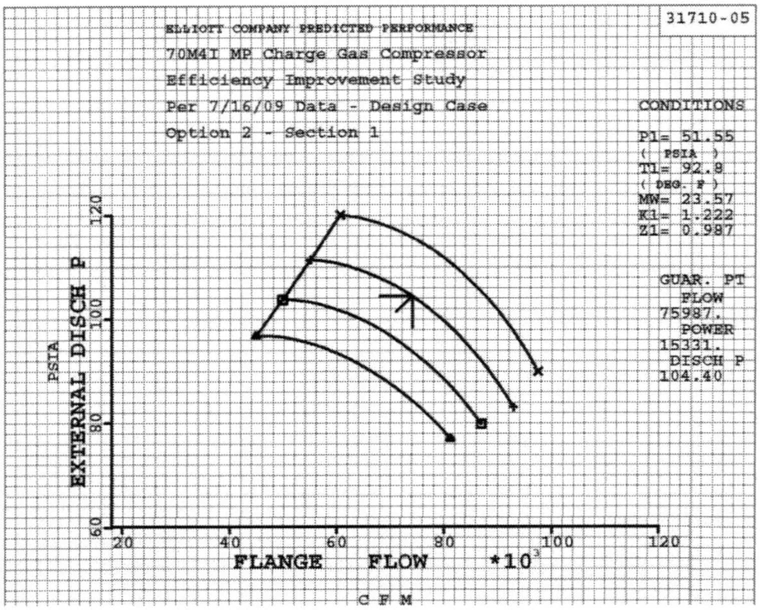

RERATE EXAMPLE

Let's say you have a…

- 1960's vintage compressor
- Wet seals
- Driven by an induction motor/speed increasing gear
- Absorbs 30,000 HP
- 78% polytropic efficiency
- Cost of power is $0.066 / kW-hr

You could rerate this compressor with new, 86% efficient aero and a gas seal upgrade. The new absorbed power is reduced 10% to 27,000 HP.

$AVING$ = $1,300,000 per year

(3000 hp x 0.7457 kW/hp x $0.066/kW-hr x 8760 hr/yr = $1.3MM)

CONCLUSION:

- Users of vintage centrifugal compressors should give serious consideration to rerating their machines to increase their efficiency.
- Tremendous power and cost savings can be realized in many cases, often with a side benefit of increased production, if desired.
- Many of the changes made while rerating the compressor will also give the machine better reliability.

ELLIOTT GROUP
EBARA CORPORATION

QUESTIONS?

ELLIOTT GROUP
EBARA CORPORATION

LISTEN.
THINK.
SOLVE.®

System-Wide Optimization and Control

Allen-Bradley · Rockwell Software

Rockwell Automation

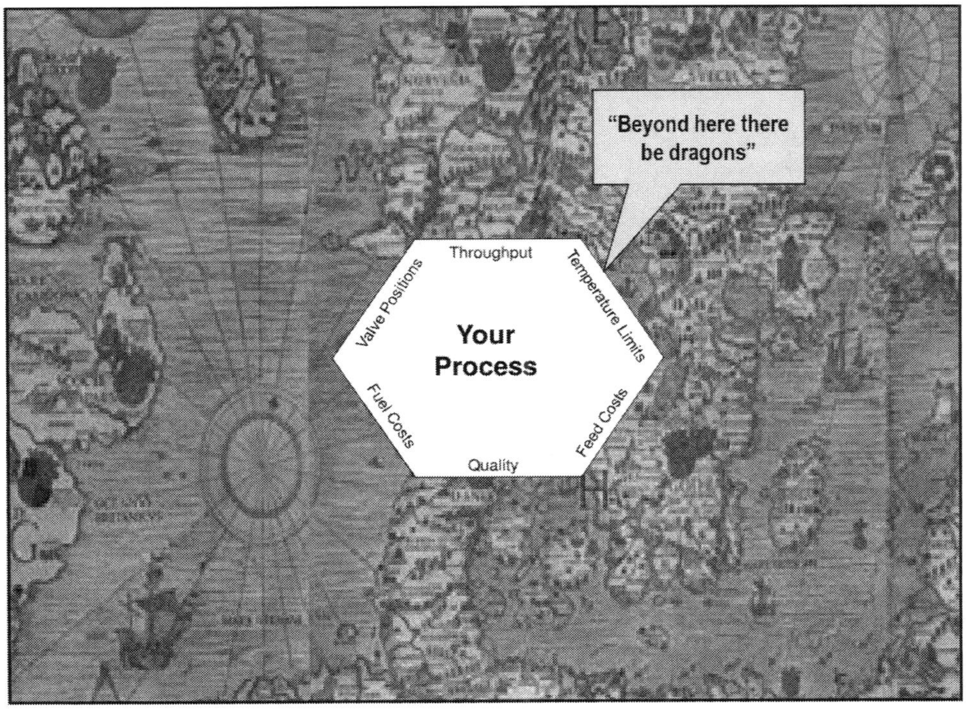

Advanced Control Objectives

Minimize operational variance

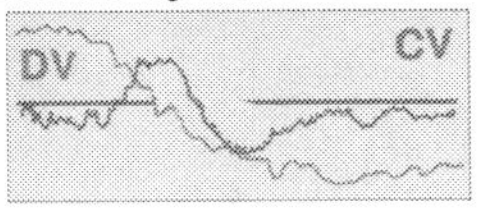

Transition effectively

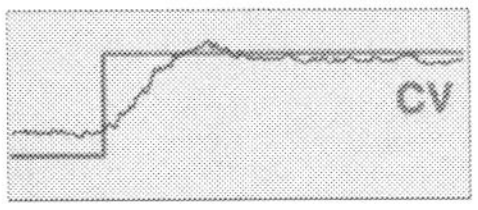

Follow operational plan

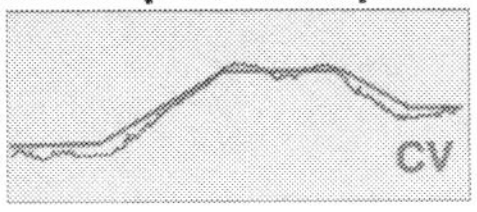

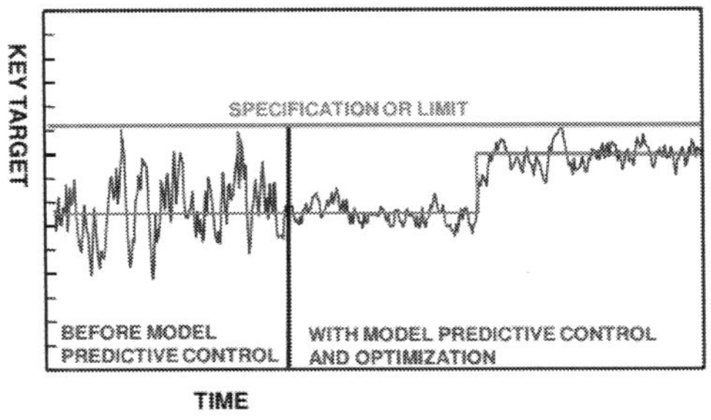

4

Regulatory Control (basic stuff)

Single Loop Control

- Single In, Single Out
- Tune Loop to Match Dynamics
- Some Extensions

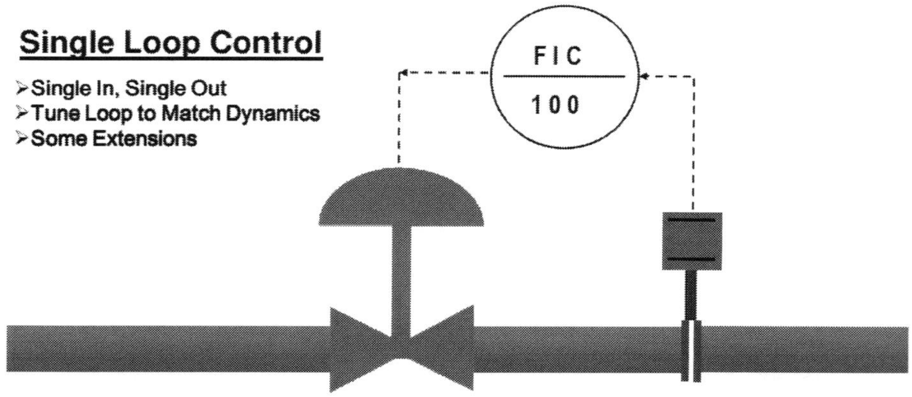

(Patented late 1800s)

5

Model Predictive Control

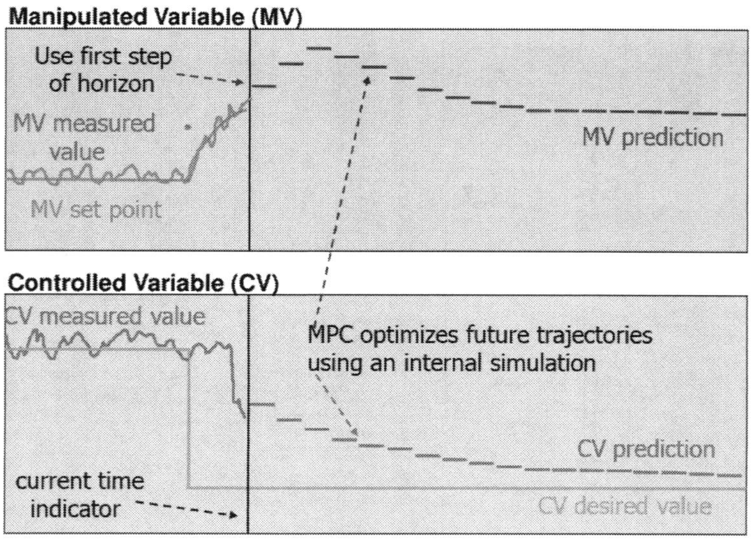

6

Multivariable Predictive Control

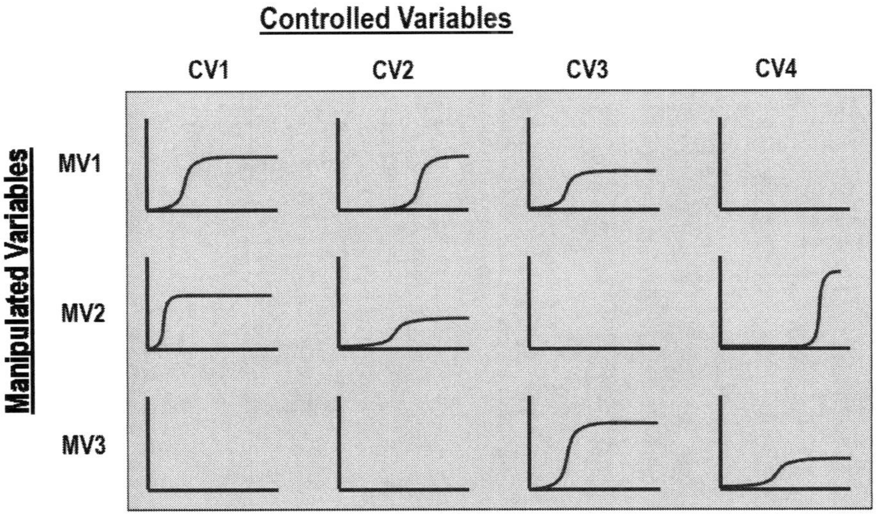

(Wide Use Since 1970s)

7

Model Predictive Control

Benefits

- Predicts Future Errors
- Handles Most Process Dynamics
- Automatic DeCoupling
- Handles Multiple Constraints

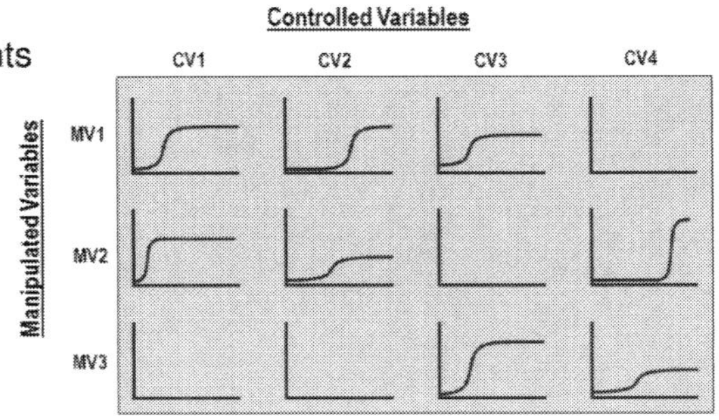

8

Models?

- Physical and Neural Models
- Linear and Non-Linear
- Fully Differentiable
- Use Same Models for Control and Optimization
- Import Existing Models

Physical

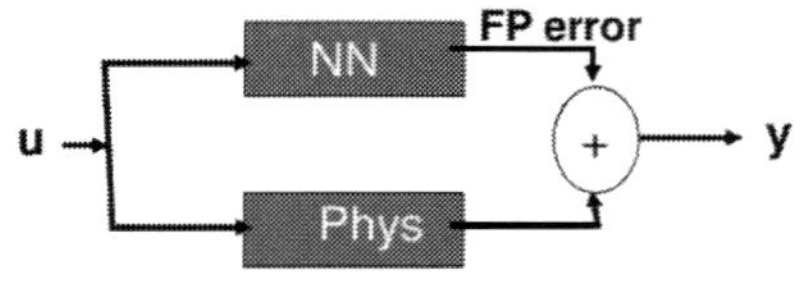

Artificial Neural Networks

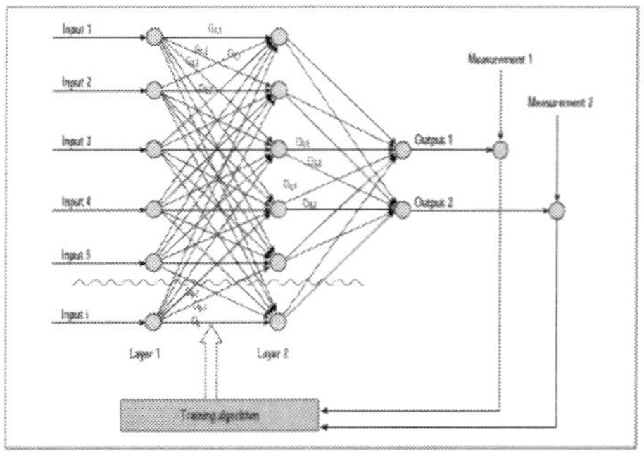

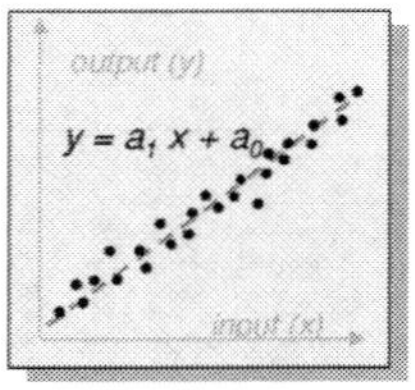

Linear

Non-Linear

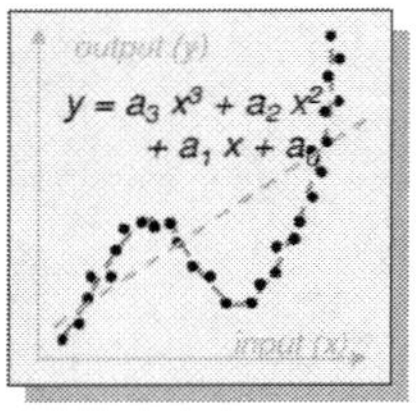

9

Predictive Load Models

Weather Based Prediction Models

Chilled Water

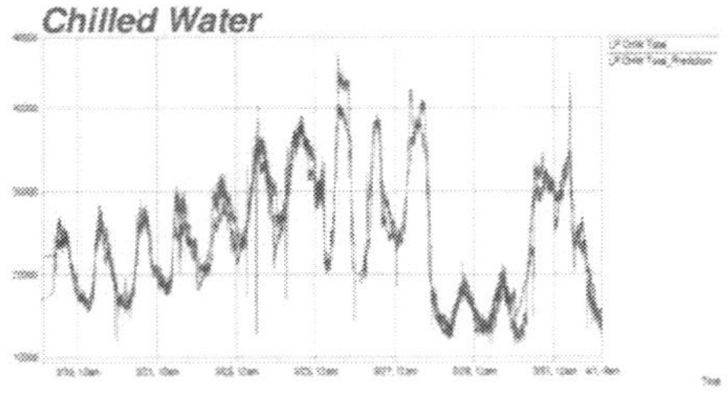

kW

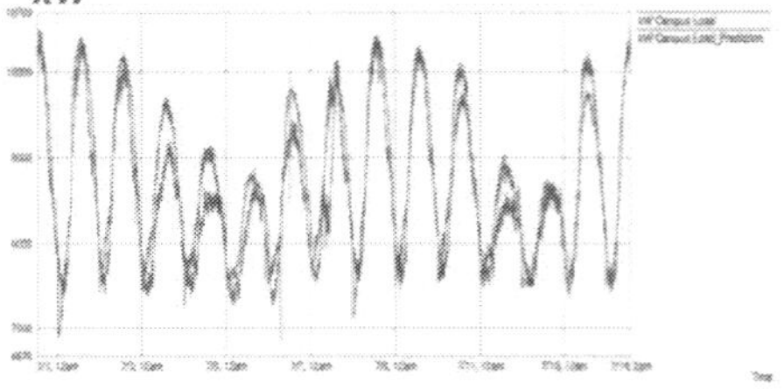

Steam

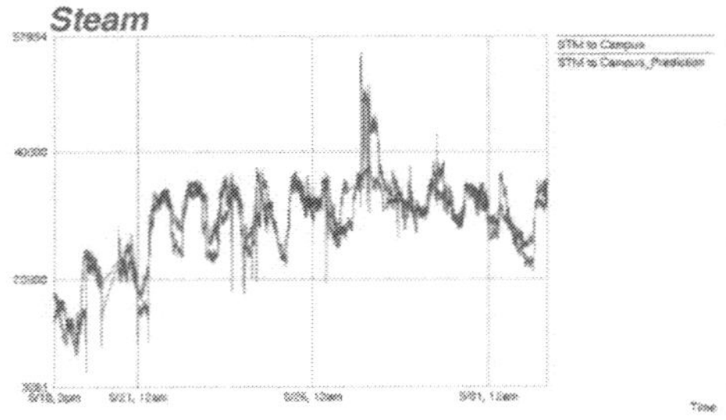

Natural Gas

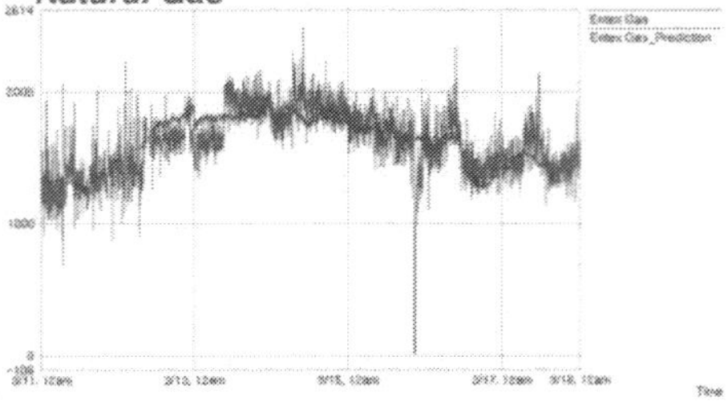

Predicted vs. Actual

10

Polymer MPC Example

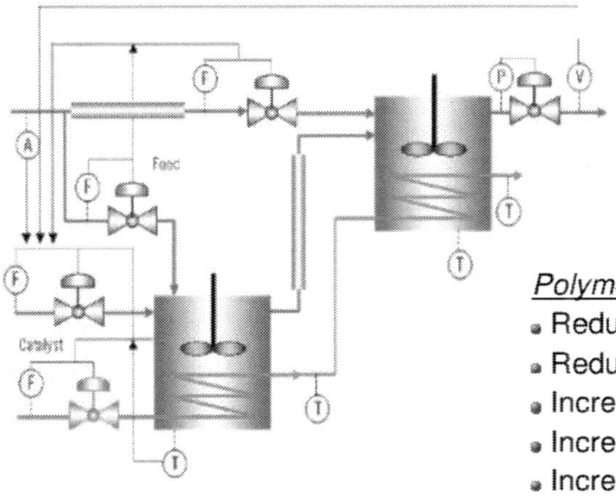

Polymer MPC Solution

- Reduced Property Variability 50%
- Reduced Transition Times 25-50%
- Increase Production 7-10%
- Increase Monomer Efficiency
- Increase Catalyst Efficiency
- Non-Linear Reaction Models

 11

Boiler MPC Example

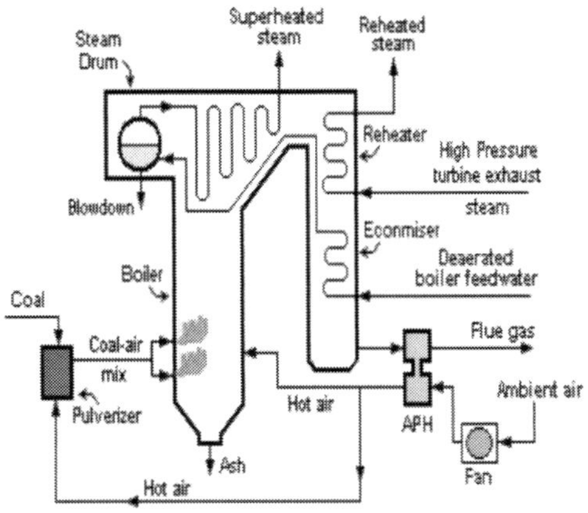

Boiler MPC Solution

- Improved Heat Rate
- Improved Ramp Rate
- Improved Reheat Temp control
- Improved Burn Control (eliminates reduction zones)
- Reduces NOx 5-35%
- Reduces CO_2 and fuel usage
- Reduces ammonia usage
- Improves SCR catalyst life

Kiln MPC Example

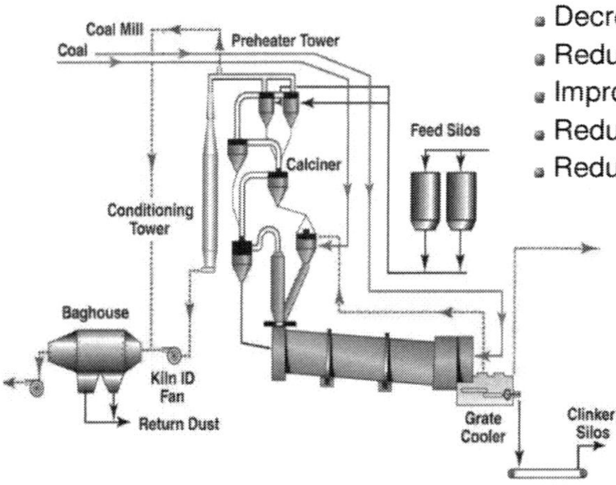

Kiln MPC Solution

- Increase Throughput 6-8%
- Decrease Variability 35-45%
- Reduce Specific Energy
- Improve Product Grade
- Reduces NOx 5-35%
- Reduces CO2 and fuel usage

Complex Systems - Mixed Integer Optimization

- Multiple Equipment Choices
- Like Equipment , Different Efficiency Curves
- Multiple Fuel Options
- Buy-Sell Power Options
- Optimally Select Equipment Based on Load Prediction
- Savings Exponentially Proportional to Number of Choices

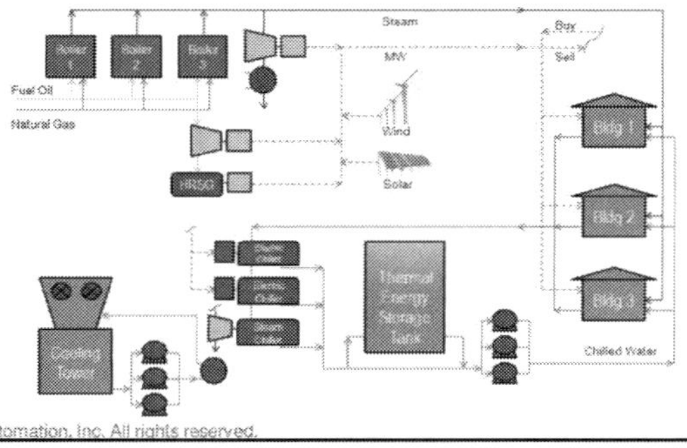

Complexity vs. Selection Choices

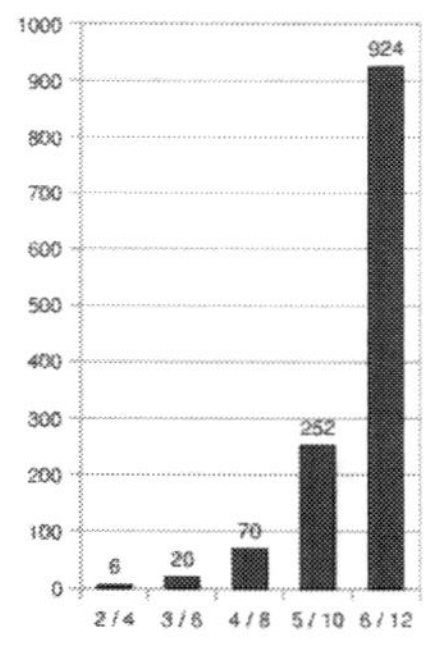

$$\frac{N!}{C! * (N-C)!}$$

Savings vs. Complexity is Exponential

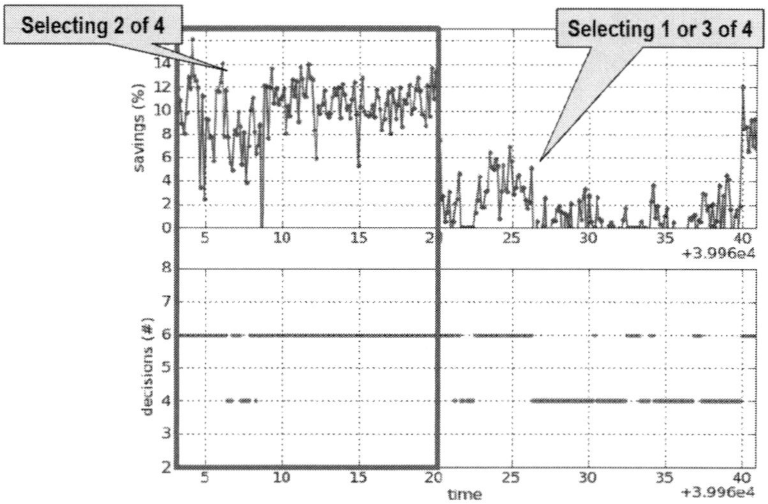

Decision complexity directly correlated to savings

Central Utility Plant Example

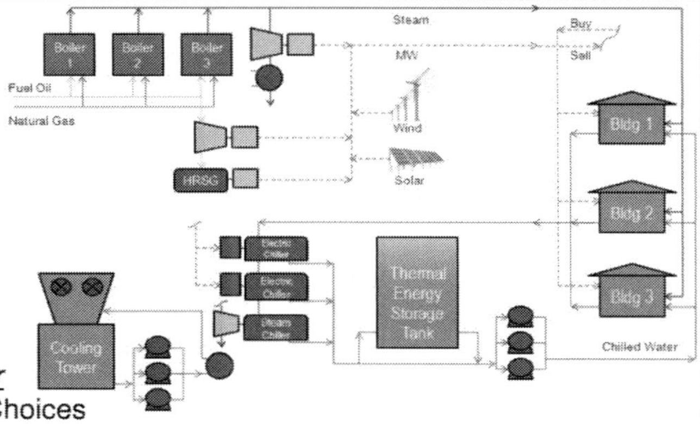

Combined Heat, Power

- Multiple Equipment Choices
- Like Equipment , Different Efficiency Curves
- Multiple Fuel Options
- Buy-Sell Power Options
- Optimally Select Equipment Based on Load Prediction
- Savings Exponentially Proportional to Number of Choices
- 10-20% Savings

Oil Recovery Steam Injection Example

Steam Injection Optimization Solution

- Hundreds of Steam Generators
- Hundreds of Turbines
- Buy-Sell Power Options
- Optimally Select Equipment based on Efficiency Curves and Load Prediction

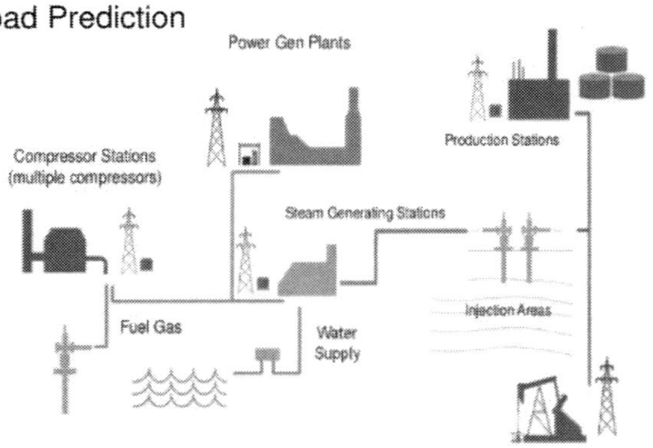

 17

Power Distribution Example

Distribution Optimization

- Consider System as a Whole
- Mixed-Integer Optimization
- Deterministic
- Exploits Efficiency Curves
- Minimzed Fuel Usage
- Minimize Coal Unit Swings
- True Economic Optimum

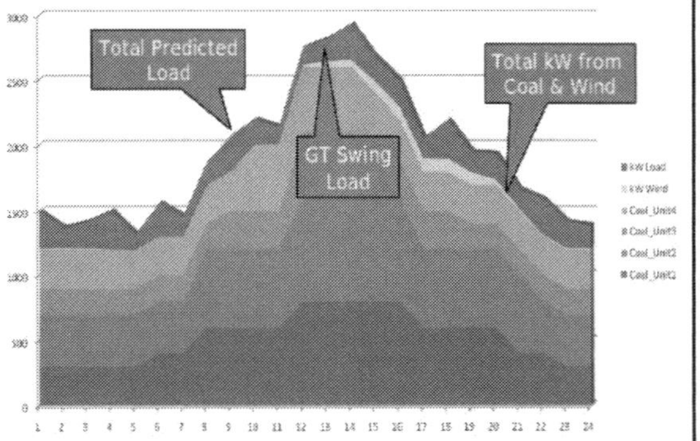

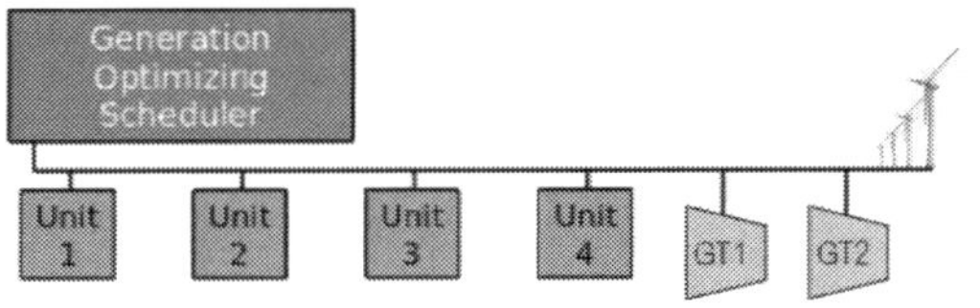

Chilled Water Systems Example

Chiller Optimization Solution

- Multiple Chillers (4-30+)
- Several Chiller Types (electric, steam)
- Partially Loaded 1/2 of the Year
- Several Energy Sources
- Strong Ambient Effect
- 10-25% Optimization Savings

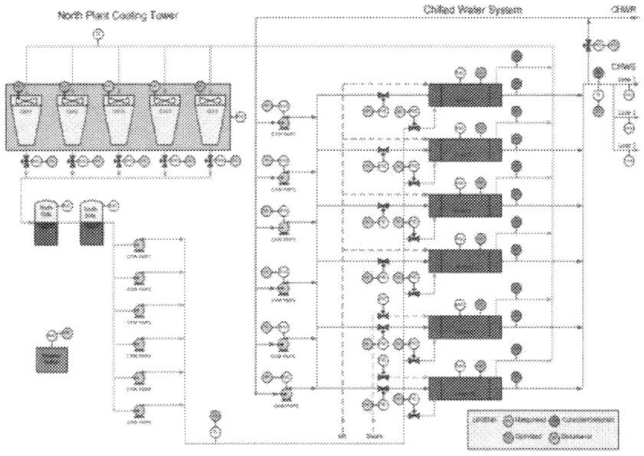

Optimization Platform Goals

System Goals

- Easier to Install
- User-Configurable
- User-Maintainable
- Flexible Framework
- Plug In Proprietary Models

Optimization System Framework

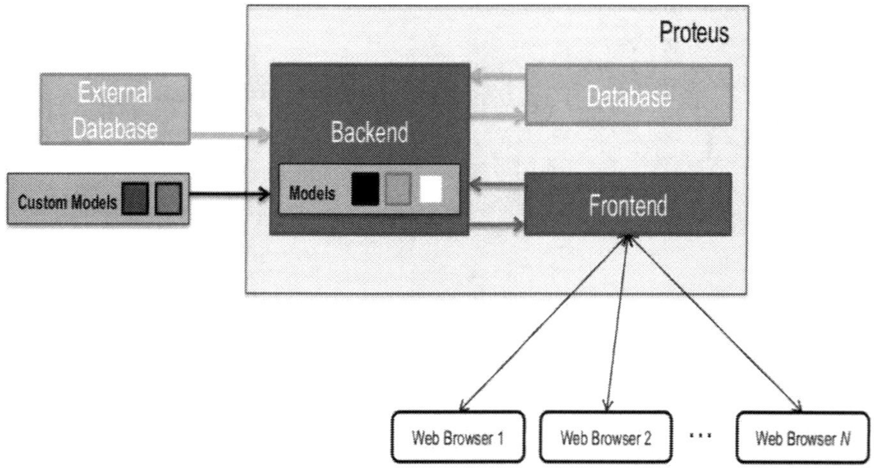

 21

Economy of Optimization

- **Optimization Models**
  - **Compromise between model detail and tractability**
  - **Non-homogenous model detail: focus on critical aspects**
- **Optimization Problem Limits**
  - **LP: millions of variables**
  - **MILP, NLP: thousands of binary variables**
  - **MINLP: hundreds of variables**

	n	$n \log_2 n$	n^2	n^3	1.5^n	2^n	$n!$
$n = 10$	< 1 sec	< 1 sec	< 1 sec	< 1 sec	< 1 sec	< 1 sec	4 sec
$n = 30$	< 1 sec	< 1 sec	< 1 sec	< 1 sec	< 1 sec	18 min	10^{25} years
$n = 50$	< 1 sec	< 1 sec	< 1 sec	< 1 sec	11 min	36 years	very long
$n = 100$	< 1 sec	< 1 sec	< 1 sec	1 sec	12,892 years	10^{17} years	very long
$n = 1,000$	< 1 sec	< 1 sec	1 sec	18 min	very long	very long	very long
$n = 10,000$	< 1 sec	< 1 sec	2 min	12 days	very long	very long	very long
$n = 100,000$	< 1 sec	2 sec	3 hours	32 years	very long	very long	very long
$n = 1,000,000$	1 sec	20 sec	12 days	31,710 years	very long	very long	very long

 22

Parametric Modeling Example

Heat Exchanger

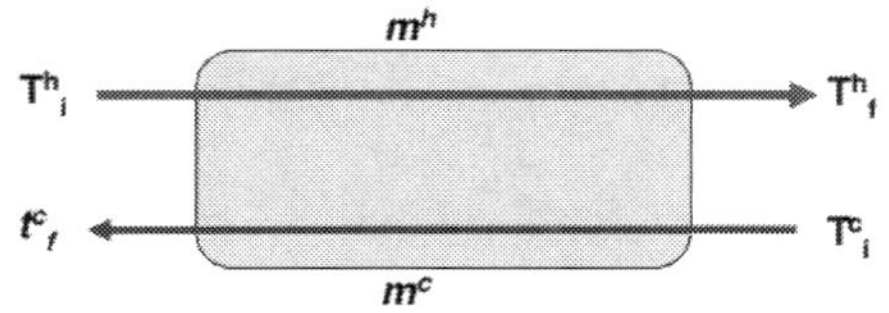

- **Thermodynamics**

$$Q = -m^h C^h_p (T^h_f - T^h_i) = m^c C^c_p (t^c_f - T^c_i)$$

- **Fourier's law**

$$Q = UA \left[\frac{(T^h_i - t^c_f) - (T^h_f - T^c_i)}{\ln \frac{T^h_i - t^c_f}{T^h_f - T^c_i}} \right]$$

 23

Heat Exchanger - First Pass Model

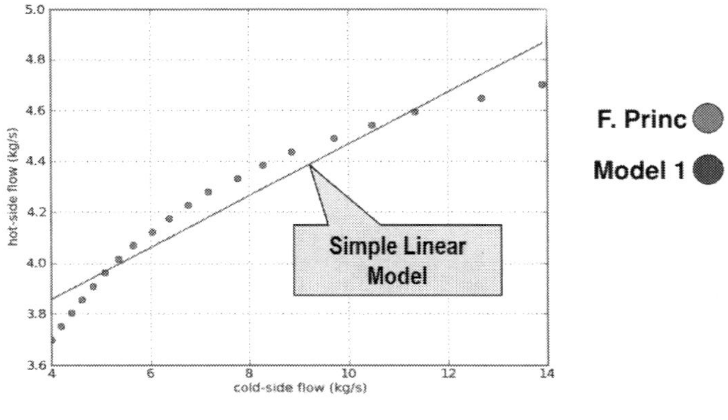

- **Model 1: Energy balance**

$$m^h = \frac{C^c_p T^c_i}{C^h_p (T^h_f - T^h_i)} m^c + \alpha$$

 24

Improved Operating Curve

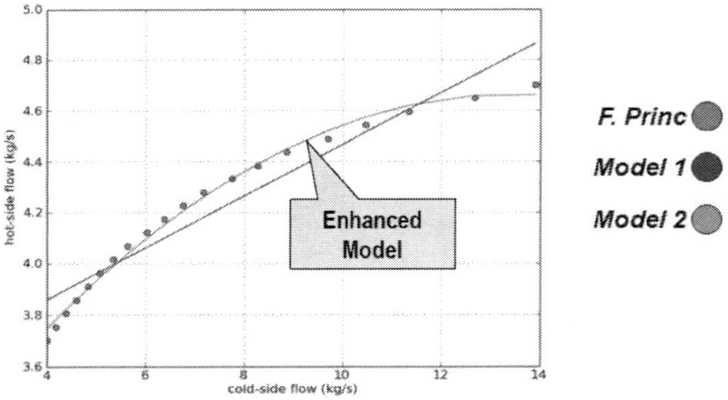

- ***Model 2: Outlet temperature approximation based on Fourier's law***

$$m^h = \frac{C_p^c T_i^c}{C_p^h (T_f^h - T_i^h)} m^c - \frac{C_p^c}{C_p^h (T_f^h - T_i^h)} m^c L_{[t_f^c]}$$

 25

Optimization Goal

GOAL

"Optimal scheduling of utility unit operations to achieve minimal cost"

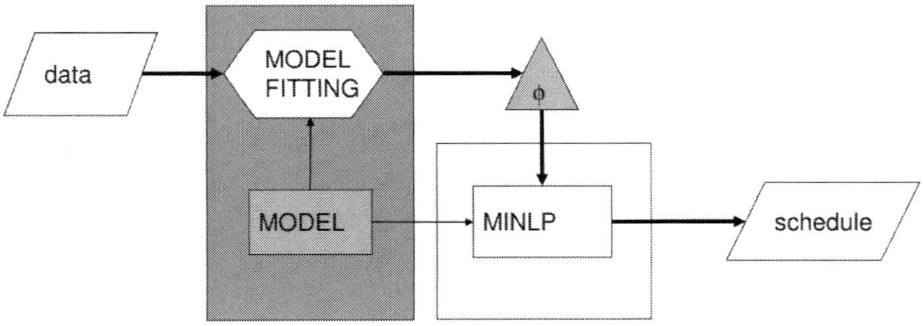

 26

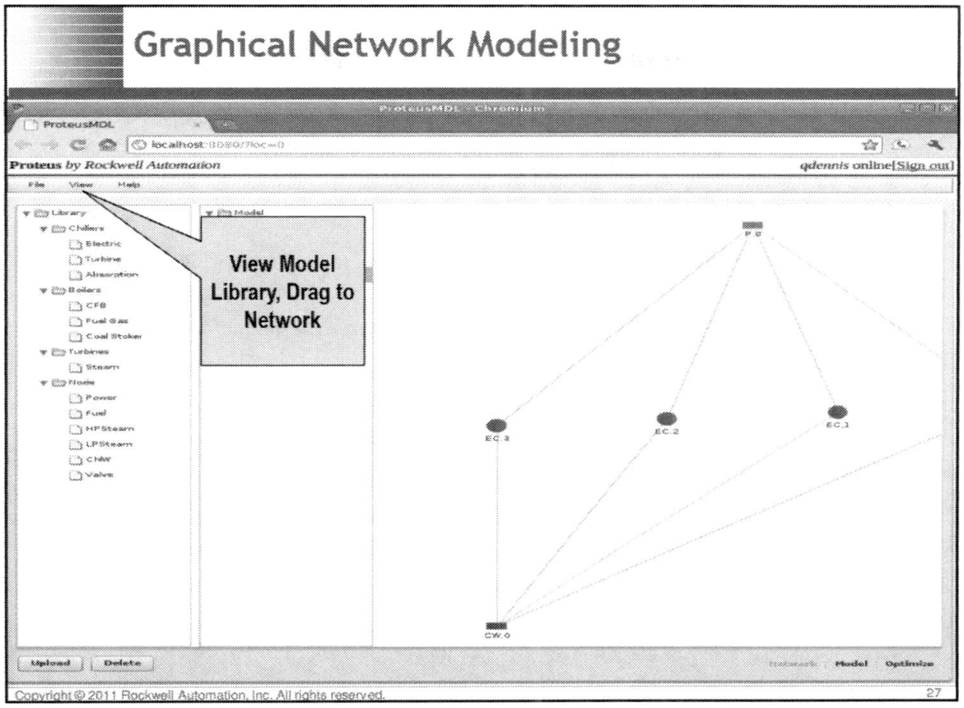
Graphical Network Modeling
ProteusMDL - Chromium
ProteusMDL
localhost
Proteus by Rockwell Automation
qdennis online[Sign out]
File
View
Help
Library
Chillers
Electric
Turbine
Absorption
Boilers
CFB
Fuel Gas
Coal Stoker
Turbines
Steam
Node
Power
Fuel
HPSteam
LPSteam
CHW
Valve
Model
View Model Library, Drag to Network
EC.3
EC.2
EC.1
CW.0
Upload
Delete
Network
Model
Optimize
Copyright © 2011 Rockwell Automation, Inc. All rights reserved.
27

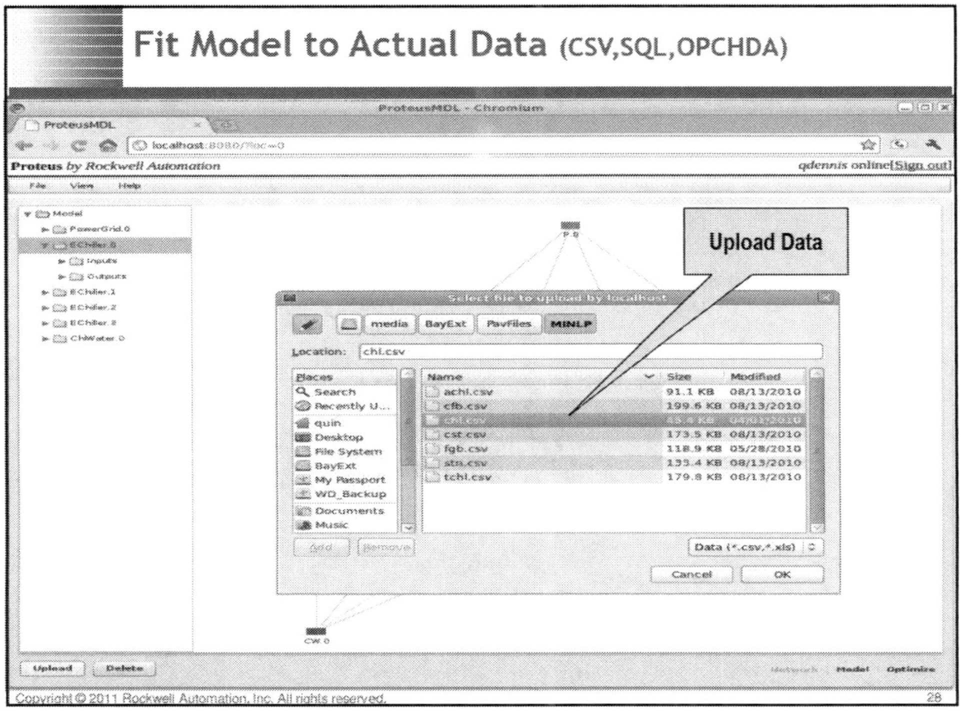
Fit Model to Actual Data (CSV,SQL,OPCHDA)
ProteusMDL - Chromium
ProteusMDL
localhost
Proteus by Rockwell Automation
qdennis online[Sign out]
File
View
Help
Model
Inputs
Outputs
Upload Data
Select file to upload by localhost
media
BayExt
PavFiles
MINLP
Location: chl.csv
Places
Search
Recently U...
quin
Desktop
File System
BayExt
My Passport
WD_Backup
Documents
Music
Add
Remove
Name
Size
Modified
achl.csv 91.1 KB 08/13/2010
cfb.csv 199.6 KB 08/13/2010
chl.csv
cst.csv 173.5 KB 08/13/2010
fgb.csv 118.9 KB 05/28/2010
sts.csv 135.4 KB 08/13/2010
tchl.csv 179.8 KB 08/13/2010
Data (*.csv,*.xls)
Cancel
OK
CW.0
Upload
Delete
Network
Model
Optimize
Copyright © 2011 Rockwell Automation, Inc. All rights reserved.
28

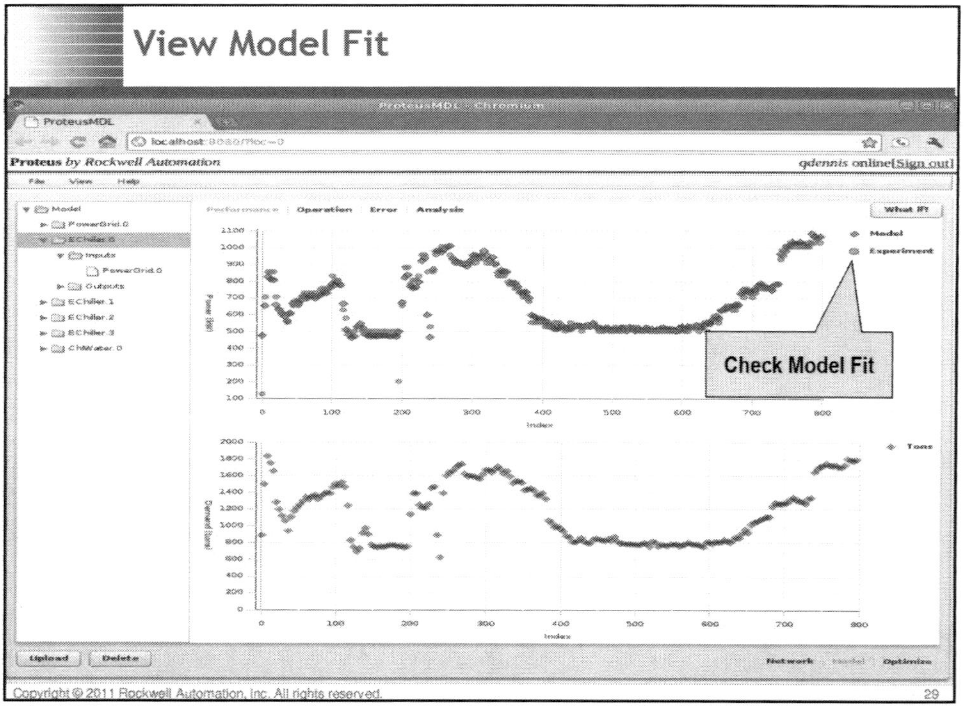
View Model Fit
ProteusMDL
localhost:8080/?loc=0
Proteus by Rockwell Automation
qdennis online[Sign out]
File View Help
Model
PowerGrid.0
EChiller.0
Inputs
PowerGrid.0
Outputs
EChiller.1
EChiller.2
EChiller.3
ChWater.0
Performance Operation Error Analysis
What If?
Model
Experiment
Power (kW)
Index
Tons
Demand (tons)
Check Model Fit
Upload
Delete
Network Model Optimize
Copyright © 2011 Rockwell Automation, Inc. All rights reserved.
29

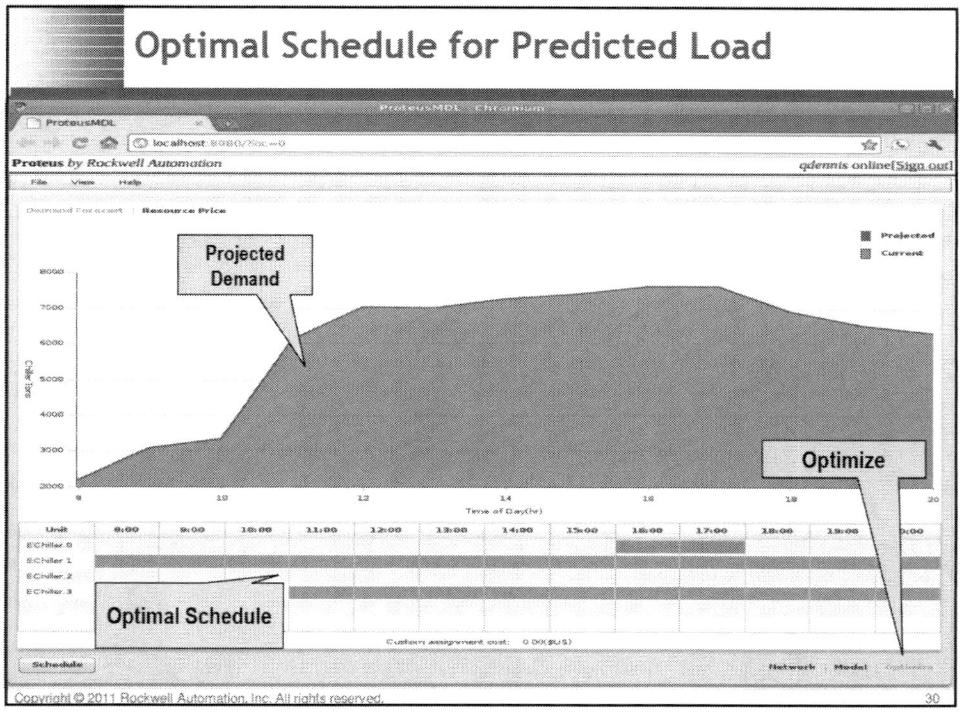
Optimal Schedule for Predicted Load
ProteusMDL
localhost:8080/?loc=0
Proteus by Rockwell Automation
qdennis online[Sign out]
File View Help
Demand Forecast Resource Price
Projected
Current
Projected Demand
Chiller Tons
Time of Day(hr)
Optimize
Unit 8:00 9:00 10:00 11:00 12:00 13:00 14:00 15:00 16:00 17:00 18:00 19:00
EChiller.0
EChiller.1
EChiller.2
EChiller.3
Optimal Schedule
Custom assignment cost: 0.00($US)
Schedule
Network Model Optimize
Copyright © 2011 Rockwell Automation, Inc. All rights reserved.
30

Operator Can Modify Schedule

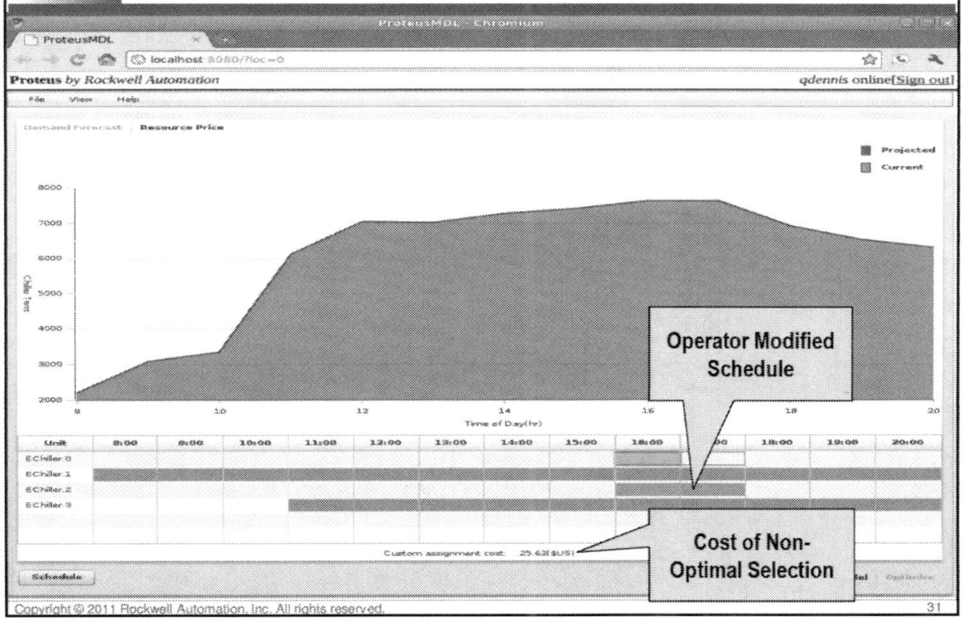

 31

Highlights

- Complete System-Wide Optimization
  - Optimize System as a Whole
  - Minimize Total System Costs
  - Optimal Equipment Sequencing (using partial load curves)
  - Robust Model Predictive Control Manages Constraints
  - Incorporate Energy Usage Projections in Production Planning
  - Fuel Selections, Buy/Sell Decisions
  - User Configurable, User Maintainable
  - KPI and Reporting System

 32

The End

33

2011 AIChE
Process Technology
Conference

Reducing Emissions in Plant Flaring Operations

Brian Duck
V.P. Flare Systems

CALLIDUS TECHNOLOGIES by Honeywell

Case Study

- 10 million ton oil refinery in Dushanzi, China

Callidus Technologies by Honeywell

Callidus.com

Facility Objectives

- Reduce emissions of CO_2, SO_2 and NOx by 130,000,000 tons/yr
  - 10 projects identified including flare systems
- Reduce smoke, light and noise to the community

3
Flare Training-0609-001.ppt

Honeywell Proprietary

Callidus Technologies by Honeywell

Callidus.com

Flare Systems

- North Flare
  - 3,500,000 lb/hr flow rate (equivalent to heating over 500,000 homes)
- South Flare
  - 2,000,000 lb/hr flow rate

4
Flare Training-0609-001.ppt

Honeywell Proprietary

Flare Gas Recovery

- North Flare System
  - Recover 60MMBtu/hr
- South Flare System
  - Recover 60 MMBtu/hr

Flare System

CALLIDUS TECHNOLOGIES by Honeywell

Callidus.com

Steam Assist

- Object is to 'quietly' produce turbulence at the flare tip exit and inspirate as much of the surrounding air as possible into the combustion zone.

7
Flare Training-0609-001.ppt

Honeywell Proprietary

CALLIDUS TECHNOLOGIES by Honeywell

Callidus.com

Steam Tip

8
Flare Training-0609-001.ppt

Honeywell Proprietary

CALLIDUS TECHNOLOGIES by Honeywell

Callidus.com

Flare in Operation

9
Flare Training-0609-001.ppt

Honeywell Proprietary

CALLIDUS TECHNOLOGIES by Honeywell

Callidus.com

CAL-IS³ Components

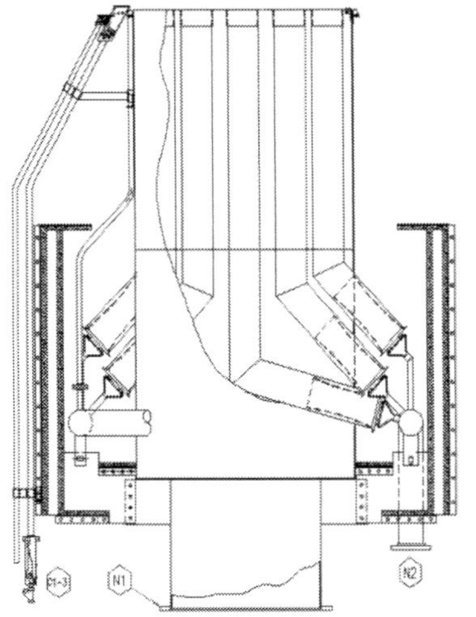

- CK-20 investment cast De Laval nozzle lower steam "spiders" allow supersonic steam flow rates

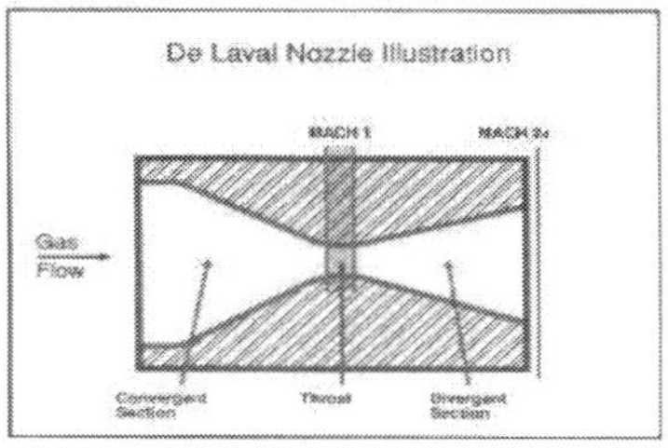

10
Flare Training-0609-001.ppt

Honeywell Proprietary

CALLIDUS TECHNOLOGIES by Honeywell

Callidus.com

CAL-IS^3 Components

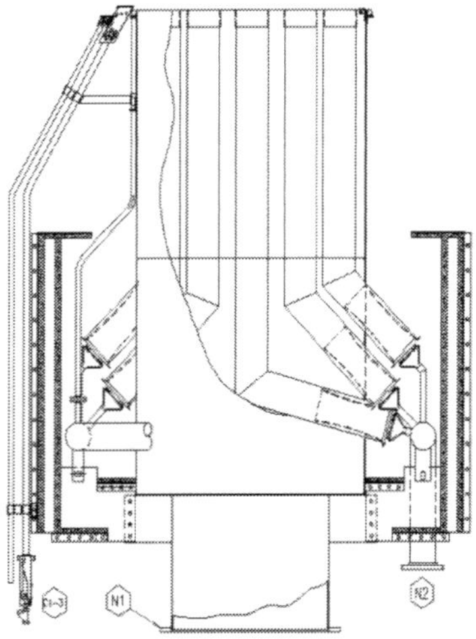

- Steam spiders are directed into an 8” venturi to maximize air induction

11
Flare Training-0609-001.ppt

Honeywell Proprietary

CALLIDUS TECHNOLOGIES by Honeywell

Callidus.com

CAL-IS^3 Components

- Steam ring placement below the flame zone allows acoustic insulation to reduce steam noise

12
Flare Training-0609-001.ppt

Honeywell Proprietary

CALLIDUS TECHNOLOGIES by Honeywell

Callidus.com

Demountable Elevated Flare

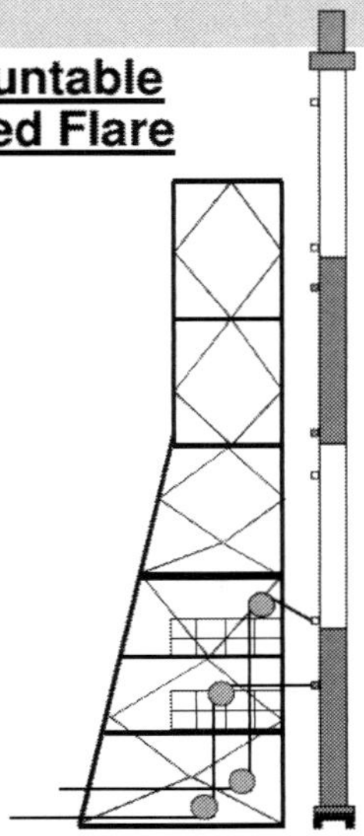

General Sequence Of Events – Demounting

1. Lowest Riser section secured.
2. Remaining sections secured at base of lowest remaining riser and unflanged from lowest section.
3. Upper section raised via winches and pulley for clearance.
4. Lowest section rotated from vertical to horizontal at tilting table for placement on transfer table.
5. Lowest section moved off transfer table, remaining sections lowered to tilting table.
6. Cables detached and attached in appropriate locations for subsequent iterations of riser relocation from stack to transfer table.
7. Repeat steps 1-6 until flare tip is moved to transfer table.

13
Flare Training-0609-001.ppt

Honeywell Proprietary

CALLIDUS TECHNOLOGIES by Honeywell

Callidus.com

Demountable Elevated Flare

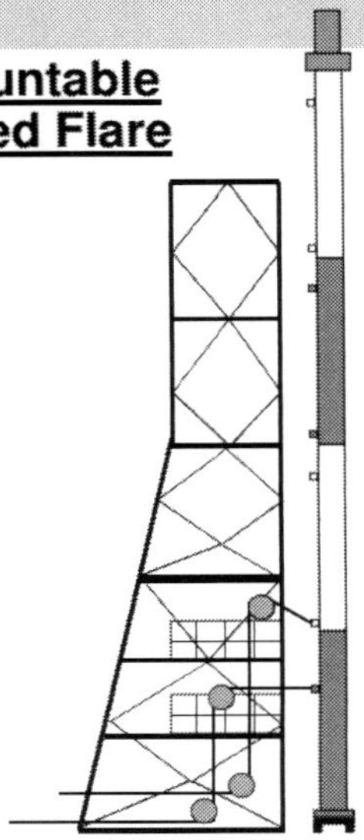

General Sequence Of Events – Demounting

1. Lowest Riser section secured.
2. Remaining sections secured at base of lowest remaining riser and unflanged from lowest section.

14
Flare Training-0609-001.ppt

Honeywell Proprietary

Demountable Elevated Flare

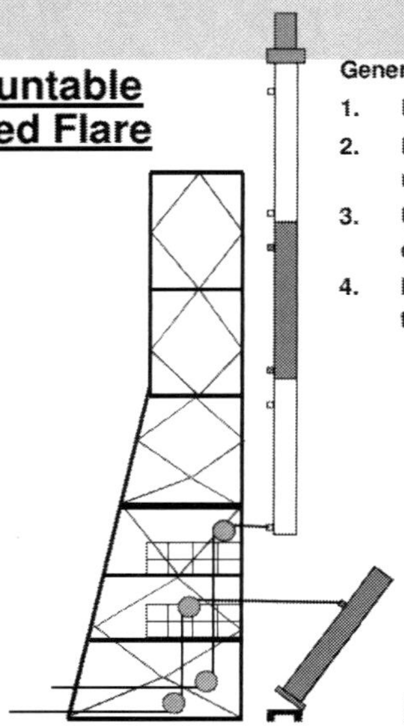

General Sequence Of Events – Demounting

1. Lowest Riser section secured.
2. Remaining sections secured at base of lowest remaining riser and unflanged from lowest section.
3. Upper section raised via winches and pulley for clearance.
4. Lowest section rotated from vertical to horizontal at tilting table for placement on transfer table.

Demountable Elevated Flare

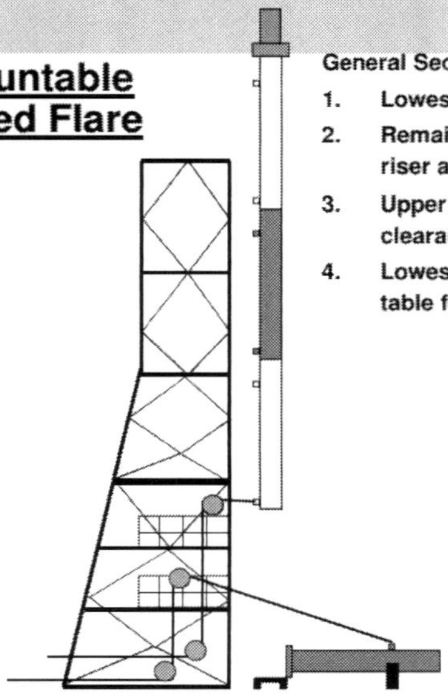

General Sequence Of Events – Demounting

1. Lowest Riser section secured.
2. Remaining sections secured at base of lowest remaining riser and unflanged from lowest section.
3. Upper section raised via winches and pulley for clearance.
4. Lowest section rotated from vertical to horizontal at tilting table for placement on transfer table.

CALLIDUS TECHNOLOGIES by Honeywell

Callidus.com

Demountable Elevated Flare

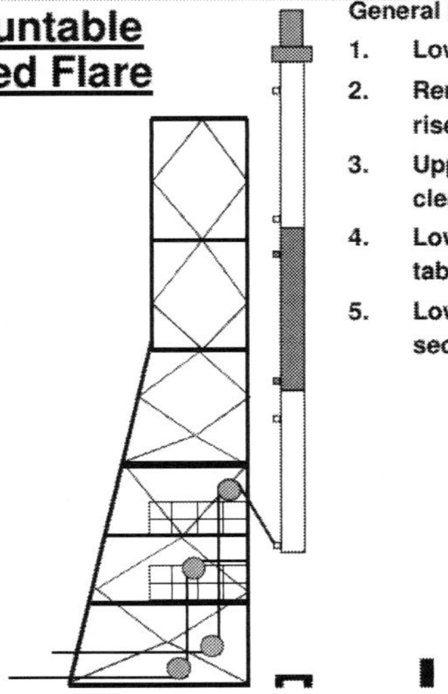

General Sequence Of Events – Demounting

1. Lowest Riser section secured.
2. Remaining sections secured at base of lowest remaining riser and unflanged from lowest section.
3. Upper section raised via winches and pulley for clearance.
4. Lowest section rotated from vertical to horizontal at tilting table for placement on transfer table.
5. Lowest section moved off transfer table, remaining sections lowered to tilting table.

17
Flare Training-0609-001.ppt

Honeywell Proprietary

CALLIDUS TECHNOLOGIES by Honeywell

Callidus.com

Demountable Elevated Flare

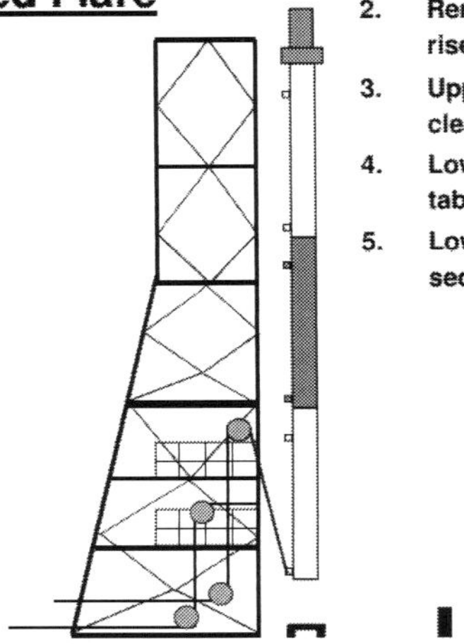

General Sequence Of Events – Demounting

1. Lowest Riser section secured.
2. Remaining sections secured at base of lowest remaining riser and unflanged from lowest section.
3. Upper section raised via winches and pulley for clearance.
4. Lowest section rotated from vertical to horizontal at tilting table for placement on transfer table.
5. Lowest section moved off transfer table, remaining sections lowered to tilting table.

18
Flare Training-0609-001.ppt

Honeywell Proprietary

Demountable Elevated Flare

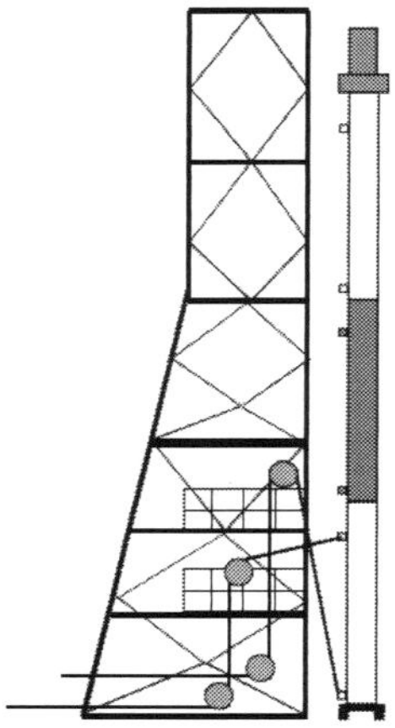

General Sequence Of Events – Demounting

1. Lowest Riser section secured.
2. Remaining sections secured at base of lowest remaining riser and unflanged from lowest section.
3. Upper section raised via winches and pulley for clearance.
4. Lowest section rotated from vertical to horizontal at tilting table for placement on transfer table.
5. Lowest section moved off transfer table, remaining sections lowered to tilting table.
6. Cables detached and attached in appropriate locations for subsequent iterations of riser relocation from stack to transfer table.

Demountable Elevated Flare

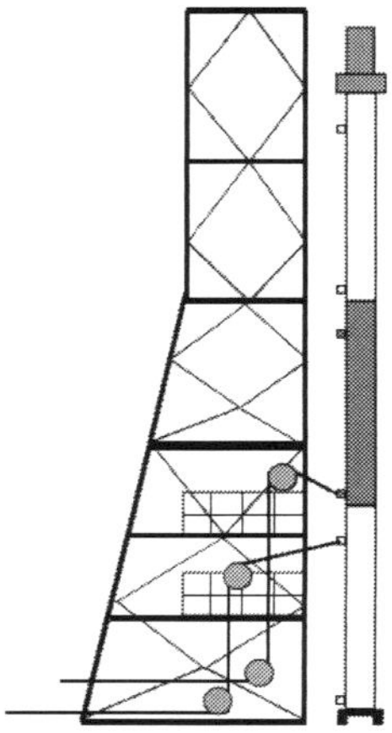

General Sequence Of Events – Demounting

1. Lowest Riser section secured.
2. Remaining sections secured at base of lowest remaining riser and unflanged from lowest section.
3. Upper section raised via winches and pulley for clearance.
4. Lowest section rotated from vertical to horizontal at tilting table for placement on transfer table.
5. Lowest section moved off transfer table, remaining sections lowered to tilting table.
6. Cables detached and attached in appropriate locations for subsequent iterations of riser relocation from stack to transfer table.
7. Repeat steps 1-6 until flare tip is moved to transfer table.

Callidus Technologies by Honeywell

Callidus.com

Demountable Elevated Flare

General Sequence Of Events – Demounting

1. Lowest Riser section secured.
2. Remaining sections secured at base of lowest remaining riser and unflanged from lowest section.
3. Upper section raised via winches and pulley for clearance.
4. Lowest section rotated from vertical to horizontal at tilting table for placement on transfer table.
5. Lowest section moved off transfer table, remaining sections lowered to tilting table.
6. Cables detached and attached in appropriate locations for subsequent iterations of riser relocation from stack to transfer table.
7. Repeat steps 1-6 until flare tip is moved to transfer table.

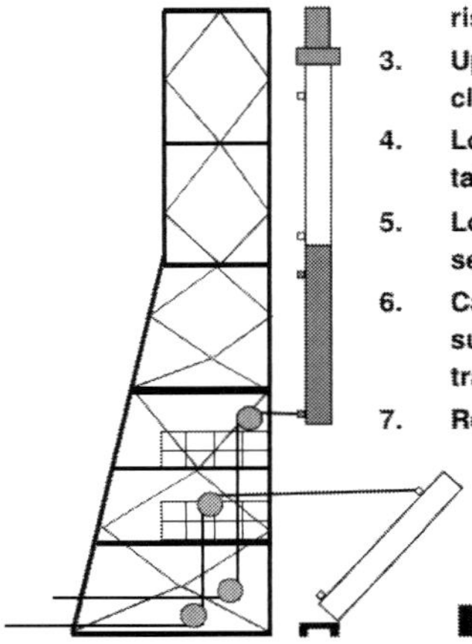

21
Flare Training-0609-001.ppt
Honeywell Proprietary

Callidus Technologies by Honeywell

Callidus.com

Demountable Elevated Flare

General Sequence Of Events – Demounting

1. Lowest Riser section secured.
2. Remaining sections secured at base of lowest remaining riser and unflanged from lowest section.
3. Upper section raised via winches and pulley for clearance.
4. Lowest section rotated from vertical to horizontal at tilting table for placement on transfer table.
5. Lowest section moved off transfer table, remaining sections lowered to tilting table.
6. Cables detached and attached in appropriate locations for subsequent iterations of riser relocation from stack to transfer table.
7. Repeat steps 1-6 until flare tip is moved to transfer table.

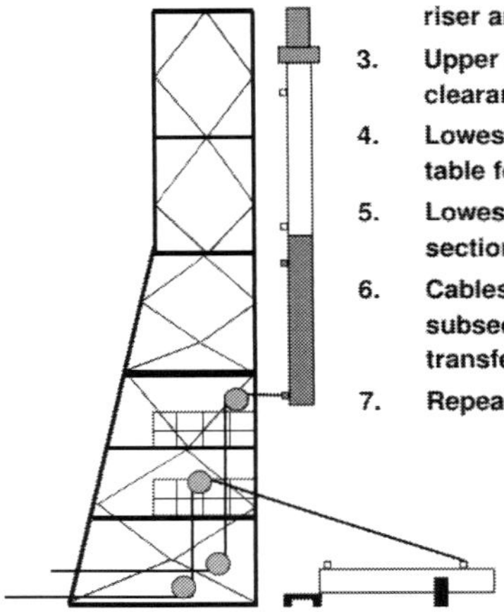

22
Flare Training-0609-001.ppt
Honeywell Proprietary

Demountable Elevated Flare

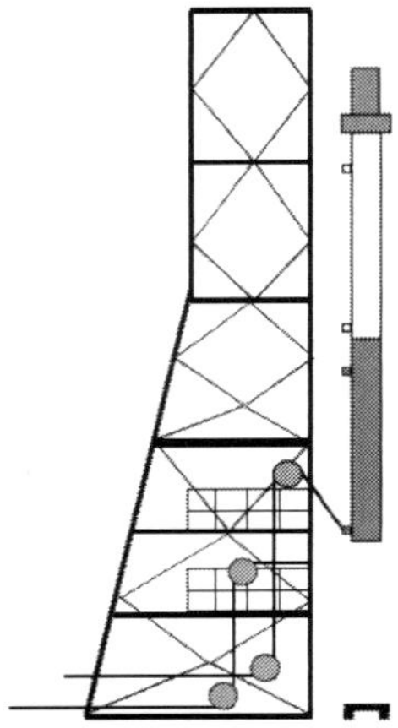

General Sequence Of Events – Demounting

1. Lowest Riser section secured.
2. Remaining sections secured at base of lowest remaining riser and unflanged from lowest section.
3. Upper section raised via winches and pulley for clearance.
4. Lowest section rotated from vertical to horizontal at tilting table for placement on transfer table.
5. Lowest section moved off transfer table, remaining sections lowered to tilting table.
6. Cables detached and attached in appropriate locations for subsequent iterations of riser relocation from stack to transfer table.
7. Repeat steps 1-6 until flare tip is moved to transfer table.

Demountable Elevated Flare

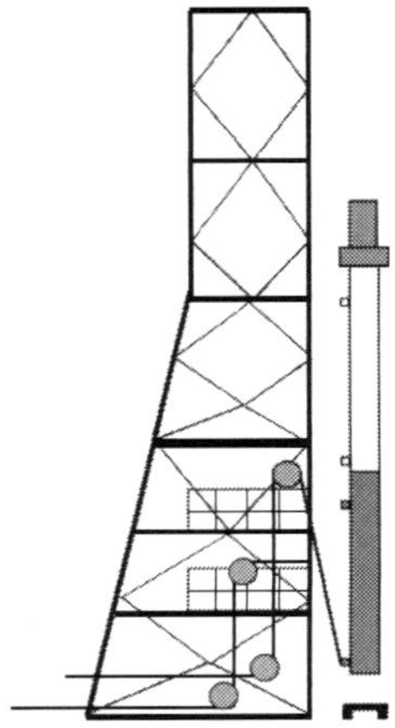

General Sequence Of Events – Demounting

1. Lowest Riser section secured.
2. Remaining sections secured at base of lowest remaining riser and unflanged from lowest section.
3. Upper section raised via winches and pulley for clearance.
4. Lowest section rotated from vertical to horizontal at tilting table for placement on transfer table.
5. Lowest section moved off transfer table, remaining sections lowered to tilting table.
6. Cables detached and attached in appropriate locations for subsequent iterations of riser relocation from stack to transfer table.
7. Repeat steps 1-6 until flare tip is moved to transfer table.

Demountable Elevated Flare

General Sequence Of Events – Demounting

1. Lowest Riser section secured.
2. Remaining sections secured at base of lowest remaining riser and unflanged from lowest section.
3. Upper section raised via winches and pulley for clearance.
4. Lowest section rotated from vertical to horizontal at tilting table for placement on transfer table.
5. Lowest section moved off transfer table, remaining sections lowered to tilting table.
6. Cables detached and attached in appropriate locations for subsequent iterations of riser relocation from stack to transfer table.
7. Repeat steps 1-6 until flare tip is moved to transfer table.

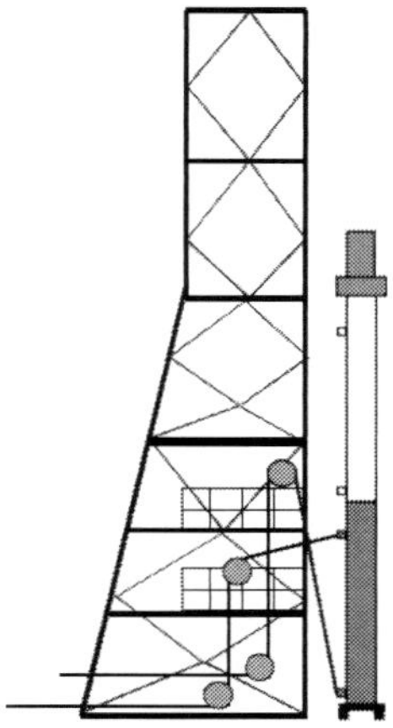

Demountable Elevated Flare

General Sequence Of Events – Demounting

1. Lowest Riser section secured.
2. Remaining sections secured at base of lowest remaining riser and unflanged from lowest section.
3. Upper section raised via winches and pulley for clearance.
4. Lowest section rotated from vertical to horizontal at tilting table for placement on transfer table.
5. Lowest section moved off transfer table, remaining sections lowered to tilting table.
6. Cables detached and attached in appropriate locations for subsequent iterations of riser relocation from stack to transfer table.
7. Repeat steps 1-6 until flare tip is moved to transfer table.

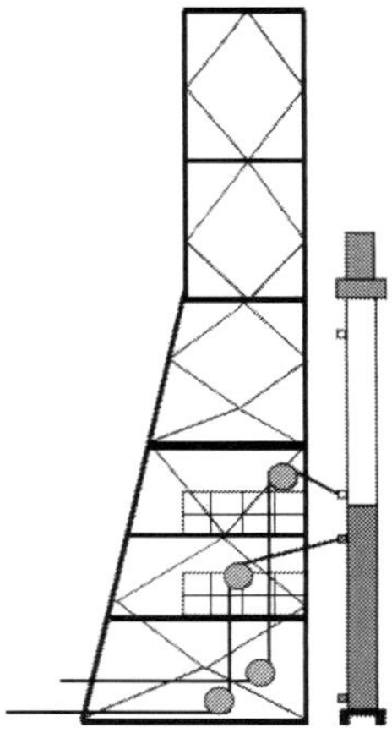

Demountable Elevated Flare

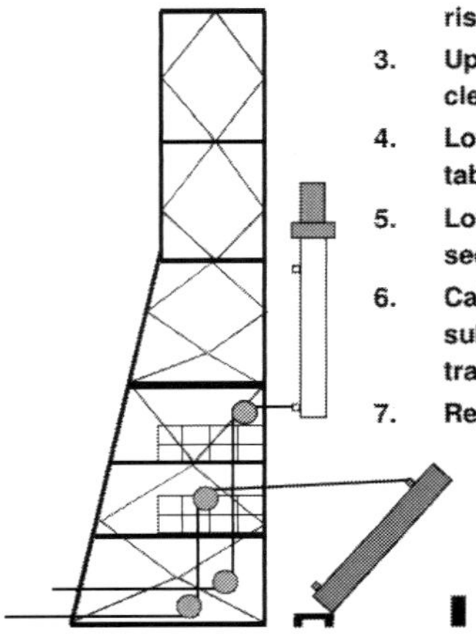

General Sequence Of Events – Demounting

1. Lowest Riser section secured.
2. Remaining sections secured at base of lowest remaining riser and unflanged from lowest section.
3. Upper section raised via winches and pulley for clearance.
4. Lowest section rotated from vertical to horizontal at tilting table for placement on transfer table.
5. Lowest section moved off transfer table, remaining sections lowered to tilting table.
6. Cables detached and attached in appropriate locations for subsequent iterations of riser relocation from stack to transfer table.
7. Repeat steps 1-6 until flare tip is moved to transfer table.

Demountable Elevated Flare

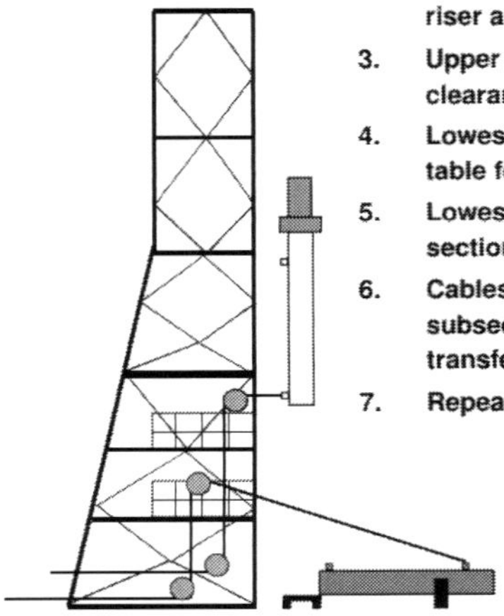

General Sequence Of Events – Demounting

1. Lowest Riser section secured.
2. Remaining sections secured at base of lowest remaining riser and unflanged from lowest section.
3. Upper section raised via winches and pulley for clearance.
4. Lowest section rotated from vertical to horizontal at tilting table for placement on transfer table.
5. Lowest section moved off transfer table, remaining sections lowered to tilting table.
6. Cables detached and attached in appropriate locations for subsequent iterations of riser relocation from stack to transfer table.
7. Repeat steps 1-6 until flare tip is moved to transfer table.

Demountable Elevated Flare

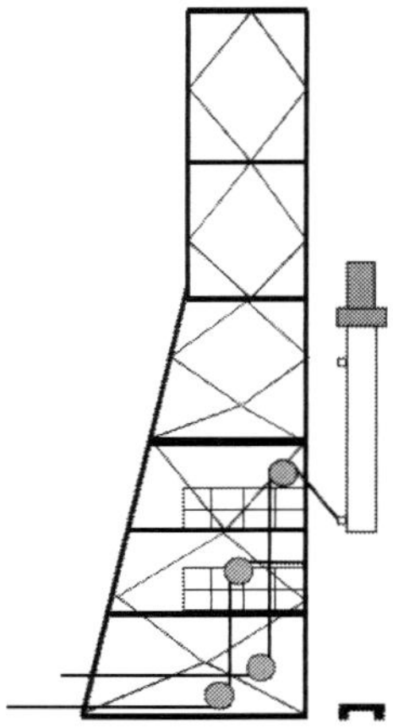

General Sequence Of Events – Demounting

1. Lowest Riser section secured.
2. Remaining sections secured at base of lowest remaining riser and unflanged from lowest section.
3. Upper section raised via winches and pulley for clearance.
4. Lowest section rotated from vertical to horizontal at tilting table for placement on transfer table.
5. Lowest section moved off transfer table, remaining sections lowered to tilting table.
6. Cables detached and attached in appropriate locations for subsequent iterations of riser relocation from stack to transfer table.
7. Repeat steps 1-6 until flare tip is moved to transfer table.

Demountable Elevated Flare

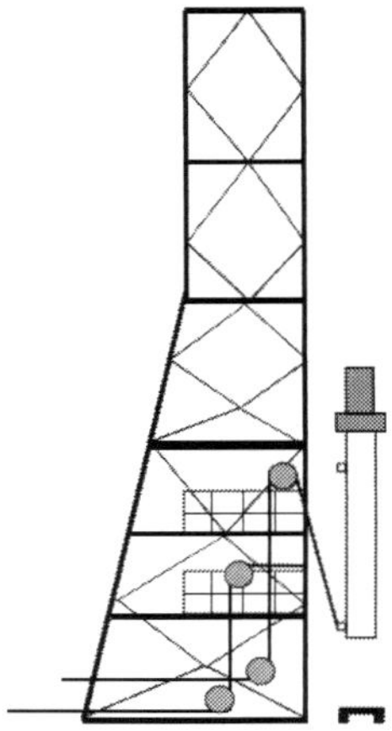

General Sequence Of Events – Demounting

1. Lowest Riser section secured.
2. Remaining sections secured at base of lowest remaining riser and unflanged from lowest section.
3. Upper section raised via winches and pulley for clearance.
4. Lowest section rotated from vertical to horizontal at tilting table for placement on transfer table.
5. Lowest section moved off transfer table, remaining sections lowered to tilting table.
6. Cables detached and attached in appropriate locations for subsequent iterations of riser relocation from stack to transfer table.
7. Repeat steps 1-6 until flare tip is moved to transfer table.

Demountable Elevated Flare

General Sequence Of Events – Demounting

1. Lowest Riser section secured.
2. Remaining sections secured at base of lowest remaining riser and unflanged from lowest section.
3. Upper section raised via winches and pulley for clearance.
4. Lowest section rotated from vertical to horizontal at tilting table for placement on transfer table.
5. Lowest section moved off transfer table, remaining sections lowered to tilting table.
6. Cables detached and attached in appropriate locations for subsequent iterations of riser relocation from stack to transfer table.
7. Repeat steps 1-6 until flare tip is moved to transfer table.

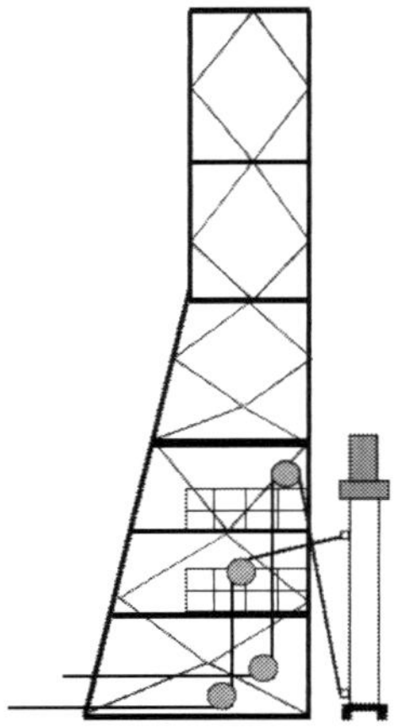

Demountable Elevated Flare

General Sequence Of Events – Demounting

1. Lowest Riser section secured.
2. Remaining sections secured at base of lowest remaining riser and unflanged from lowest section.
3. Upper section raised via winches and pulley for clearance.
4. Lowest section rotated from vertical to horizontal at tilting table for placement on transfer table.
5. Lowest section moved off transfer table, remaining sections lowered to tilting table.
6. Cables detached and attached in appropriate locations for subsequent iterations of riser relocation from stack to transfer table.
7. Repeat steps 1-6 until flare tip is moved to transfer table.

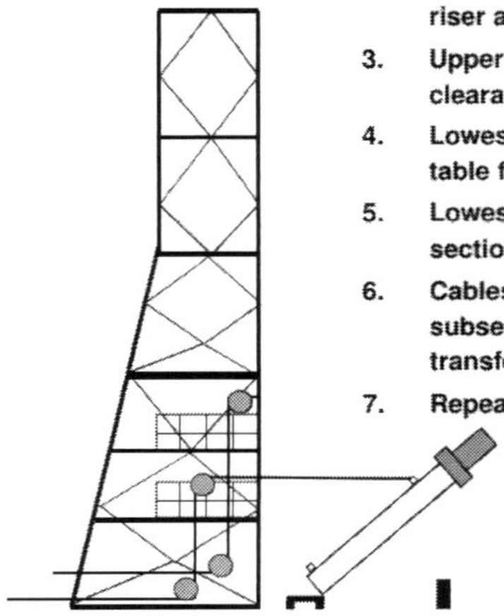

Demountable Elevated Flare

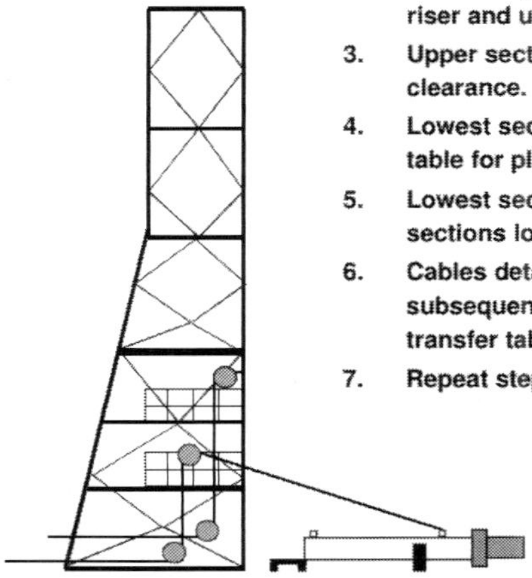

General Sequence Of Events – Demounting

1. Lowest Riser section secured.
2. Remaining sections secured at base of lowest remaining riser and unflanged from lowest section.
3. Upper section raised via winches and pulley for clearance.
4. Lowest section rotated from vertical to horizontal at tilting table for placement on transfer table.
5. Lowest section moved off transfer table, remaining sections lowered to tilting table.
6. Cables detached and attached in appropriate locations for subsequent iterations of riser relocation from stack to transfer table.
7. Repeat steps 1-6 until flare tip is moved to transfer table.

Demountable Flare

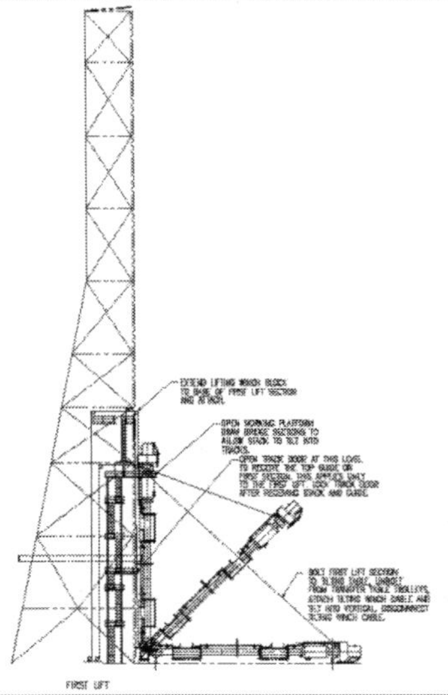

CALLIDUS TECHNOLOGIES by Honeywell

Callidus.com

Demountable Flare

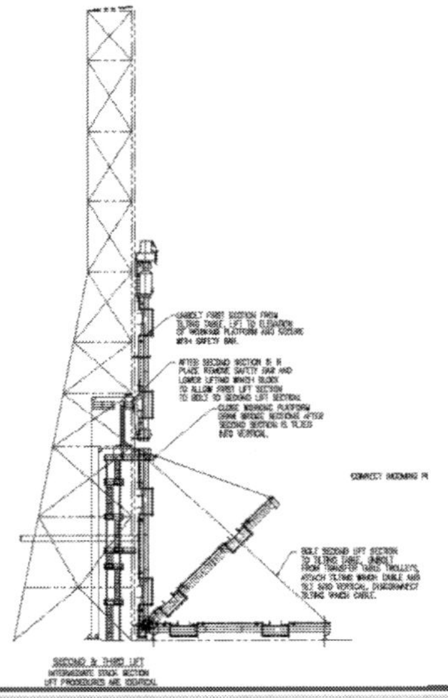

35

Flare Training-0609-001.ppt

Honeywell Proprietary

CALLIDUS TECHNOLOGIES by Honeywell

Callidus.com

Demountable Flare

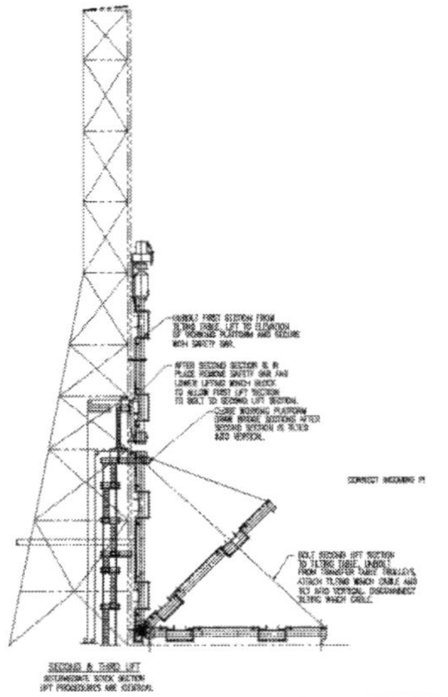

36

Flare Training-0609-001.ppt

Honeywell Proprietary

Flare Gas Recovery

Vapor Recovery

Objective: Control displaced hydrocarbon vapors for improved air quality through combustion/destruction and recovery systems.

Vapor Recovery Typical Applications

- Refining
- Petrochemical
- Storage Tanks

Why Do We Have Flare Gas Recovery?

- Reduced facility emissions
- Improved public relations
- Reduced purchase gas requirements and/or increased salable
- Zero impact on safety relief system
- Extended flare tip life
- Reduced steam consumption

Callidus Technologies by Honeywell

Callidus.com

FGRU Schematic

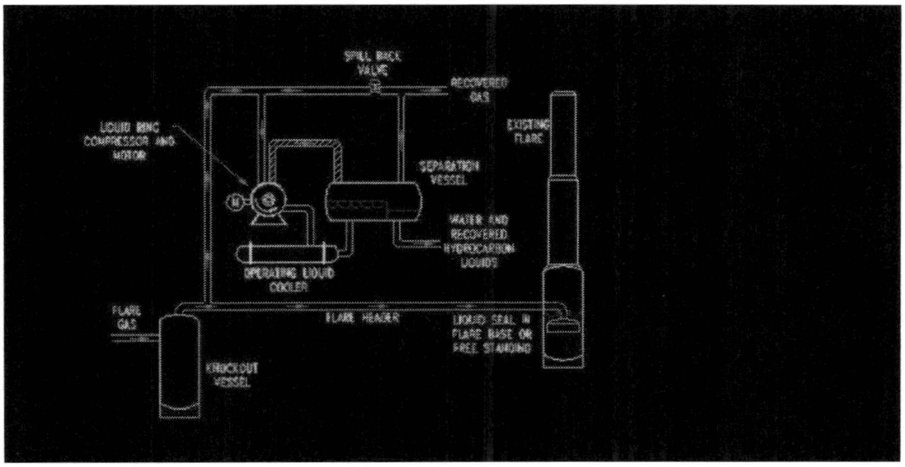

41
Flare Training-0809-001.ppt

Honeywell Proprietary

Callidus Technologies by Honeywell

Callidus.com

Start-Up of FGRS

42
Flare Training-0809-001.ppt

Honeywell Proprietary

CALLIDUS TECHNOLOGIES by Honeywell

Callidus.com

Project Results

- Reduced Emissions
  - Up to 32.5 tons/yr of NOx
  - Up to 176.8 tons/yr of CO
  - Up to 67,000 tons/hr of CO_2
- Smokeless capacity of 1,000,000 lb/hr
- Reduced noise
- Reduced light
- Payback of FGRS in nine (9) months

43
Flare Training-0609-001.ppt

Honeywell Proprietary

CALLIDUS TECHNOLOGIES by Honeywell

Callidus.com

TULSA WORLD Sunday, April 13, 2008 E3

FYI BUSINESS | NEWS, TRENDS AND EVENTS IN TULSA BUSINESS

Callidus builds steam flare tip for Petro-China

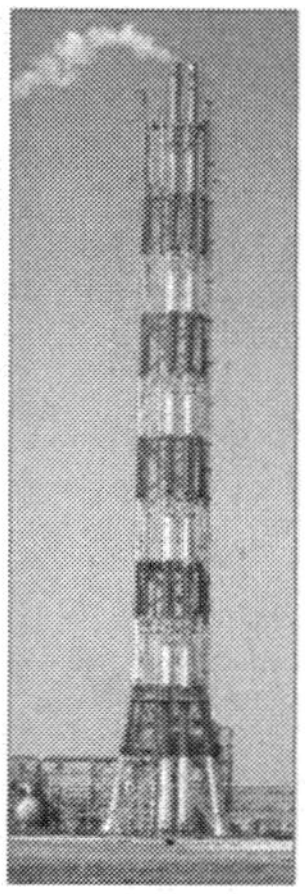

Callidus Technologies LLC is exporting an internal steam flare tip to China which is one of the largest tips ever made.

The flare tip is part of a complete flare system sold to Petro-China, and is one of the largest elevated flare systems in the world.

When the internal steam tip, which is 80 inches in diameter, is positioned on top of the system, the flare will stand 510 feet — just 138 feet shy of the height of the CityPlex Towers in Tulsa.

"Winning this order, . . . destined for the largest petrochemical facility in China, further confirms Callidus as a leader in combustion/environmental technologies worldwide," said William P. Bartlett, CEO and founder of Callidus.

Building the tip was a crew at Callidus' fabrication and manufacturing facility in Beggs. The company has been working on the flare tip — which includes the derrick, piping, stacks and tip — for 14 weeks.

The tip is now crated and will be trucked to a port in Houston, then shipped to a port on the northern coast of China.

At that time, the tip will be trucked to Dushanzi, which is in the mountains in northwestern China. The flare tip is but a portion of the entire facility, which has more than 60,000 workers.

According to Brian Duck, partner and vice president of the flare division, the flare system will have the combined energy generated at maximum continuous flow equivalent to heating approximately 500,000 average-size homes or 75 billion Btu per hour.

Callidus engineers designed the system, which they said will reduce the emissions to the lowest rate possible.

Petro-China is one of the largest energy companies in the world and is engaged in a broad range of activities related to petroleum, natural gas, exploration, production, refining, transportation and marketing of oil and gas.

Callidus has offices in China and India, and was given the 2007 Governor's Award for Excellence in Exporting.

Courtesy

Above, Callidus Technologies employees gather around the internal steam tip, which measures 80 inches in diameter, they manufactured at their Beggs facility. Left, an example of the tip the company built for Petro-China, which will stand 510 feet tall.

44
Flare Training-0609-001.ppt

Honeywell Proprietary

CALLIDUS TECHNOLOGIES by Honeywell

Tulsa World — THURSDAY, APRIL 26, 2007

Callidus.com

Callidus wins export award

▶ The company makes combustion and environmental systems for the refinery, petrochemical and chemical industries.

By Jason Womack
World Staff Writer

Tulsa-based Callidus Technologies LLC has received the 2007 Governor's Award for Excellence in Exporting in recognition of the company's growth in international markets.

The award was presented Wednesday to Bill Bartlett, chairman and CEO of Callidus, at the World Trade Conference in Oklahoma City.

"We're excited about winning the award and we're excited about winning it in an area that has played such a dynamic role in growing our company," Bartlett said in an interview.

The company manufactures combustion and environmental systems for the refinery, petrochemical and chemical industries. It has more than doubled in size over the last three years.

The outfit currently exports about 65 percent of its products to six different continents.

Bartlett said he would like to see that percentage continue to increase.

"We don't think we've topped out yet," he said. "We can see our overseas business being at 70 or 80 percent."

Due to growing international business, the size of the Callidus work force has increased to 320. And the business is approaching annual revenue of $100 million.

"It's impressive that an Oklahoma company which began in 1989 is already doing business in 24-plus countries," Gov. Brad Henry said in a prepared statement. "Callidus has truly become a world leader in its industry."

Bartlett

The company, which partnered with the Oklahoma Department of Commerce, began targeting foreign markets in 2001. The state assisted Callidus with making contacts in China, leading to the company establishing an office there.

Dennis Apostolova, director of international offices for the department, said Callidus has set the "gold standard" for globalizing its business model.

"They exemplify what it takes for an Oklahoma company to become globally savvy," she said. "And they are reaping the rewards."

Three other Tulsa-based companies — Qualtron Inc., Midwestern Manufacturing Co. and Melton Truck Lines Inc. — were also recognized during the awards ceremony. They received export achievement certificates from the U.S. Department of Commerce.

Jason Womack 581-8380
jason.womack@tulsaworld.com

Courtesy

A demountable flair unit manufactured by Tulsa-based Callidus Technologies LLC operates at a petrochemical plant in China.

45
Flare Training-0609-001.ppt — Honeywell Proprietary

CALLIDUS TECHNOLOGIES by Honeywell

Callidus.com

UNITED STATES DEPARTMENT OF COMMERCE
U.S. Commercial Service

EXPORT ACHIEVEMENT CERTIFICATE

Presented to

Callidus Technologies

for Recent Accomplishments
in the Global Marketplace

US COMMERCIAL SERVICE
United States of America
Department of Commerce

Assistant Secretary and Director General

46
Flare Training-0609-001.ppt — Honeywell Proprietary

Questions?

CALLIDUS
TECHNOLOGIES
by Honeywell

BAYER

Science For A Better Life

Modular MicroReactor Technology -

A Novel Approach for Process Intensification from Lab to Production

*Bharat Marwaha, Maria Krusche**

Bayer Technology Services, *Ehrfeld Mikrotechnik BTS

AIChE Process Technology Conference Galveston, TX

Track 2: Catalysis and Reaction Engineering

Galveston, TX [Oct 6-7, 2011]

Agenda/ Content

- Process Intensification
  - Definition
  - Drivers for Process Intensification
- Multi-Functional Reactors / Catalyst for Process Intensification.
- How can Microreaction Technology support Process Intensification
- Importance of a.) Scale-up Concepts b.) Modularity
- Ehrfeld Microstructured Mixers, Reactors & Heat Exchangers
  - **Overview & Scale-up Concept**
  - **Heterogenous Catalysis Microreactor Modules**
  - **Miprowa Reactors for Production of Fine & Specialty Chemicals**
- Application Examples for Process Intensification

Process Intensification Definition

AD 1556*

"Smaller..
Faster..
Cheaper..
Safer..
Cleaner"

AD 2002*

Process intensification consists of the development of novel apparatus and / or techniques that, are expected to bring _dramatic_ improvements in manufacturing and processing:

- decrease equipment-size for a production-capacity.
- increase production capacity for an equipment size.
- decrease energy consumption
- decrease waste production

} *sustainable*

* *A. I. Stankiewicz et al, 2000, Chemical Engineering Progress*

page 3

Drivers for Process Intensification

Drivers

- Increasing importance of specialty chemicals.
- Rapidly changing market demand due to globalisation.
- Strive for more sustainable [eco-friendly] processes.
- Manage Hazardous chemistries more safely.

Impacts

- Distributed Scale or on-demand business model.
- Faster time-to-market [scale-up] with reduced CAPEX and OPEX.
- Push for Batch to Continuous / Flow Chemistry.
- Reduce resource consumption, waste

page 4

Multi-Functional Reactors / Catalyst for Process Intensification

- Reactor is the core of a Process
  - Large Influence on Downstream / Separation requirements.

MultiFunctional Reactors → Combine "separation" unit operation [ex. membrane reactors] / Improve Yield, Selectivity by Heat Integration and Catalyst Improvement → Intensify Process by Reducing Downstream footprint

page 5

How can Microreactor Technology Support Process Intensification

	$A/V = m^2/m^3$
Microreactor:	2,000 – 20,000
Millireactor:	1,000 – 2,000
Tubular reactor:	100 - 500
Stirred vessel:	4 - 40

Micro structured channels + Continuous flow

- Rapid mixing (without back-mixing)
- Rapid heat exchange (no "hot spots")
- Defined residence time with narrow distribution
- Increased safety due to low hold-up

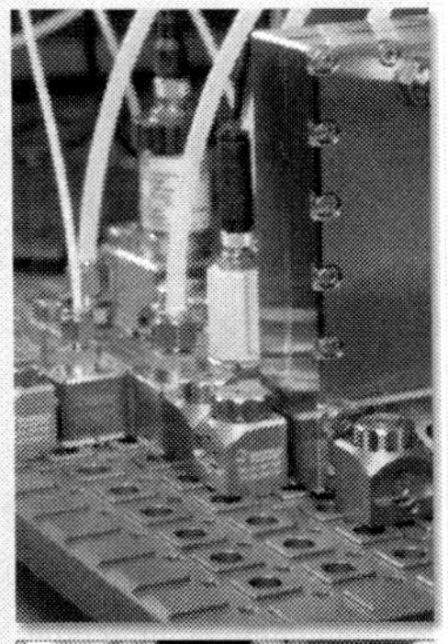

page 6

How can Microreactor Technology Support Process Intensification

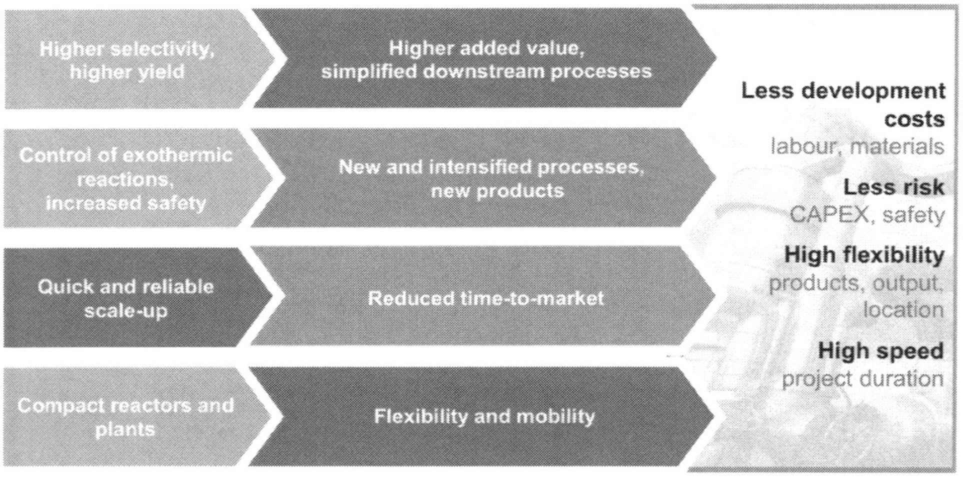

Importance of a.)Scale-up Concepts and b.) Modularity

a.) Scale-up Concept:

- Fast and reliable scale-up critical to quickly meet market demand.
- Example numbering-up of micro-channels.

b.) Modularity:

- Enables to quickly adapt process to different products.
- Enables to quickly adapt pilot plant equipments to various processes.
- Example multi-purpose microchannel modular unit.

Ehrfeld Modular MicroReaction Technology Overview

Modular Micro - Reaction System for R&D	• For development of materials & synthesis routes • For process development & intensification • Small scale production
Equipment for Production of Chemicals	• Scalable to high throughput • For chemical production & multi-purpose applications
Equipment for Production of Pharmaceuticals	• For preclinical, clinical and production purpose • Easy to scale-up and to handle
Support & Training	• Maintenance & after-sales-support • Installation support & application training
Service	• Proof of principle studies • Recommendation of plant configuration

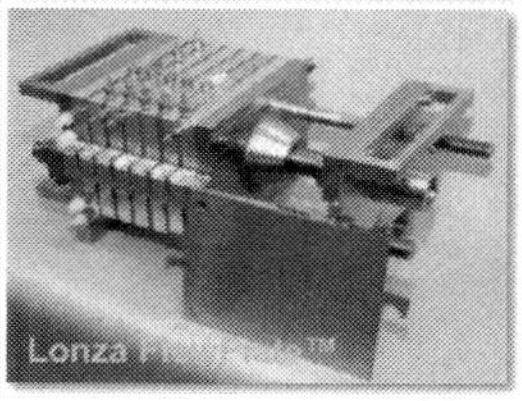

Page 9

Fast and Flexible Assembly of Microchannel Modules

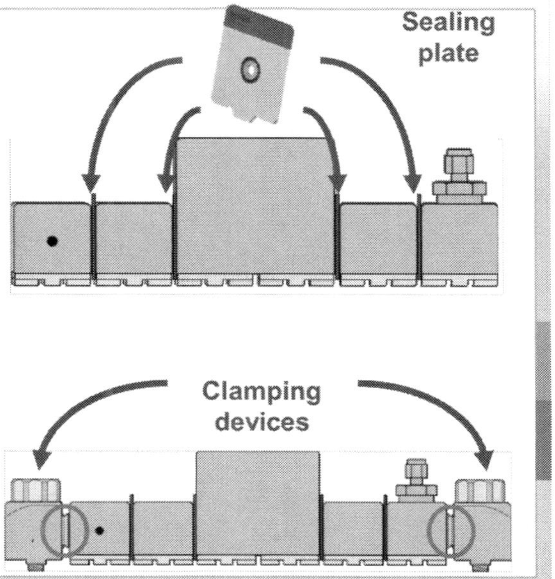

page 10

Modular MicroReaction System Maximized Flexibility

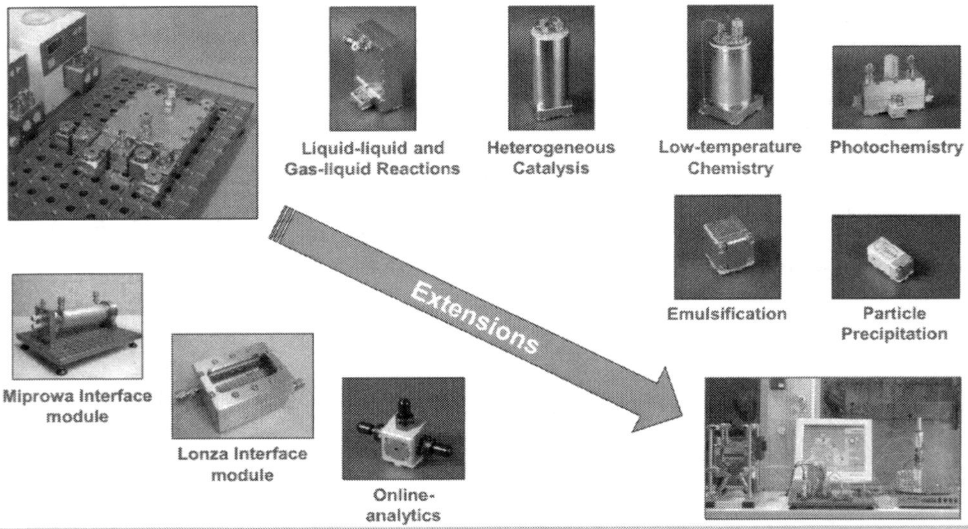

Heterogenous Microchannel Catalysis Modules

Module	Max. Temp [C]	Max. Press [bar]	Inner channel diameter [mm]	Volume for Catalyst [mL]	Recommended catalyst size [μm]
Fixed bed Meander Reactor	200	35	4	24.5	150 - 500
Cartridge Reactor XL	240	100	10	5	50 - 1000
Cartridge Reactor	200	30	8	0.7	50 - 500
Cartridge Reactor [High Temp]	600	100	4.5	0.8	50 - 500
Gas-Phase Reactor [washcoat]	200	20	0.5		

Fixed-bed Meander Reactor

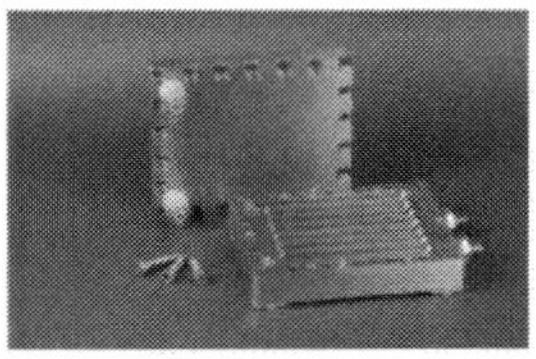

- Fluidic cooling / heating
- Dismountable and cleanable
- Integrated Temperature sensors [2], process inlet and outlet

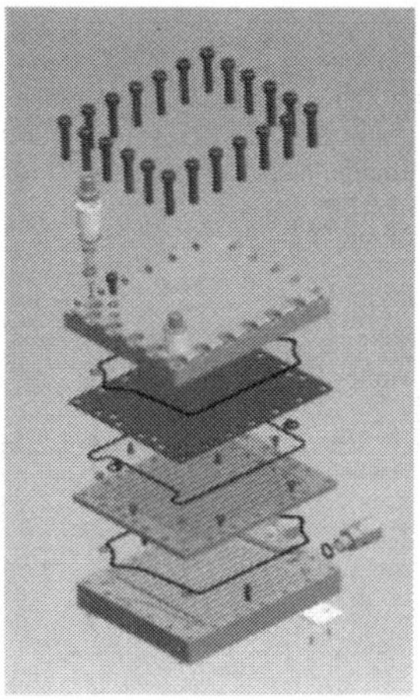

page 13

Cartridge Reactor XL

MMRS base plate compatible

- Process volume*: 5 ml
- Dismountable and cleanable
- 2 x 100 W heating power (or 2 x 200 W)
- Integrated pre-heater
- Integrated temperature sensors
- Working range (others on request)
    - T: -20 … 240 °C
    - P: up to 100 bar

Easy remove of cartridge

page 14

Gas Phase Reactor [washcoat]

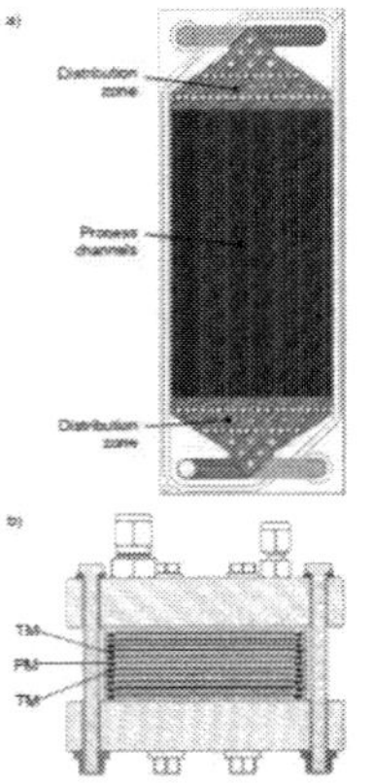

Fig. 1: Conduction of fluids within the gas phase reactor in a top and front cross view respectively (red/PM: process fluid; blue/TM: heat transfer fluid).

Parameters	
Allowable fluid temperature range	-20 ... 200 °C
Allowable ambient temperature range	-20 ... 200 °C
Max. allowable pressure (process medium)	20 bar
Max. allowable pressure (heat transfer medium)	20 bar
(Pressure values are valid for 200 °C)	
Fluid volume (process medium), total	22,4 mL
Fluid volume (process layer)	4,2 mL/PL*
Fluid volume (process channel zone)	2,4 mL/PL*
Channel dimensions (L x H x W)	130 x 0,5 x 0,5 mm³
Maximum diameter at narrowest position	500 µm
Number of process channel layers	1 ... 5
Number of process channels per layer	93
Recommended for viscosities up to	100 mPas
Recommended minimum coolant flow	1,5 L/min/CL*
Possibility to sterilize	yes
(* "PL" = process layer; "CL" = cooling layer)	

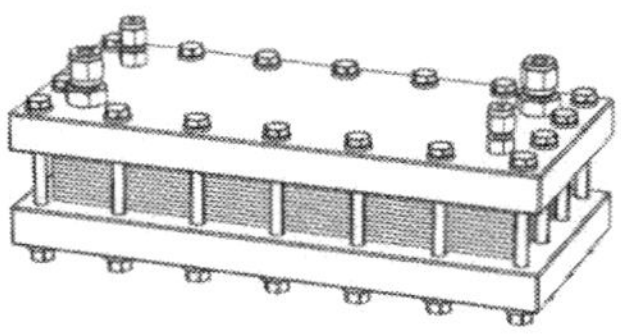

page 15

Modular MicroReaction System Speeds up your R&D

Modular MicroReaction System

Page 16

Scale-up Strategy

R&D and kilo lab
0.05 – 10 L/h

Pilot scale
10 L/h – 100 L/h

Production Scale
40 L/h – 10,000 L/h

Page 17

Miprowa® Reactors
High Performance Technology

- Rectangular channels (18 x 3 mm) combined like tube bundle
- Exchangeable mixing inserts generate microstructure
- Pressure drop and mixing efficiency adaptable
- Fluidic heat transfer

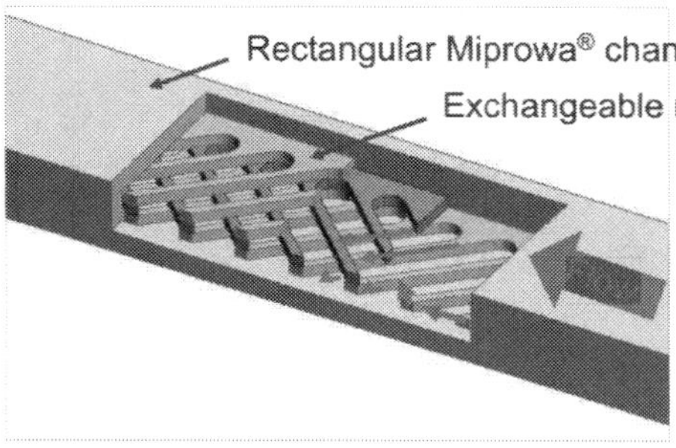

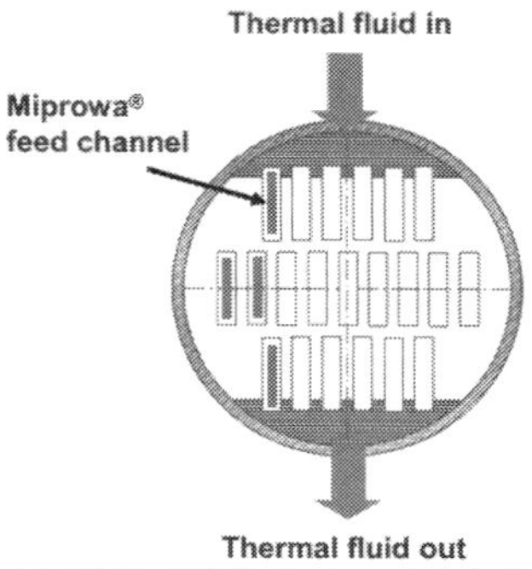

Page 18

EMB : Miprowa Scale-up concept

Lab scale:
No. of channels: 1-8
Channel size: 12 x 1.5 mm
Throughput: 3-10 L/h

Pilot scale:
No. of channels: 3-100
Channel size:
12 x 1.5 mm & 18 x 3 mm
Throughput: 10-300 L/h

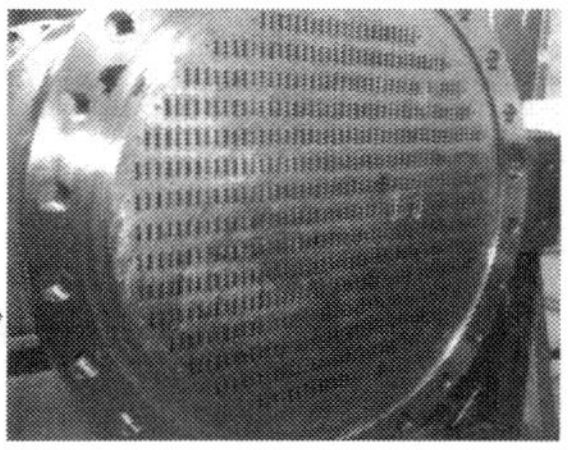

Production scale:
No. of channels: 10-1000
Channel size: 18 x 3...120 x 3 mm
Throughput: 300-10,000 L/h

Peroxide Synthesis via Direct Oxidation

[Distributed scale model; Manage Hazardous chemistry]

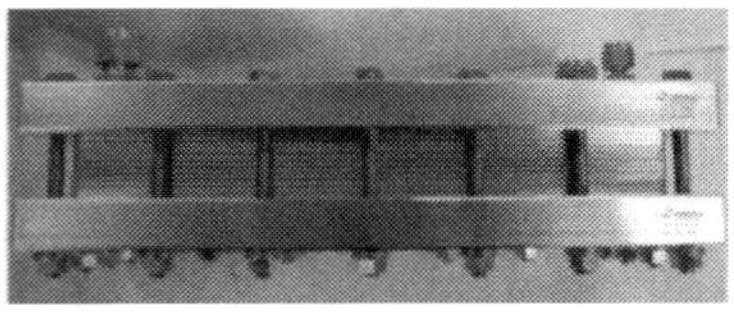

10-plate micro-channel reactor from BTS for peroxide synthesis

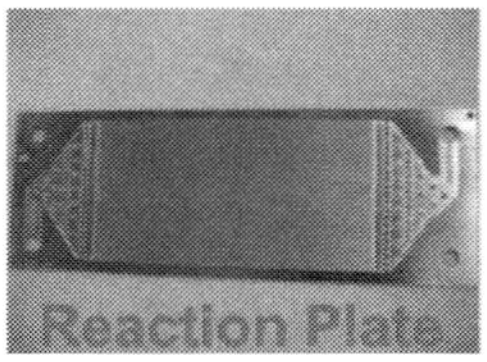

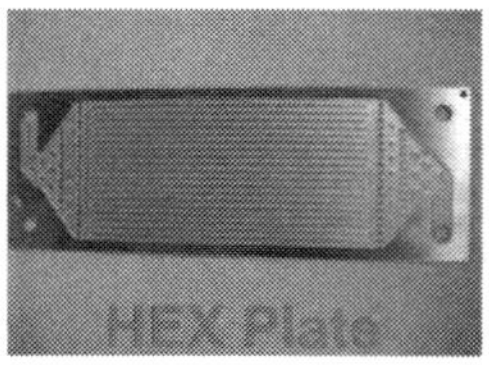

- TDA is developing a compact portable system that will generate liquid H2O2 on-site
- The system uses a very active, highly selective long life catalyst for direct oxidation of H2 and O2 to H2O2
- TDA's system requires only water, air and electricity
- H2 mixes with the O2 stream in the microchannels in contact with TDA's direct oxidation catalyst to produce H2O2
- H2O flows in adjacent channels to the reaction zones to remove the heat generated in the exothermic oxidation reaction (ΔH = -32.4 kcal/mol of H2O2 produced)

Peroxide Synthesis: Catalyst Preparation in Microchannels

Empty Channels

Washcoated Channels

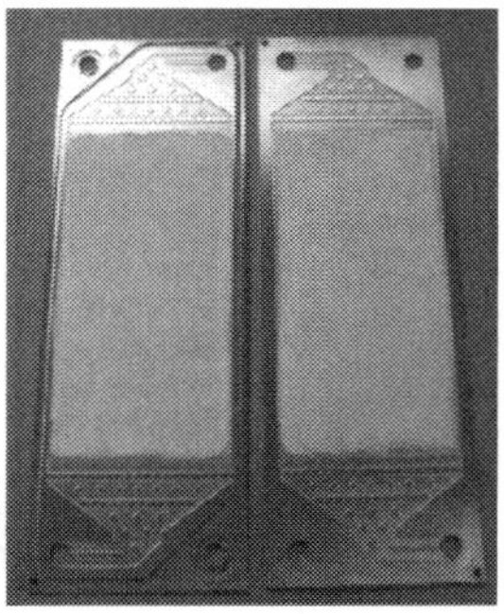

Catalyst coated Channels

- **BTS microchannel plates were washcoated with a catalyst support**
- **TDA then introduced the active catalyst phase on to the washcoat following their proprietary technique**
- **These catalyst loaded microchannel modules are being tested at TDA for direct peroxide synthesis**

page 21

Pilot Plant for Selective Hydrogenation

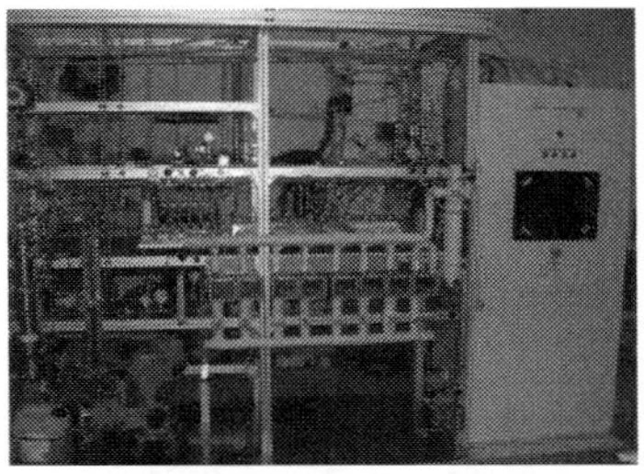

xytel
CONCEPT TO COMMERCIAL

EHRFELD
Mikrotechnik BTS
A Bayer Technology Services Company

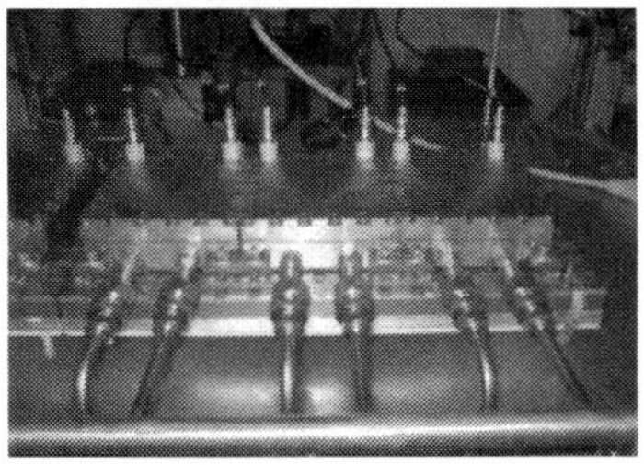

page 22

Solid-Catalyzed Liquid Phase Synthesis [Phenol-Derivative]

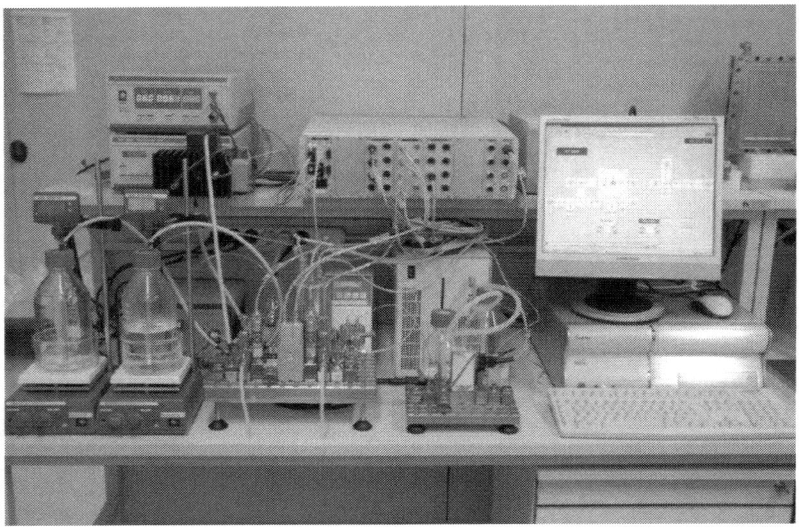

page 23

Low Temperature Organometallic Chemistry

[Batch-to-Continous; Improve Hazardous chemistry]

Low-Temperature Organometallic Chemistry

Reactand —(1. Lithiation; 2. Electrophile)→ Product

- Throughput: 20 kg/h
- Continuous production of 100 kg isolated product

Benefits:

- Increased selectivity from 86 % to 94 %
- Fastened reaction rate by temperature increase: -70 °C to -40 °C
- Increased safety: hold-up reduced from 400 L to 2 L

page 24

Continuous Metal Halogen Exchange Reaction [Batch – Continuous]

Bing-Shiou Yang, Matthew Yakush, Jon Lorenz and Daniel Fandrick
Boehringer Ingelheim Pharmaceutical Inc.

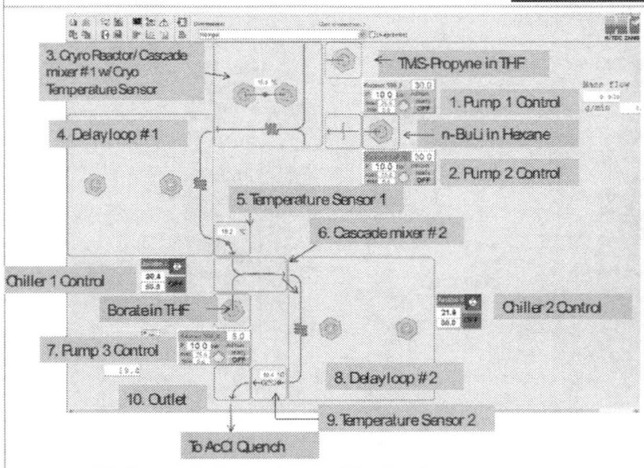

- Challenges in the scale-up of the Batch process:
  - Variable yield with different addition time (impurity formation with extended process time)
  - Decreased cooling efficiency with increasing batch size
- Continuous process was able to **overcome the scale up challenges** observed in the batch reaction

Simplify Downstream Process using Ehrfeld Microreactor

[Safer, Cleaner, Cheaper]

- Potential Benefits
  - Yield
    - **Reduced generation of by-products**
      → shorter residence time
      → increased reaction rate
  - Downstream processing
    - **Lowered/avoided use of solvent**
  - Safety concept & plant layout
    - **Smaller holdup**
    - **Reduced backmixing (vs. tank)**
- Key Requirements
  - **Efficient heat removal during reaction by microstructured devices**

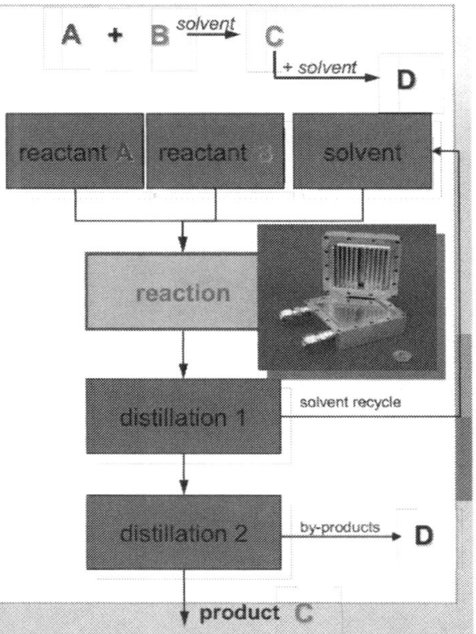

Simplify Downstream Process using Ehrfeld Microreactor

[Safer, Cleaner, Cheaper]

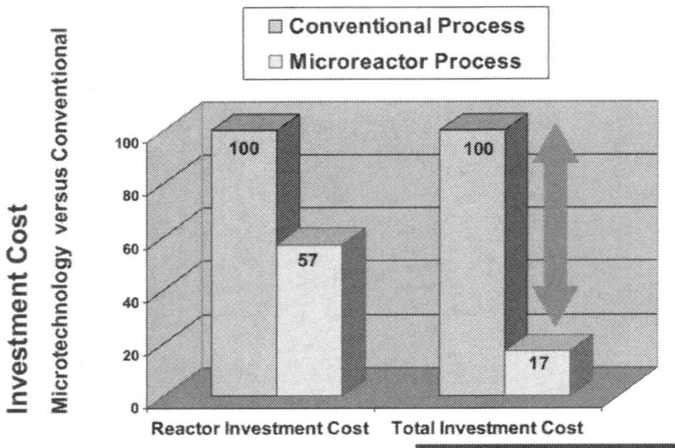

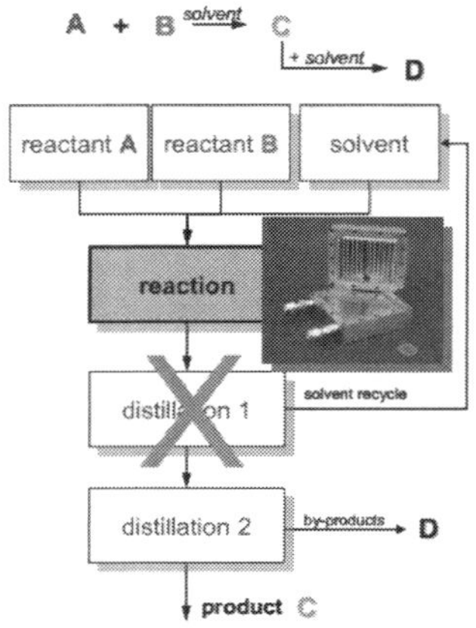

Microtechnology enables a complete redesign of the process to optimize economics and increase process safety!

page 27

Nanoparticle production using microreactor

[Batch-to-Continous; Safer, faster, cleaner]

Valve assisted mixer, scale-up options:

- different slit widths (mixing plates)
- different number of slits (mixing plates)

- Total product amount: about 50 m^3 suspension (~ 1.5 t solid material)
- 2 modules in alternate operation: change over when pressure is exeeded, followed by flushing the blocked module
- 24 h operation, 3 months

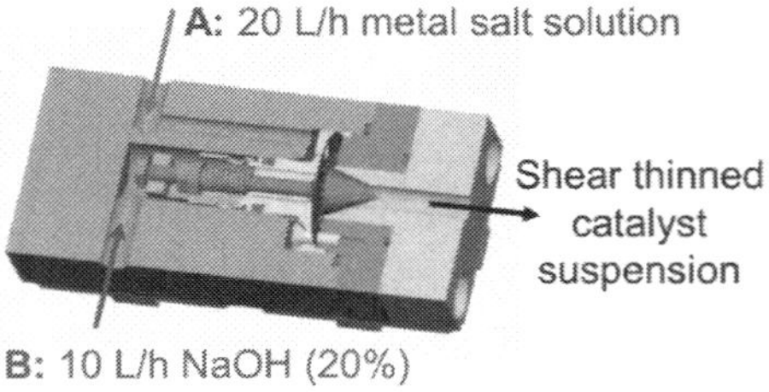

Result

Safe, fast and reliable production of high-quality nano-scale particles; mixer easy to maintain and clean

page 28

Nanocatalyst production using microreactor

[Batch-to-Continous; improved catalyst]

Valve assisted mixer, scale-up options:

- different slit widths (mixing plates)
- different number of slits (mixing plates)

Result

Replacement of the conventional batch process by innovative synthesis routes for catalyst particles; the improvement of catalyst quality using micromixers is relevant for industrial applications

- Synthesis of $Cu/ZnO/Al_2O_3$ catalysts
- Synthesis of highly cristalline Hydroxycarbonate precursor predominantly with the highly active Malachite and Rosasite phases
- Result: catalysts with high specific Cu surface and high activity in MeOH synthesis

Source: 44. Jahrestreffen Dt. Katalytiker u. Jahrestreffen Reaktionstechnik, Weimar, 17.03.2011, Dr. Georg Simson, Kontinuierliche Fällung von Cu/ZnO/Al2O3-Katalysatorsystemen in Mikroreaktoren, Catalysis Research Center und Department Chemie, TU München

Miprowa® Reactors Application Examples

Reactor with different temperature zones for process development

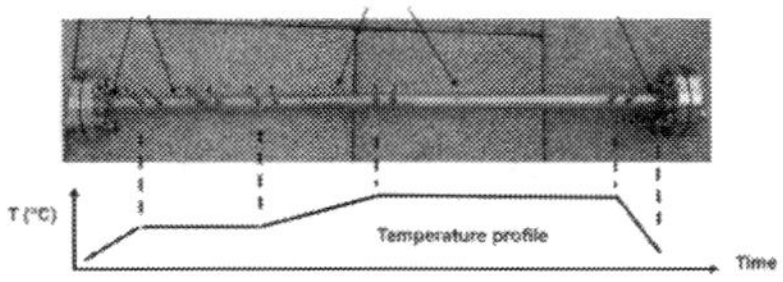

Reactor with pre-mixer for fast and highly exothermic gas-liquid reaction

Pilot-scale reactor for polymerisation

Heat exchanger for total evaporation of temperature-sensitive fluid

Summary

- Quick & Reliable Scale-up
- Simplified Process Intensification
  - New synthesis routes
  - Lower energy
  - Lower raw material usage
- Safety - less potential for hazard
- Easy product switches
- Flexibility & Mobility
- Feasibility Studies… Confirm Suitability

We had a small problem with the scale-up out of the laboratory....

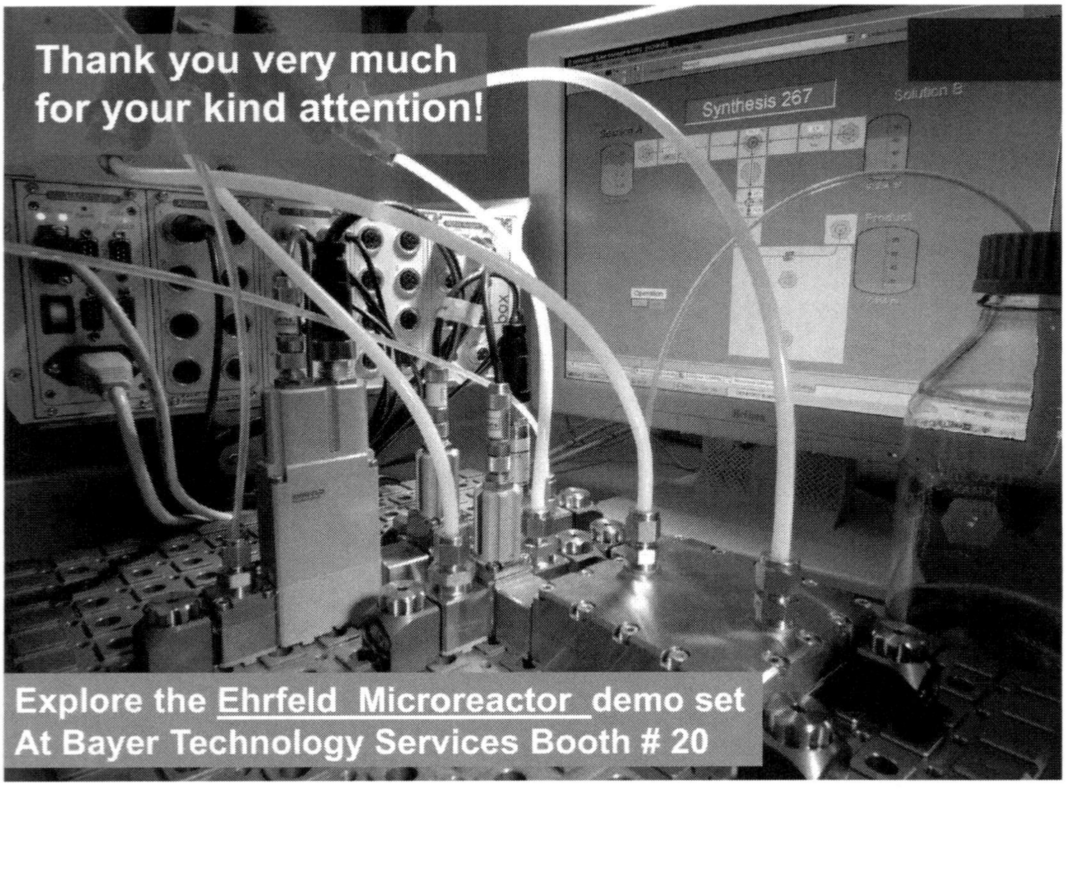

The State of the Working Catalyst on Alkali-Promoted Pt/Al2O3 and Pt/TiO2 for the Water Gas Shift Reaction

Pazmino, J., Purdue University
Wang, J., Purdue University
Shekhar, M., Purdue University
Williams, W. D., Purdue University
Miller, J. T., Argonne National Laboratory
Akatay, M. C., Purdue University
Stach, E. A., Purdue University
Delgass, W. N., Purdue University
Ribeiro, F. H., Purdue University
Originally presented on: 10/6/2011 0:25:00 - 0:50:00

The search for active and stable low-temperature catalysts for the production of hydrogen has been subject of investigation due to the increasing interest in commercializing portable proton exchange membranes fuel cells (PEM) for power generation. The water gas shift (WGS) reaction ($H_2O + CO \rightarrow H_2 + CO_2$) on noble metal (Pt, Au, Pd) catalysts is a suitable process for such technology as it offers greater stability during start-up and shut-down operations. However, these catalysts require developing formulations that can surpass the performance of the commercially available Cu-based catalysts.

Here, we report on the promotional effect of alkali additives (Na, Li, K) and we examined the nature of such promotion by combining kinetic analysis, x-ray absorption (XAS), transmission electron microscopy (TEM) and steady state isotopic transient methods (SSITKA). Among the alkali, Na showed the most exceptional enhancement of the turnover rate (TOR, 250 °C) up to 107 times that of Pt/Al_2O_3 and up to 4 times the TOR of Pt/P25. The alkali-promoted catalysts can therefore exhibit rates of reaction higher than the conventional $Cu/ZnO/Al_2O_3$ under fuel cell operating conditions. We found that the reaction kinetics for either support were identical upon the addition of Na, suggesting the creation of the same active sites. In situ XAS experiments under WGS conditions indicate that Pt remained metallic under WGS conditions in most of the catalysts. On samples where PtO was present no correlation was found between the rate and the fraction of Pt oxides. Transmission infrared experiments done in operando mode revealed important changes in the adsorption of CO on Pt, characterized by the appearance of low frequency bands. Additional bands were observed in the formate region, which we attributed to sodium formates. SSITKA measurements show that these formate species exchange 13C at the same rate as the product $13CO_2$, invoking a formate-type mechanism. Lastly, the estimation of the size of the "carbon pool" by SSITKA suggests that the role of Pt is to provide CO to other intermediates in the alkali-modified support where the rate determining step occurs. As the amount of alkali increases, the probability of Pt being surrounded by alkali increases and the size of the carbon pool largely exceeds the available surface area of Pt. Therefore, the promotion by alkali can be regarded simply as a support-driven effect.

Accurate Estimation of Si Accumulation in Coker Naphtha Hydrotreaters Using an Improved Direct Injection ICP Analysis and Hot Loop Feed Sampling

Thienan D. Tran, Patrick J. Gripka, & Lawrence (Larry) S. Kraus

CRITERION
CATALYSTS & TECHNOLOGIES
The Clean Fuels Company

2

Sources of Silicon (Si)

Coker Naphtha
Silicone oil used in the delayed coking process

SR Naphtha
Silicone oil used at the well

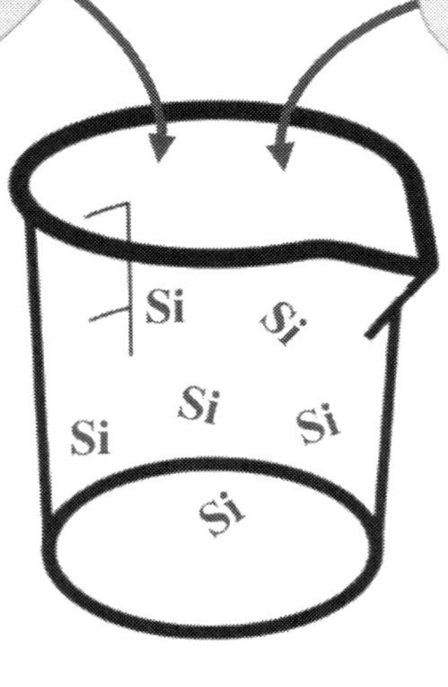

3

Siloxane

Polydimethylsiloxane (PDMS)

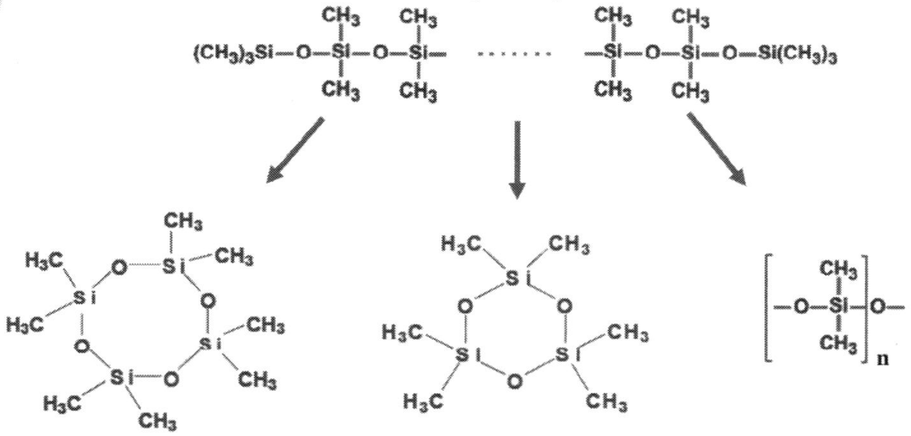

Octamethylcyclotetrasiloxane **Hexamethylcyclotrisiloxane** **Dimers and Trimers of Dimethylsiloxane**

4

Silicon Poisoning

- **The silica fragments in the feed quickly adsorb on the catalyst surface. No chemical bonds to the alumina or metal sulfides are formed.**
- **The deposition of Si is similar to the deposition of coke.**
- **Catalyst deactivation is caused by Si adsorption on catalyst surface covering the active sites and restricting catalyst pores.**
- **Si poisoning is irreversible.**
- **HDN activity is affected by Si deposition more than HDS activity.**
- **Nitrogen and silicon are poisons to reforming catalyst.**

5

Effect of Si Deposition on NHT Cycle Length

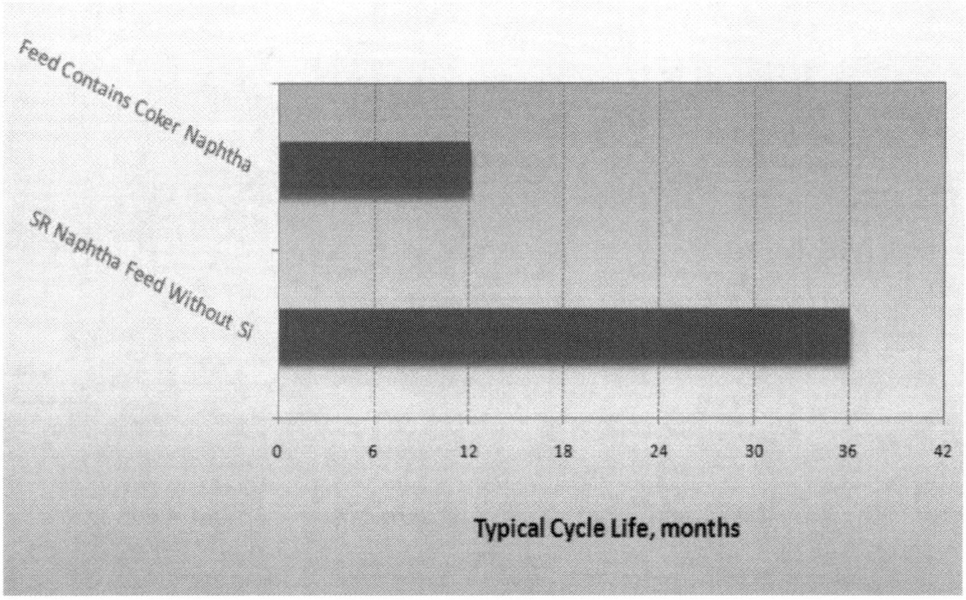

6

Cycle Length of NHT

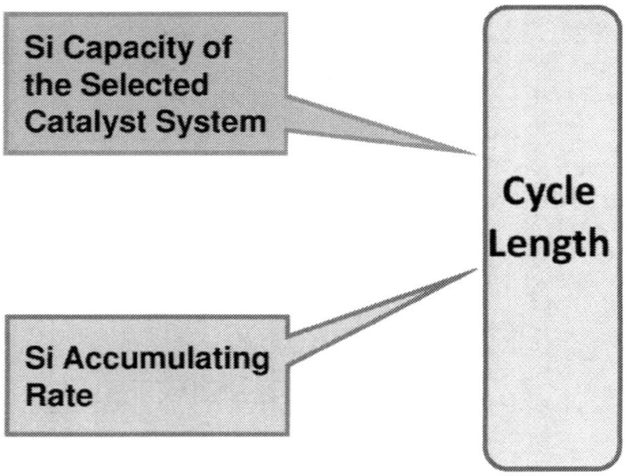

7

Cycle Length of NHT

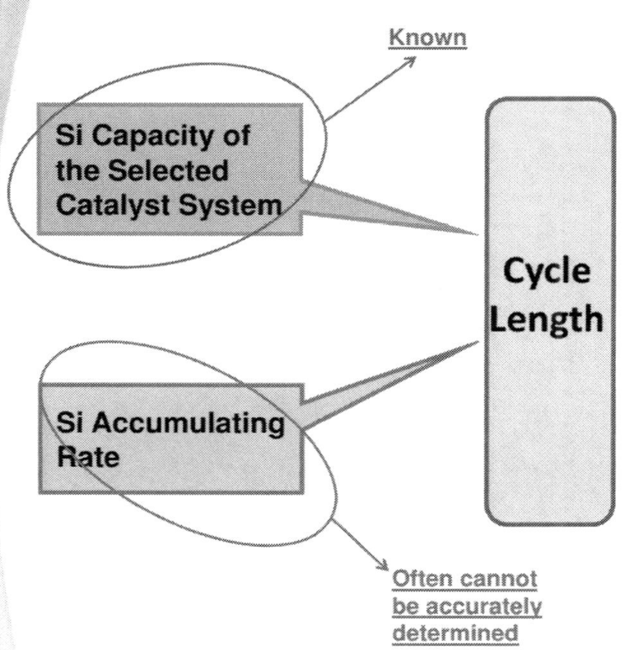

8

Problems Causing Inaccurate Prediction of Si Deposition Rate

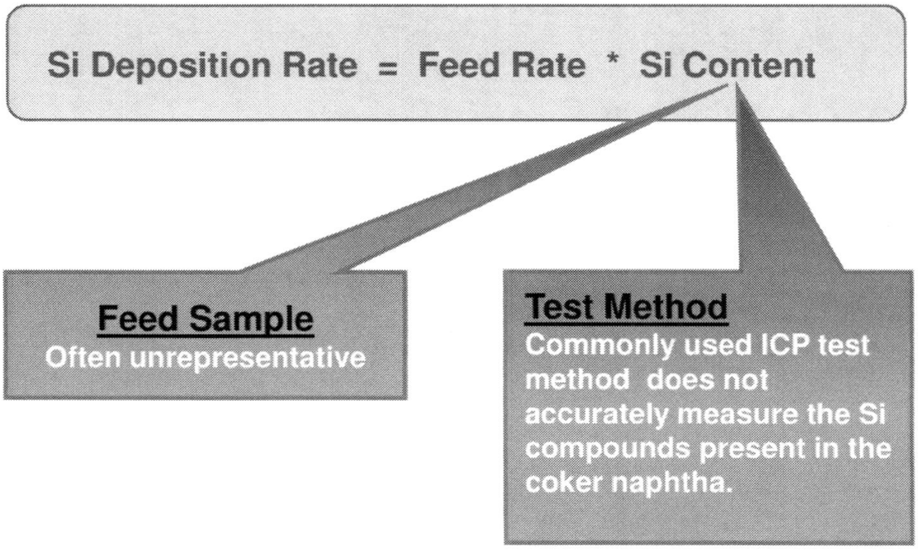

9

Source of Errors in Standard ICP Analysis

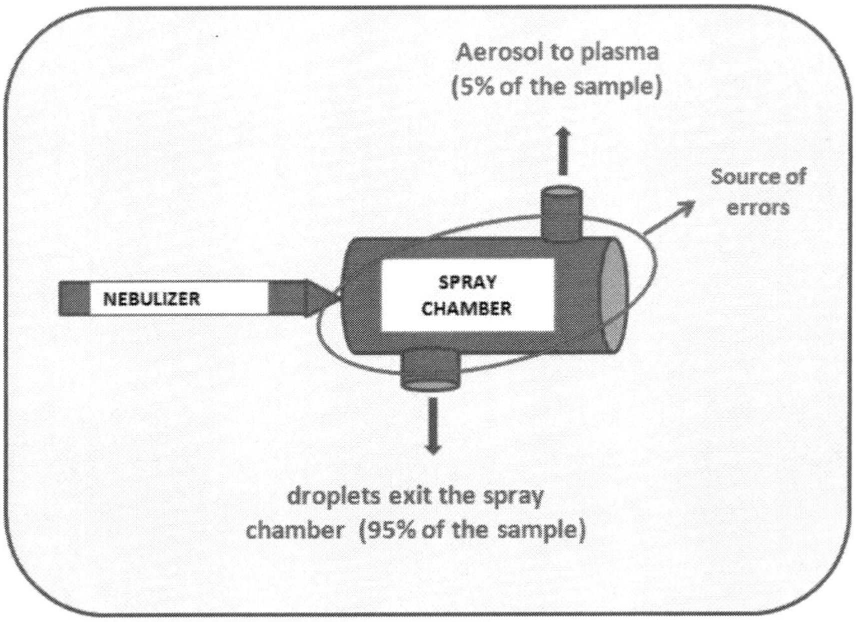

10

Cycle Life of NHT

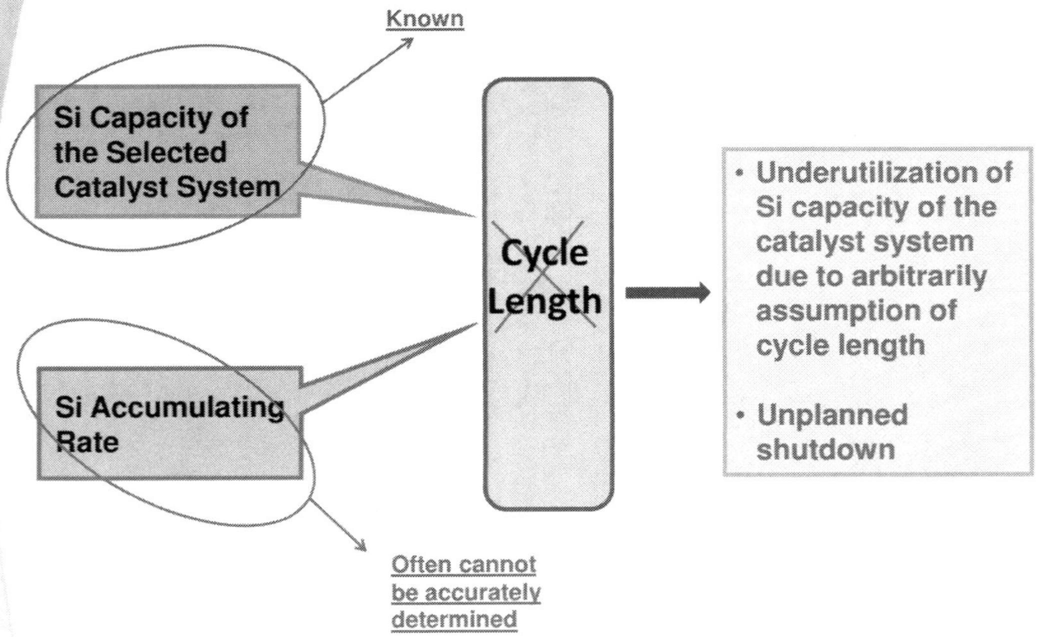

Case Study

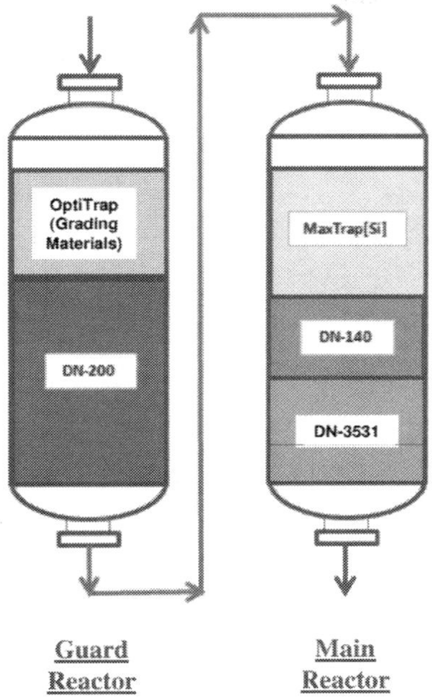

Unit Description

Feed	
Composition	18 vol% Coker Naphtha
	82 vol% SR Naphtha
Sulfur, wppm	580
Nitrogen, wppm	18
Operating Conditions	
WABT, °F	> 600
LHSV in the Main Reactors	4.1
Product Properties	
Sulfur, ppmw	0.2
Nitrogen, ppmw	< 0.2

Si breakthrough capacity	8,750 lbs
Scheduled run length	12 months

Action 1 – Installed a Hot Loop Composite Sampler

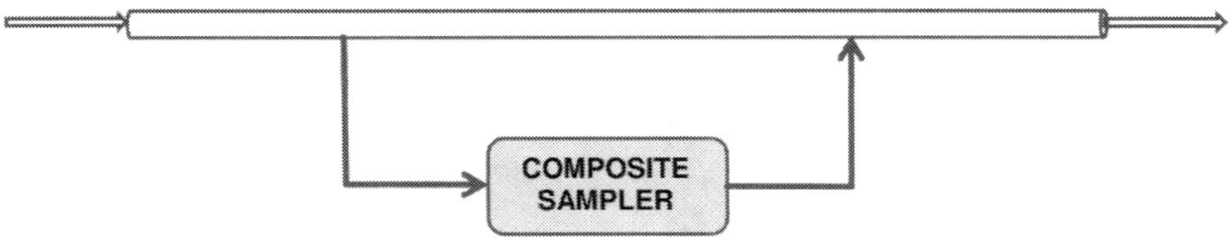

Hot loop feed sampling system

- The composite sampler is designed to collect a defined amount of feed at a fixed time interval.
- Composite sample was collected once/week for Si analysis.

Action 2 – Selected Shell Proprietary ICPDIN Analysis

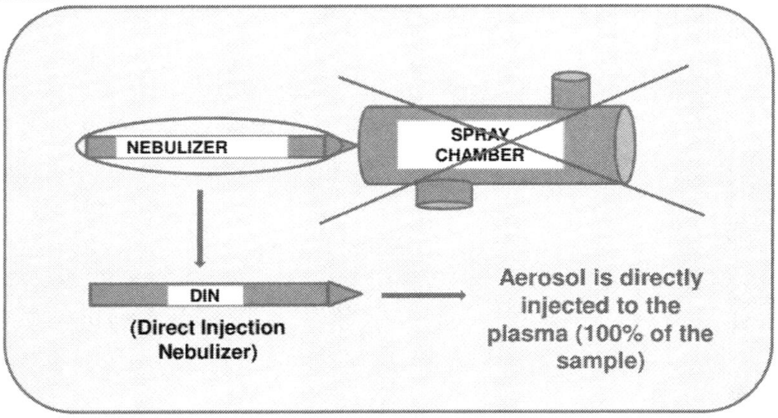

Shell proprietary Direct Injection ICP (ICPDIN) analysis

- ICPDIN analysis is designed to accurately measure the concentration of volatile analytes.
- A custom design direct injection nebulizer (DIN) is used to introduce the aerosol into the ICP.
- Spray chamber is eliminated
- Maximum margin of error is 10%

Weekly Average Si Content in Feed

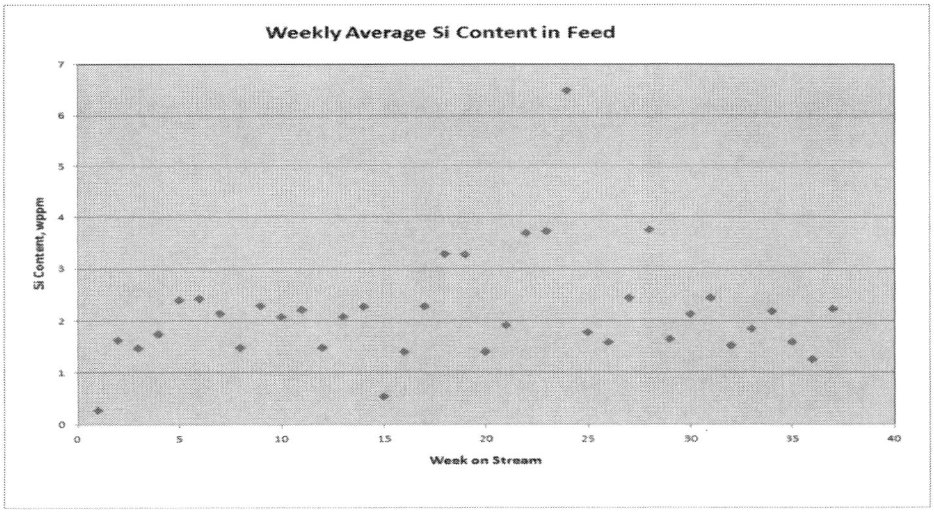

- Si content in the feed varied from 0.26 to 6.5 wppm.
- 62% of the data population were in the range of 1.5 to 2.5 wppm.
- At the time the unit was shutdown, the estimated Si accumulation was 4,770 pounds (55% of the Si capacity of the catalyst system).

Spent Catalyst Silicon Profile

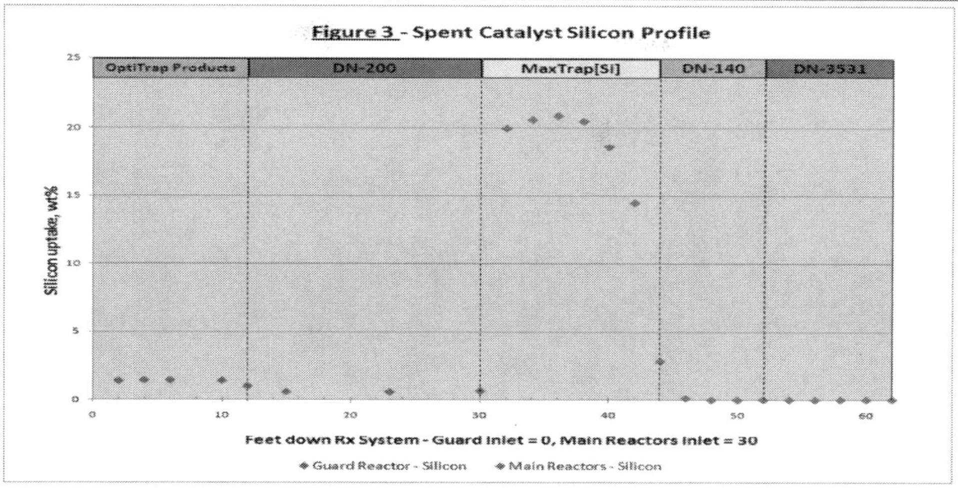

- Guard reactor picked up minimal amount of Si.
- MaxTrap[Si] picked up 20+% Si. Si had not slipped into the DN-140 and DN-3531 layer.
- Si accumulated on the spent catalyst was 5,220 pounds.
- The actual Si accumulated on the catalyst was 9% higher than the estimated number.

Silicon Accumulation – Estimated vs. Actual

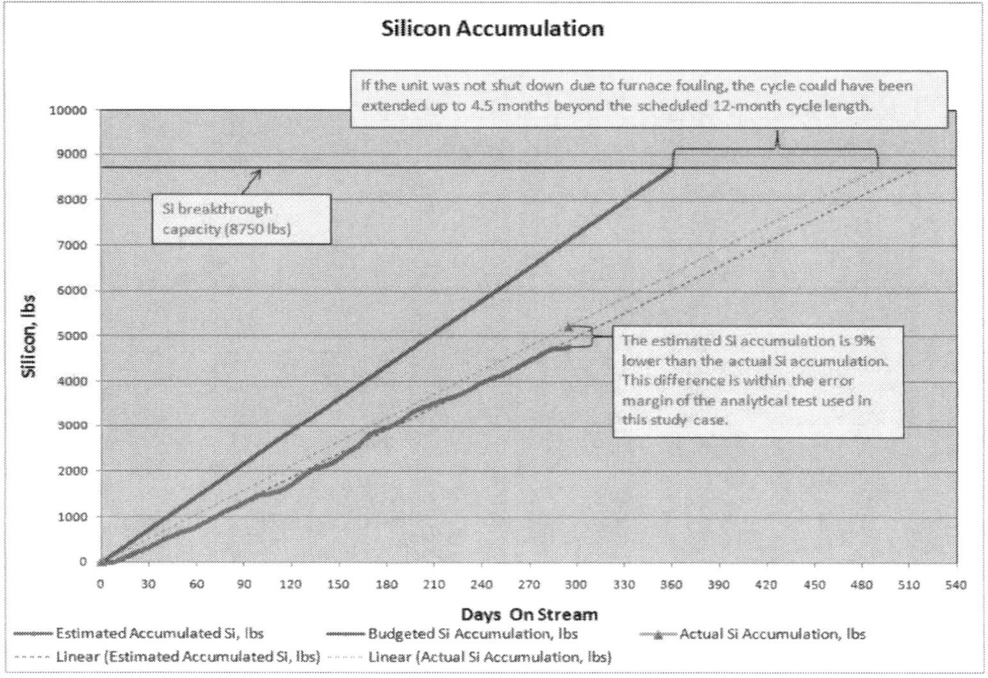

17

Conclusions

- Quality of the estimation of Si deposition depends on:
  - Quality of the feed sampling program
  - Accuracy of the analytical method
- Hot loop sampling system can be used to obtain composite samples which capture the changes in the Si content of the feed.
- Shell proprietary ICPDIN analysis accurately measures the Si content in coker naphtha.
- Combination of good feed sampling program and accurate Si analysis results in a more reliable estimation of silicon deposition.
  - Cycle life of the unit in this study case could have been extended up to 4.5 months beyond the scheduled 12 months.
  - This same unit achieved a 16 month cycle life in the cycle following the cycle in this case study.

SECARB ED

Southeast Regional Carbon Sequestration Technology Training Program

Track 3: Carbon Capture and Storage Workshop

CCS Overview, Carbon Capture & Pipeline Transportation

Presented by:

Gerald R. Hill, Ph.D.

Southern States Energy Board

2011 AIChE Regional Process Technology Conference

October 6, 2011

Moody Gardens Hotel

Galveston, TX

Acknowledgements

- SECARB-Ed is supported by the U.S. Department of Energy's (DOE) National Energy Technology Laboratory as part of the American Recovery and Reinvestment Act of 2009 under DE-FE0001930.
  - www.netl.doe.gov/technologies/carbon_seq/arra/training.html
- The Southern States Energy Board is the Principal Contractor of SECARB-Ed to DOE.
  - http://www.secarb-ed.org/

2

SECARB ED

Southeast Regional Carbon Sequestration Technology Training Program

Outline

- Overview – Hill
- CO_2 Capture & Transportation – Hill
- Storage Site Characterization – Peters
- Break
- CO_2 Injection Well Design – Peters
- CO_2 Monitoring, Verification & Auditing - Young

SOUTHERN STATES ENERGY BOARD

NETL

3

E-Alerts, Newsletters & Webpage

SECARB ED

Southeast Regional Carbon Sequestration Technology Training Program

Quarterly Training Newsletter and E-Alert:

- Summary of upcoming training opportunities
- Material of a technical nature relevant to the SECARB-Ed region
- Results from on-going CCS research
- Public policy updates related to CCS
- Alerts to forthcoming R&D solicitations

Webpage

- www.SECARB-Ed.org

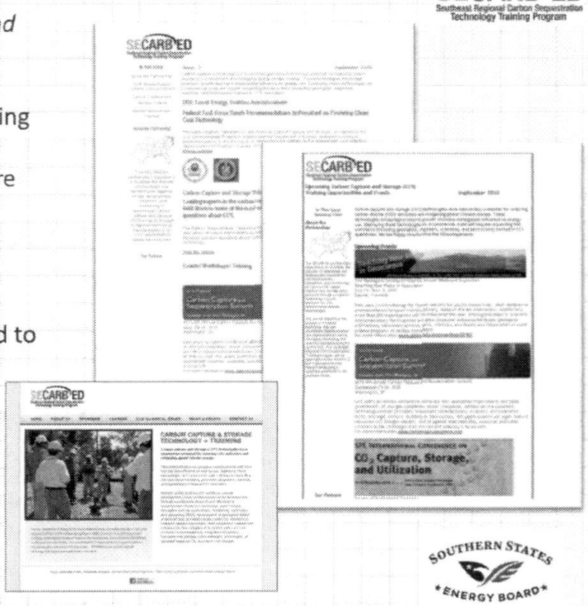

SOUTHERN STATES ENERGY BOARD

4

Regional Carbon Sequestration Partnerships

Developing the Infrastructure for Wide Scale Deployment

Seven Regional Partnerships

400+ distinct organizations, 43 states, 4 Canadian Provinces

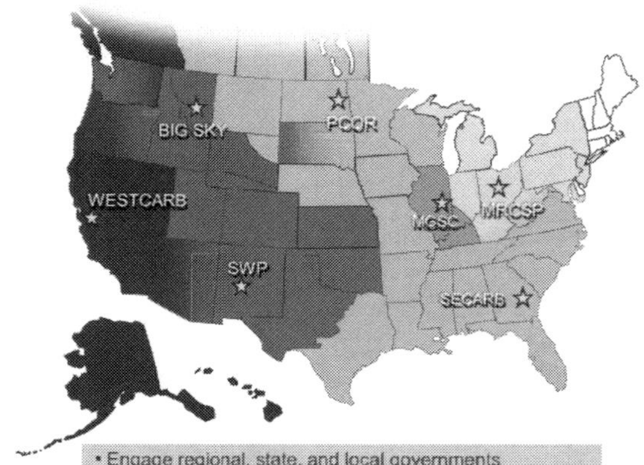

- Engage regional, state, and local governments
- Determine regional sequestration benefits
- Baseline region for sources and sinks
- Establish monitoring and verification protocols
- Address regulatory, environmental, and outreach issues
- Validate sequestration technology and infrastructure

Characterization Phase (2003-2005)

Search of potential storage locations and CO_2 sources	Found potential for 100's of years of storage

Validation Phase (2005-2011+)

19 injection tests in saline formations, depleted oil, unmineable coal seams, and basalt

Development Phase (2008-2018+)

Large scale injections	Commercial scale understanding	Regulatory, liability, ownership issues

5

NATIONAL ENERGY TECHNOLOGY LABORATORY

RCSP Phase III: Development Phase

Large-Scale Geologic Tests

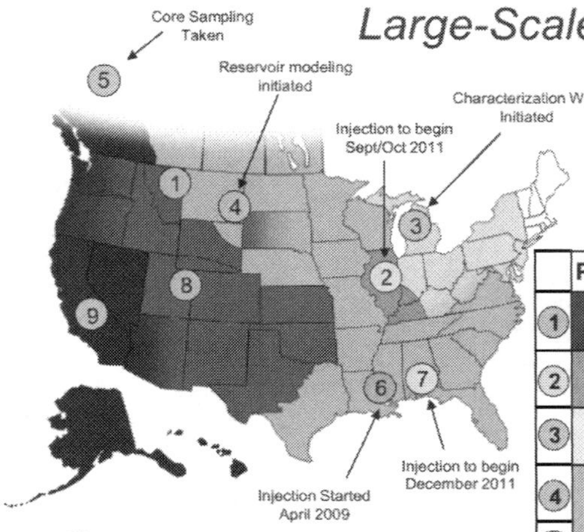

- Injection Ongoing
- 2011 Injection Scheduled
- Injection Scheduled 2012-2015

Note: Some locations presented on map may differ from final injection location

- ✓ *Injection Targets -minimum planned volumes*
- ✓ *One injection commenced April 2009*
- ✓ *Remaining injections scheduled 2011-2015*

	Partnership	Geologic Province	Storage Type
1	Big Sky	Sweetgrass Arch-Duperow Formation	Saline
2	MGSC	Illinois Basin-Mt. Simon Sandstone	Saline
3	MRCSP	Michigan Basin-St Peter SS or Niagaran Reef	Saline/Oil
4	PCOR	Powder River Basin-Muddy Formation	Oil Bearing
5	PCOR	Alberta Basin-Sulphur Point Formation	Saline
6	SECARB	Interior Salt Basin-Tuscaloosa Formation	Oil/Saline
7	SECARB	Interior Salt Basin-Paluxy Formation	Saline
8	SWP	Wasatch Plateau-Navajo Sandstone	Saline
9	WESTCARB	Regional Characterization	TBD

6

Project Integration

Carbon Capture and Storage (CCS)

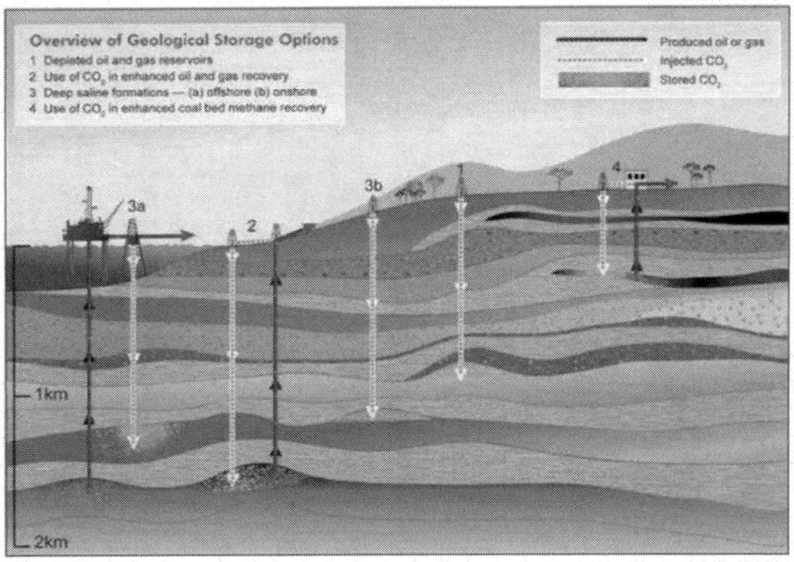

Source: IPCC, 2005

7

Integrating CO_2 Utilization and CO_2 Storage (CCUS)

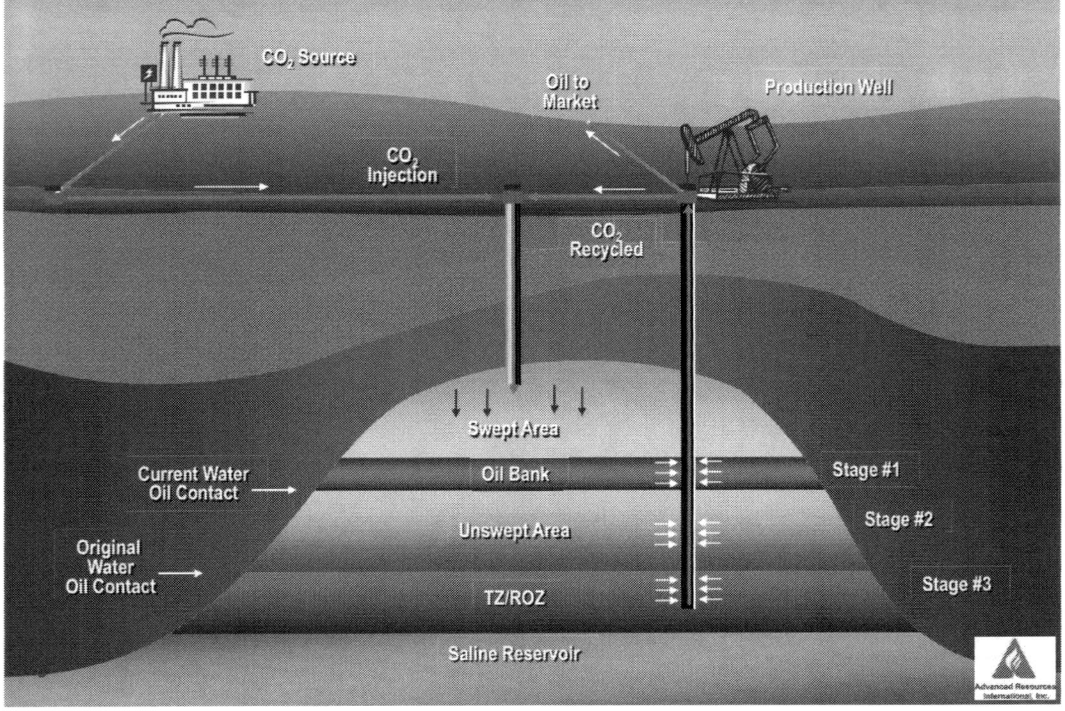

CCUS Targets: Opportunities for "Early Movers" in Anthropogenic Carbon Capture

SECARB ED
Southeast Regional Carbon Sequestration Technology Training Program

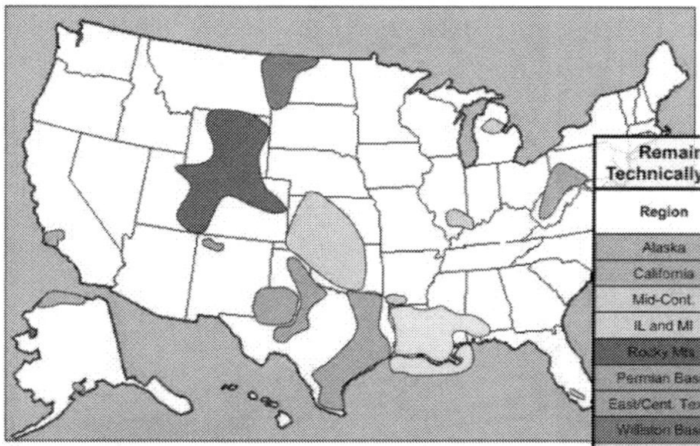

Source: ARI, February 2009

Remaining Oil in Place and Technically Recoverable Oil (BBls)

Region	ROIP*	Technically Recoverable
Alaska	45.0	12.4
California	57.3	6.3
Mid-Cont.	65.6	10.6
IL and MI	11.5	1.2
Rocky Mts.	22.5	3.9
Permian Basin	61.7	15.9
East/Cent. Texas	73.6	17.6
Williston Basin	9.4	2.5
Gulf Coast	27.5	7.0
LA Offshore	15.7	5.8
Appalachia	10.1	1.6
Total	**400**	**84.8**

* Remaining Oil in Place

Producible if costs, oil price and risks justify investment

U.S. CO_2-EOR Activity
Oil Fields and CO_2 Sources

SECARB ED
Southeast Regional Carbon Sequestration Technology Training Program

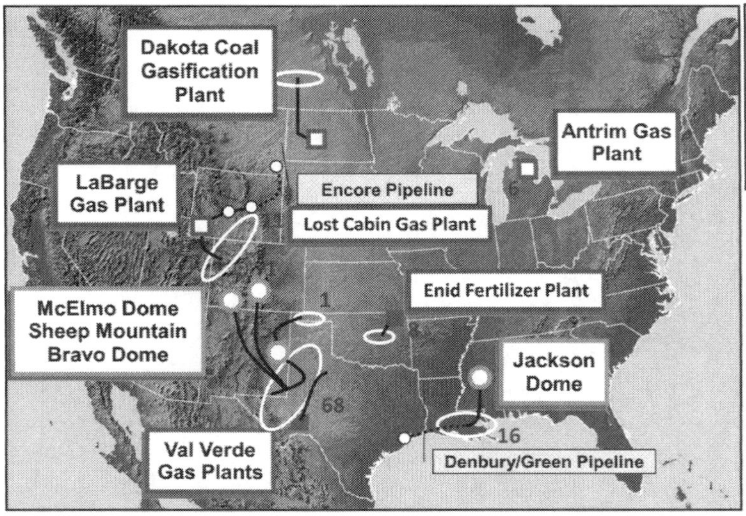

114	Number of CO_2-EOR Projects
○	Natural CO_2 Source
□	Industrial CO_2 Source
———	Existing CO_2 Pipeline
········	CO_2 Pipeline Under Development

- Most CO_2-EOR activity currently in North America
- Currently, 114 projects provide 281,000 B/D in U.S.
- **Additional projects planned to help increase CO_2 supply to CO_2-EOR**
  - **Natural & anthropogenic**
  - **New CO_2 pipelines**

Source: Advanced Resources International, Inc., based on Oil and Gas Journal, 2010 and other sources.

Advanced Resources International, Inc.

Oil Recovery Performance
Permian Basin: San Andres Formation

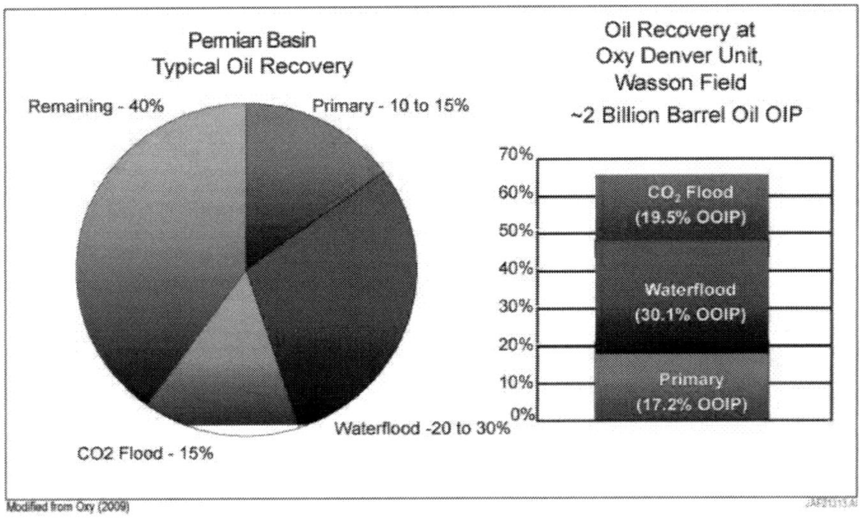

Modified from Oxy (2009)

11

Advanced Resources International, Inc.

Another CCUS Option: Enhanced Coal Bed Methane (ECBM) Production Combined with CO_2 Storage

Major U.S. Coal Basins
Coal capacity ~65-128 Gt CO_2
Shale capacity being assessed

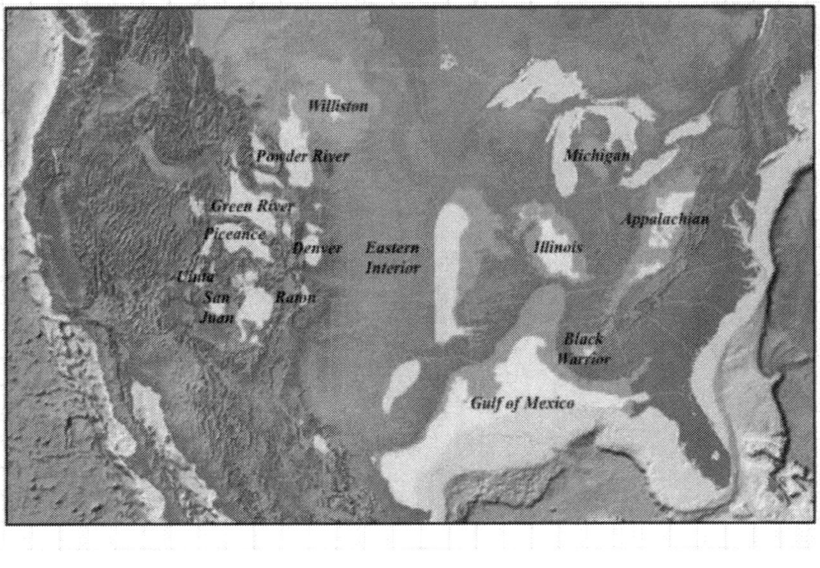

12

Dual Porosity

Butt Cleats and Face Cleats Facilitate Flow in ECBM Fields

SECARB ED
Southeast Regional Carbon Sequestration Technology Training Program

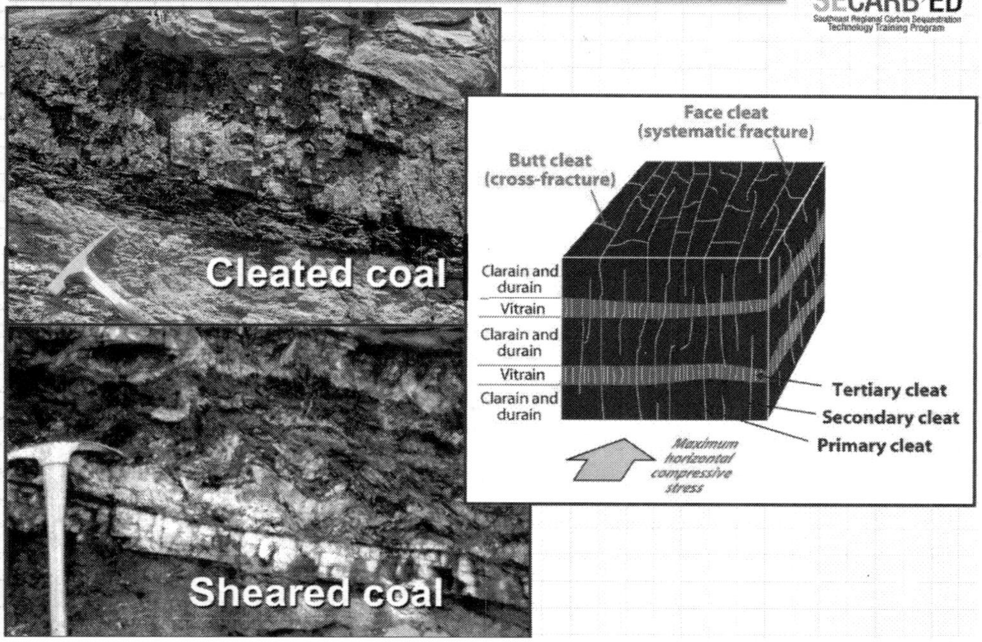

13

CO_2 – CH_4 Exchange: Injected CO_2 flows via Cleat Systems, Adsorbing to Coal and Desorbing Methane

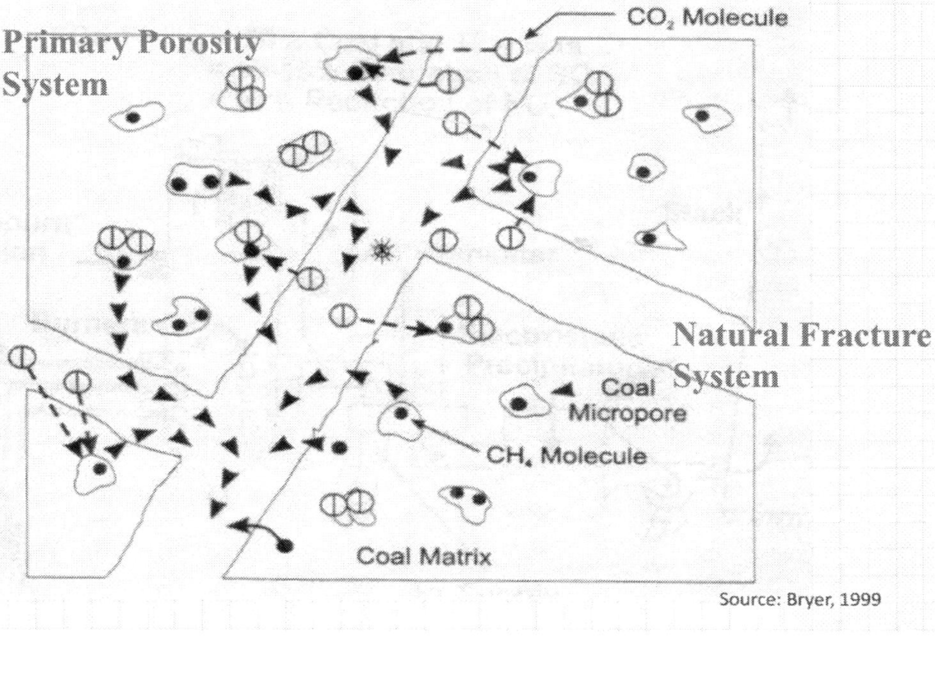

Source: Bryer, 1999

CO_2 Phase Relationships: CO_2 is in its Gas Phase when Adsorbed to Coal

SECARB-ED
Southeast Regional Carbon Sequestration Technology Training Program

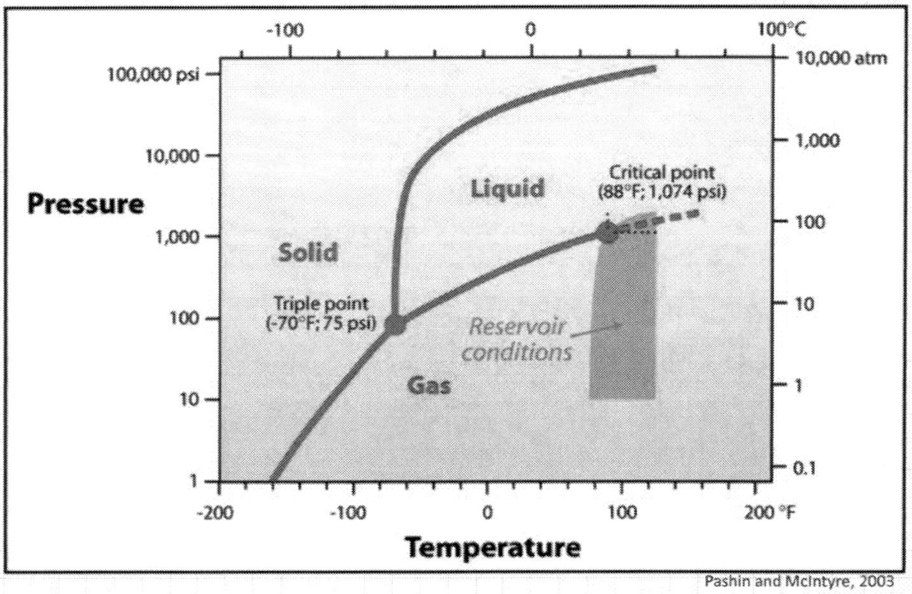

Pashin and McIntyre, 2003

15

CO_2 Capture

Three General Categories

SECARB-ED
Southeast Regional Carbon Sequestration Technology Training Program

- Install on new fossil plants or retrofit to existing plants
- Integrated Gas Combined Cycle (IGCC) gasification with carbon capture
- Oxygen enrichment results in concentrated CO_2 stream

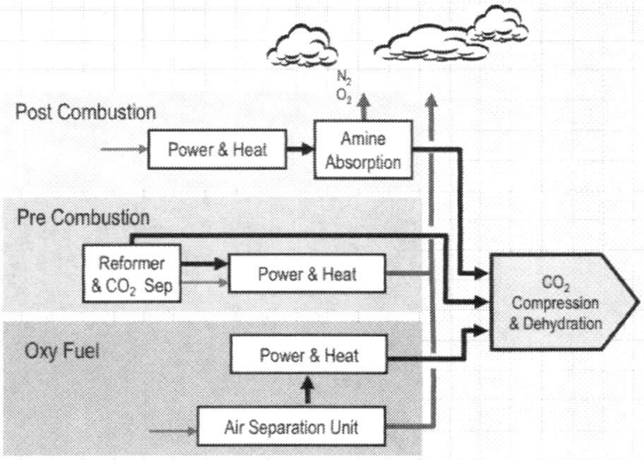

SOUTHERN STATES ENERGY BOARD · NETL · CLEMSON UNIVERSITY · JACKSON · VirginiaTech

16

Post Combustion Capture

CO_2 Scrubbing Process (Amine)

SECARB-ED
Southeast Regional Carbon Sequestration Technology Training Program

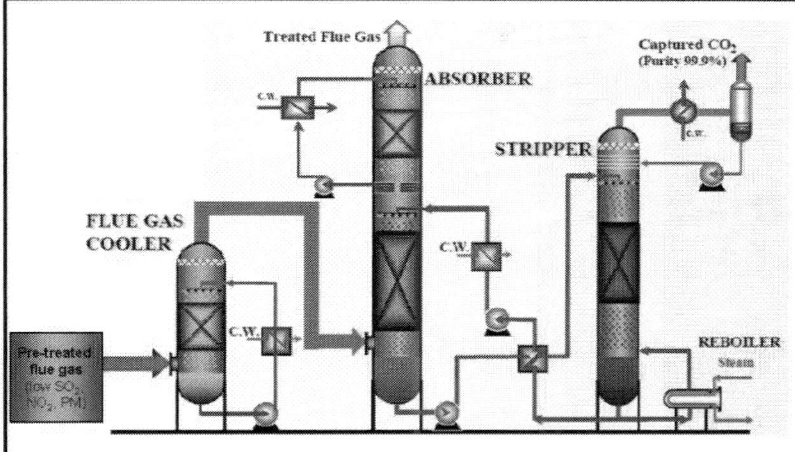

Key Points: needs >99% SO_2 removal; consumes process steam

CLEMSON UNIVERSITY
VirginiaTech

17

Pre Combustion Capture:

Integrated Gasification Combined Cycle (IGCC) with Carbon Capture

SECARB-ED
Southeast Regional Carbon Sequestration Technology Training Program

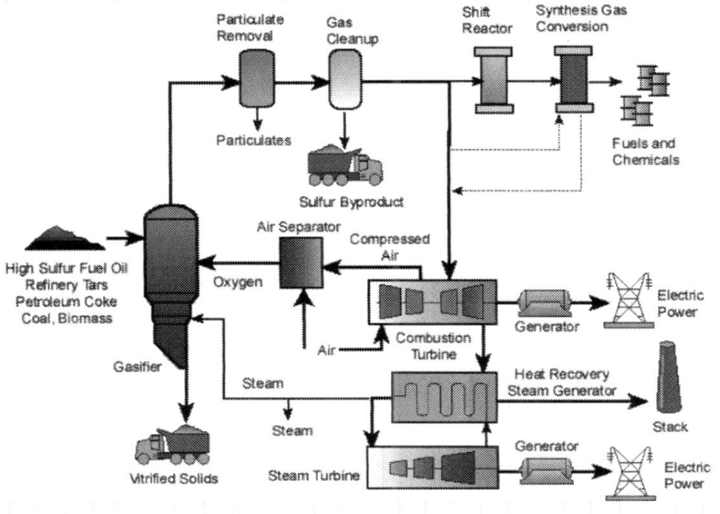

CLEMSON UNIVERSITY
VirginiaTech

18

Carbon Capture Capable (post combustion example)

Dominion's Virginia City, VA plant currently under construction

SECARB-ED Southeast Regional Carbon Sequestration Technology Training Program

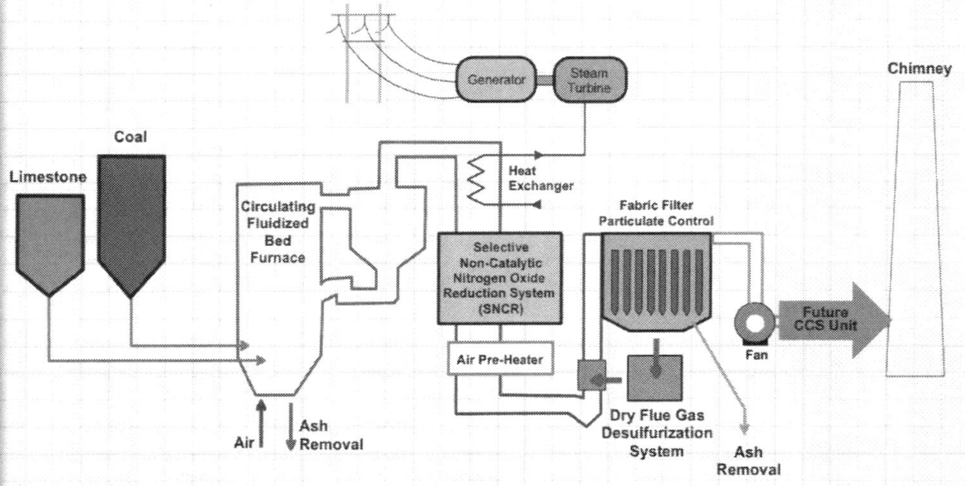

19

CCS Increases Water Use & Water Consumption

Water Use With and Without Carbon Capture

SECARB-ED Southeast Regional Carbon Sequestration Technology Training Program

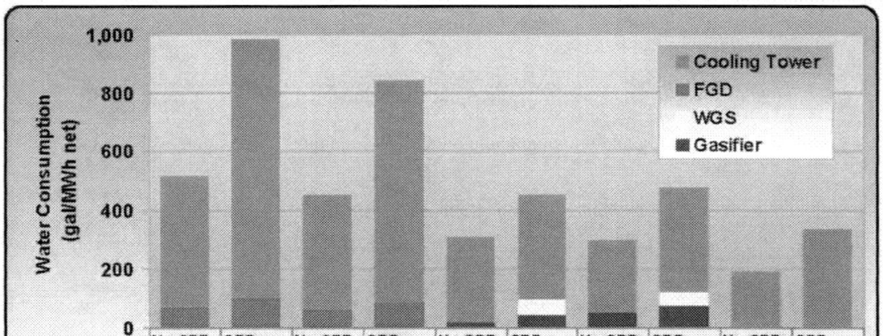

	Subcritical PC		Supercritical PC		IGCC (slurry fed)		IGCC (dry fed)		NGCC	
	No CDR	CDR	No CDR	CDR	No CDR	CDR	No CDR	CDR	No CDR	CDR
Total	517	985	452	844	309	452	295	475	192	338
Cooling Tower	449	884	392	759	290	355	243	355	192	338
FGD	68	101	59	86						
WGS						52		49		
Gasifier					19	45	53	71		

Comparison of water consumption factors with and without carbon capture for greenfield plants using wet recirculating cooling towers - net power basis. (Note: FGD refers to flue gas desulfurization; WGS refers to water gas shift; and CDR refers to carbon dioxide recovery.)

Source: Gerdes 2008, DOE/NETL, Section 4

20

Water Reuse: Water extraction may be needed for reservoir pressure management in some geologic settings

Research on water treatment and reuse is underway

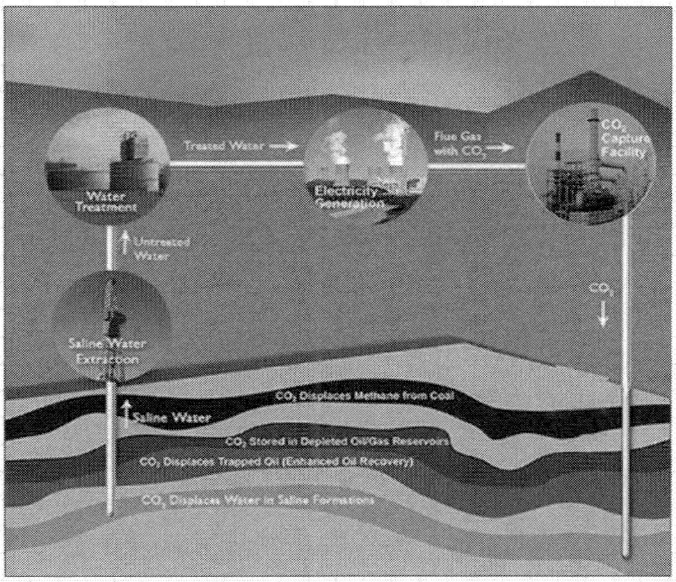

Source: Sandia, 2009

21

CO_2 Transportation: Pipeline networks currently exist in support of EOR. Billions will be needed to construct integrated CCUS transportation systems

SECARB ED
Southeast Regional Carbon Sequestration Technology Training Program

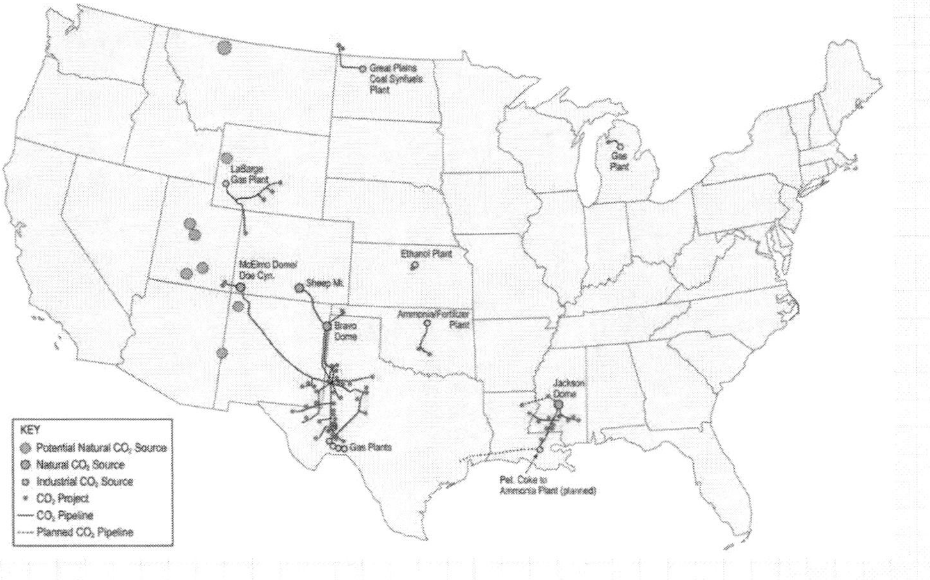

22

Example of CO_2 Phase Diagram used to calculate wellhead pressures needed to achieve injectivity and maintain supercritical state for deep injections (>800 meters)

SECARB ED
Southeast Regional Carbon Sequestration Technology Training Program

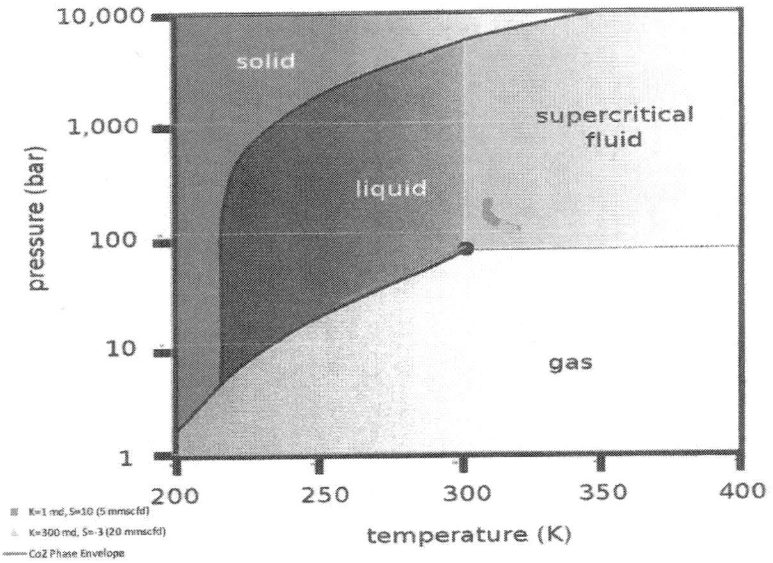

23

Integrated CCS Pilot Project: Southeast Region

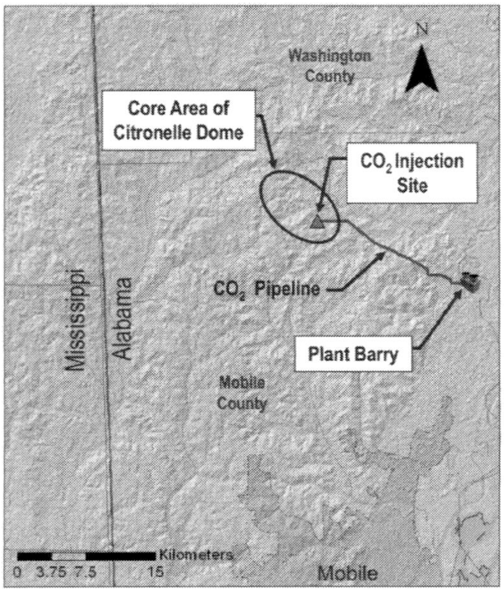

SECARB Anthropogenic Test

The 25 MW CO_2 capture unit at Alabama Power's (Southern Co.) Plant Barry became operational June 2011.

A newly built 12 mile CO_2 pipeline (4Q 2011) will transport CO_2 from Plant Barry to the Citronelle Dome.

From 100 to 300 thousand metric tons of CO_2 will be injected into the Paluxy saline formation for up to 3 years.

Advanced Resources and supporting researchers will conduct 3 years of monitoring after CO_2 injection and then close the site.

24 SECARB NETL EPRI SOUTHERN COMPANY

Anthropogenic Test: Plant Barry & Citronelle Field

Capture Facility at Alabama Power's Plant Barry (July 2010 – Present)

Characterization Well D9-8 #2 at Citronelle Field - Drilled (Dec. 2010/Jan. 2011)

- Convert to Obs/Monitoring Well
- 11,800' TD
- Whole core (98 feet in two intervals)
- 45 sidewall cores
- Full set of logs (Triple Combo, MRI, Mineralogy, Dipole Sonic, CBL, etc.).

Risk Management Framework: Integrated CCS Project

	CONSEQUENCE				LIKELIHOOD/FREQUENCY				
					A	B	C	D	E
					Remote:	Unlikely:	Possible:	Probable:	Frequent:
CONSEQUENCE SEVERITY	Environment	Cost	Reputation	Schedule to start-up of operations	Very unlikely to occur during life-time of project	Unlikely to occur during lifetime of project	50-50 chance of occuring during lifetime of project	Likely to occur at least once during lifetime of project	Expected to occur several times during lifetime of project
Very High	Persistent severe environmental damage. Extensive remediation required. Environment restored > 5 years.	More than $10 million	National or international media attention. Regulators shut down operations.	More than 12 months	M	M	H	H	H
High	Severe environmental damage. Remediation measures required. Environment restored < 5 years	$1 to $10 million	Regional media attention. Regulatory or legal action taken	6-12 months	L	M	M	H	H
Medium	Limited environmental damage managed by Company response teams. Environment restored < 2 years	$100 to $1000 k	Local media attention. Regulatory or legal action likely	3-6 months	L	L	M	M	H
Low	Minor environmental damage, but no lasting effect	$10 to $100 k	Public awareness may exist, but there is no public concern	1-3 months	L	L	L	M	M
Very Low	Slight environmental damage contained within the premises	Less than $10 k	On-site communications	Less than 1 month	L	L	L	L	M

MANAGING RISK

SECARB ED

Southeast Regional Carbon Sequestration Technology Training Program

Track 3: Carbon Capture and Storage Workshop

CCS Overview, Carbon Capture & Pipeline Transportation

QUESTIONS?

http://www.secarb-ed.org/

hill@sseb.org

2011 AIChE Regional Process Technology Conference

October 6, 2011

Moody Gardens Hotel

Galveston, TX

SOUTHERN STATES ENERGY BOARD

NETL

Schlumberger
Carbon Services

Commercial CO_2 Sequestration

Dwight Peters
North America Business Manager

2011 AIChE Regional Process Technology Conference
Moody Gardens Hotel
Galveston, TX

Oct 6, 2011

Acknowledgements

- Some graphics in this material is based upon work supported by the U.S. Department of Energy (DOE) National Energy Technology Laboratory (NETL). This work is managed and administered by the Regional Carbon Sequestration Partnerships and funded by DOE/NETL and cost-sharing partners.
- SECARB-Ed is supported by the U.S. Department of Energy's (DOE) National Energy Technology Laboratory as part of the American Recovery and Reinvestment Act of 2009 under DE-FE0001930.
  - www.netl.doe.gov/technologies/carbon_seq/arra/training.html
- The Southern States Energy Board is the Principal Contractor of SECARB-Ed to DOE.
  - www.secarb-ed.org/

2

The Commercial CO2 Storage Workflow

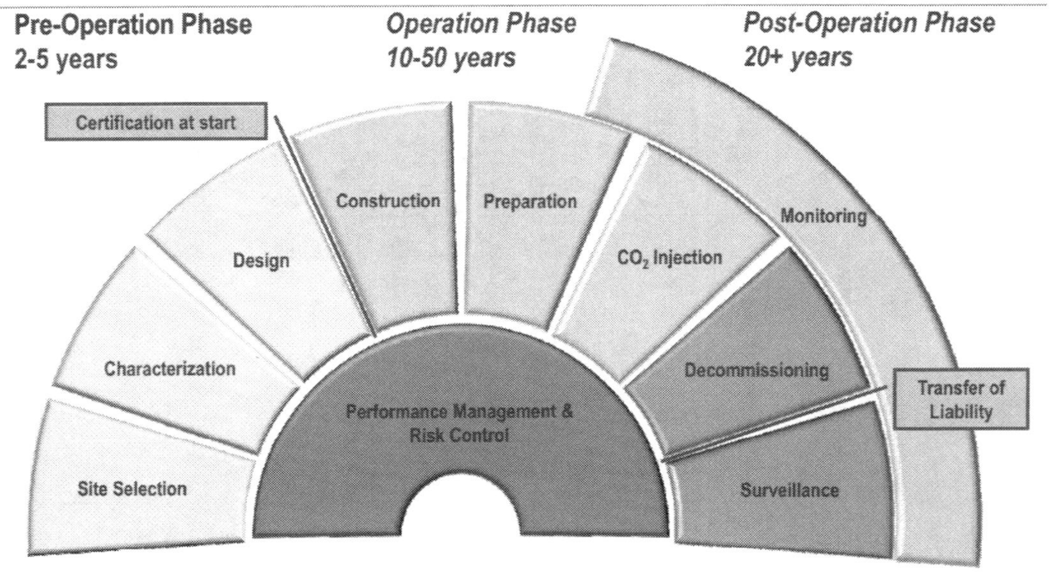

Current Knowledge

- Scientific work has identified many potential storage sites in the US and rest of world
- Only some of these sites can provide low risk, low cost commercial storage
- Injection pilots build acceptance but leave many unknowns about commerciality
- Today' s best practices manuals have been derived from small scale experiences
- Many important pilot project experiences are not completely understood
- Scale-up will require commercial processes adapted from the oil and gas industry.

Potential Storage Reservoirs

Schlumberger
Carbon Services

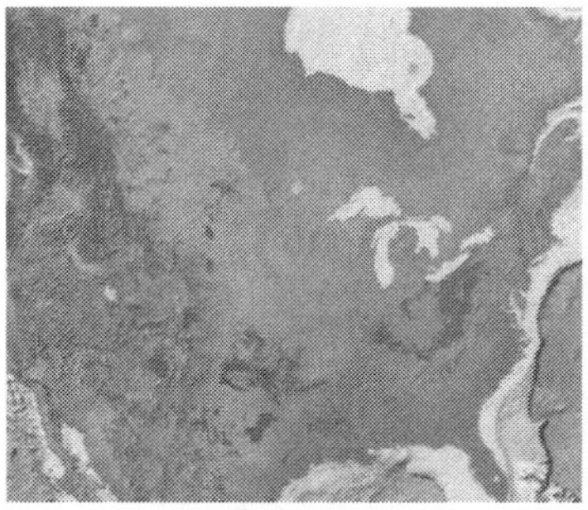

Depleted oil & gas
84 Gt

Unmineable coal
156-184 Gt

Basalts
?? Gt

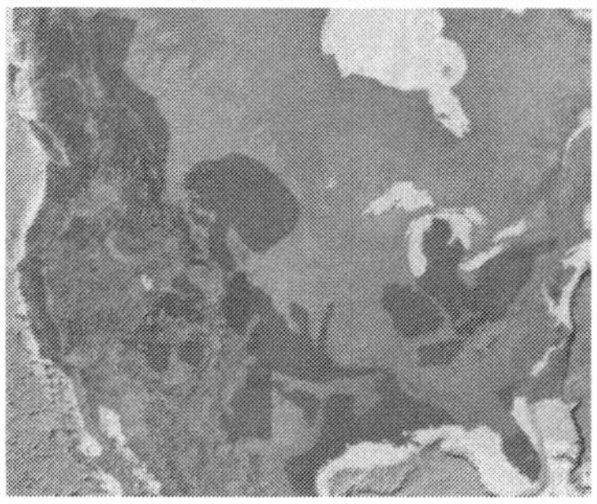

Saline aquifers
919-3378 Gt

Potential Saline Storage Basins

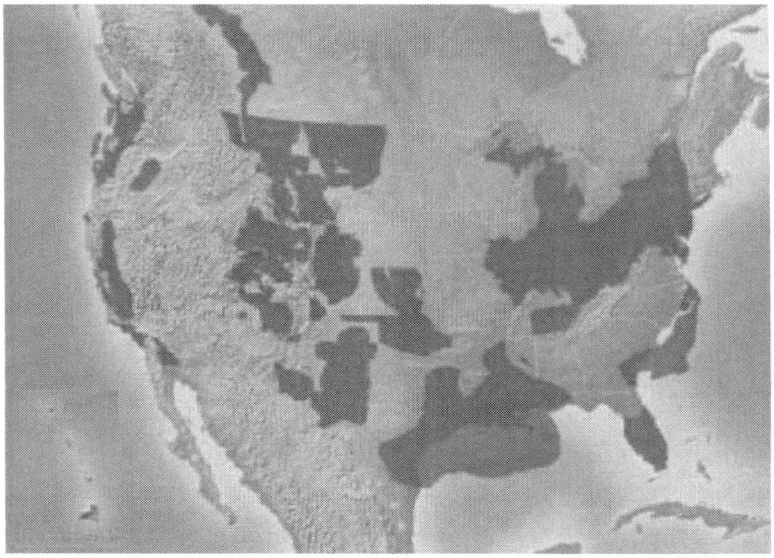

Saline Formation Storage Fundamentals

Schlumberger
Carbon Services

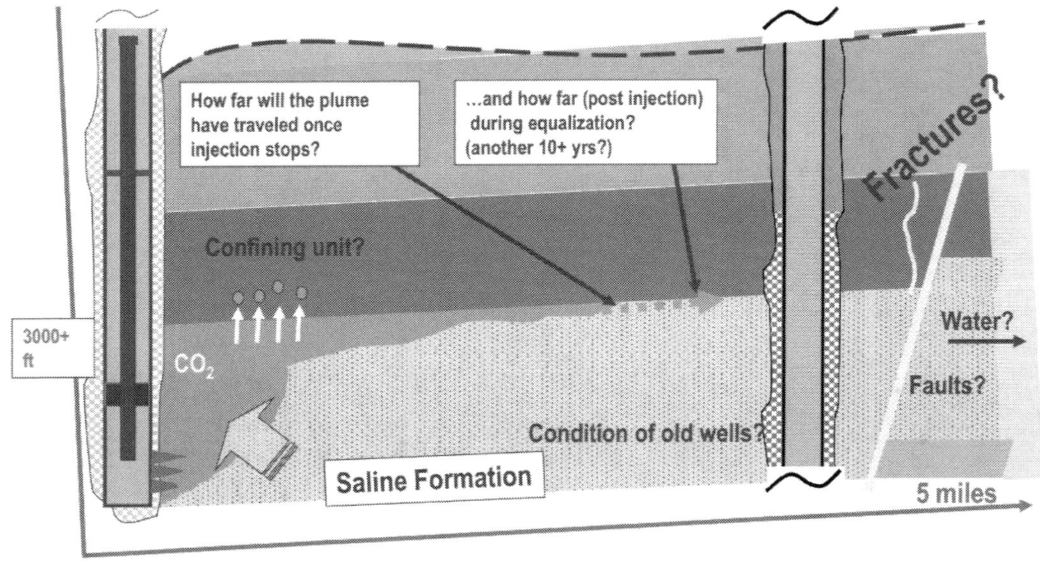

Commercial Geologic Storage Project Timeline

Schlumberger
Carbon Services

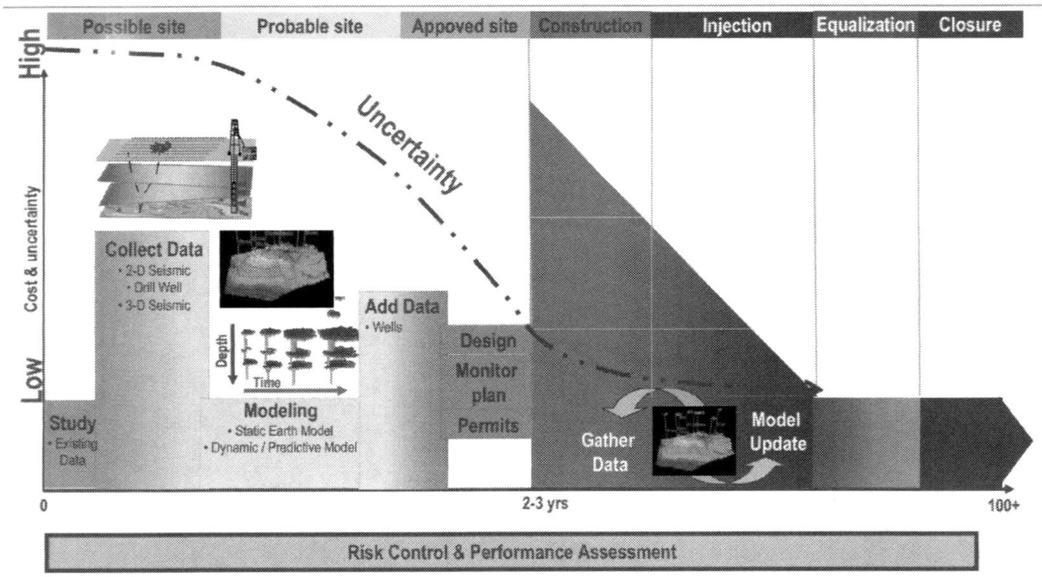

The Need for Data

Schlumberger
Carbon Services

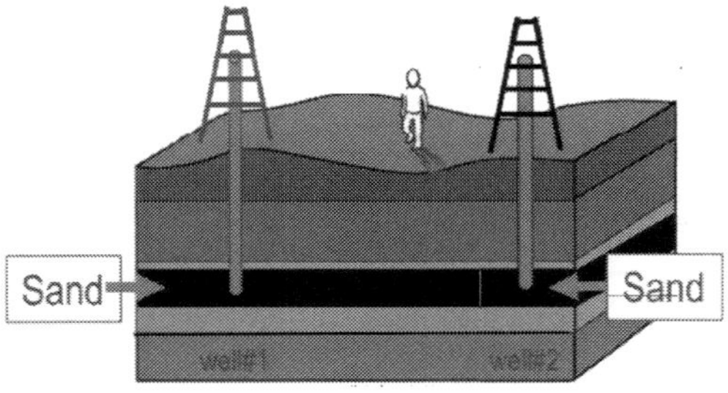

possible
scenario

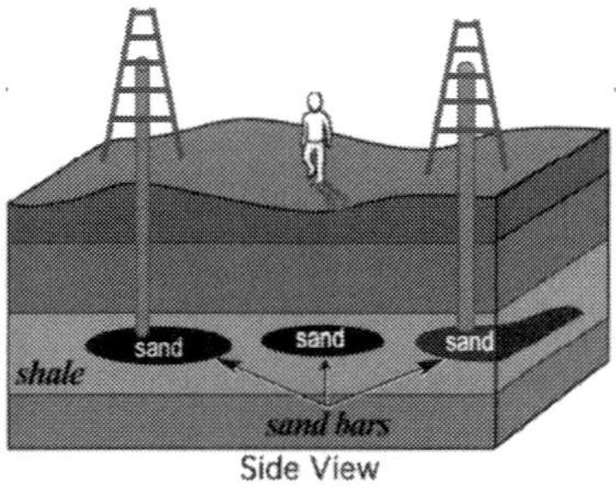

correct
scenario

reality - top view

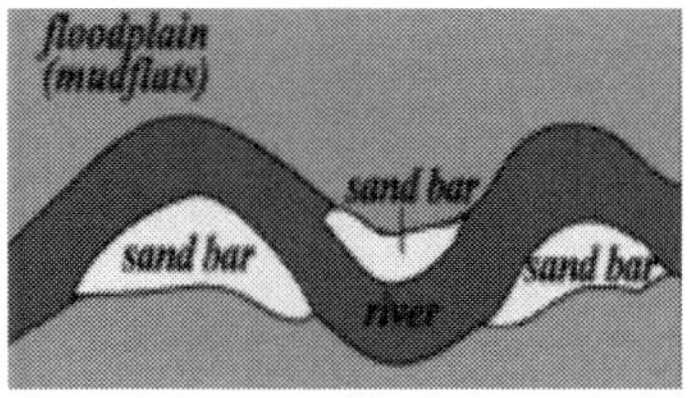

Location of Geological Uncertainties

Schlumberger
Carbon Services

Porosity and Permeability are Key

Schlumberger
Carbon Services

These have the same porosity (40%) but different permeability

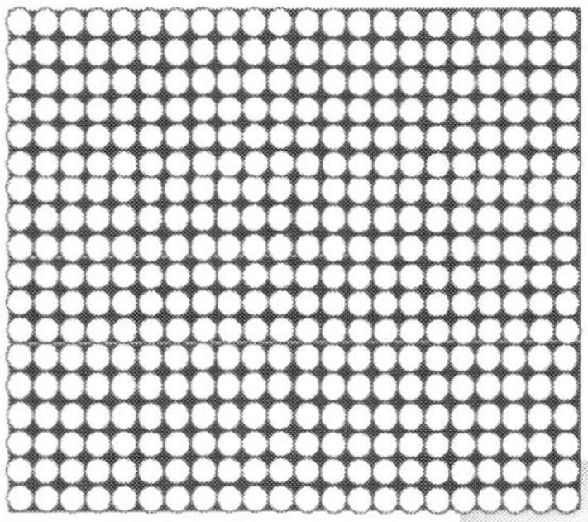

Big Grains
= high permeability

Small Grains
= low permeability

Data and Reservoir Scales

Schlumberger
Carbon Services

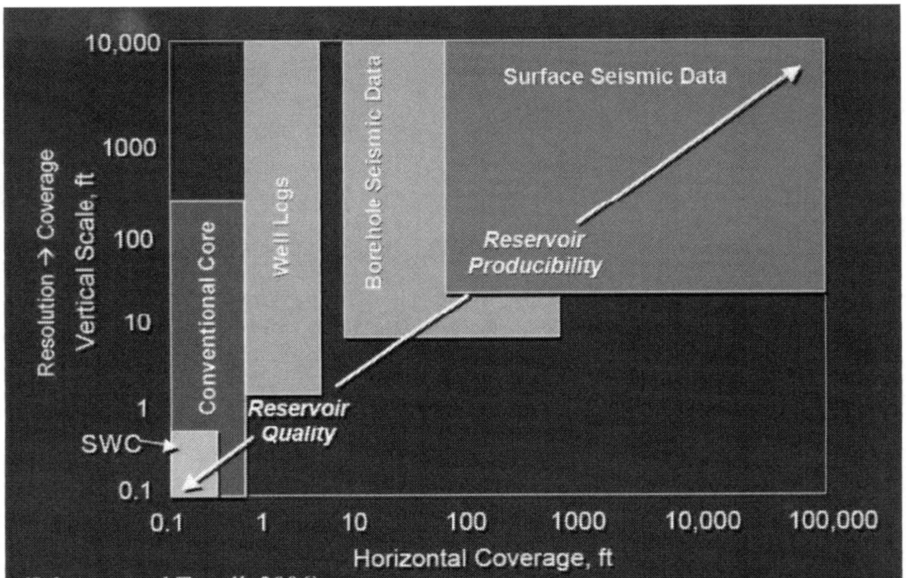

(Johnston and Terrell, 2006)

Conventional Data vs Hi-Resolution Data

Schlumberger
Carbon Services

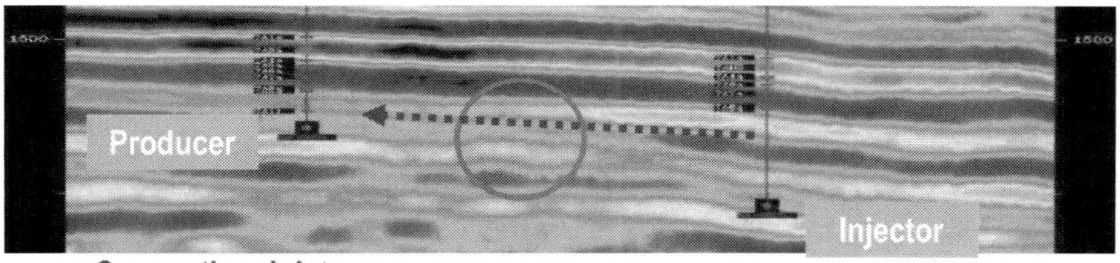

Conventional data

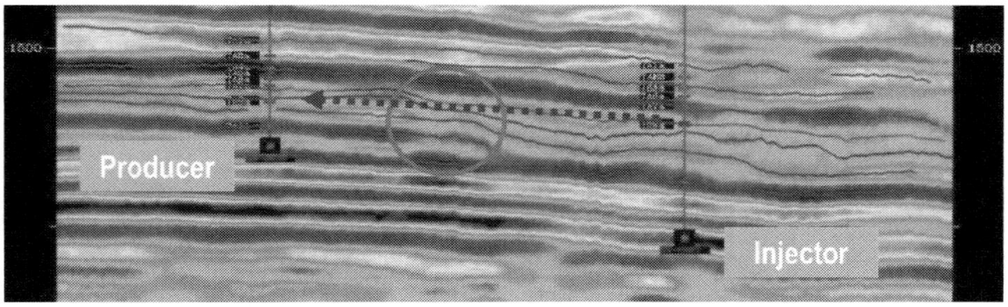

Single-sensor data

CCS Full System Integration

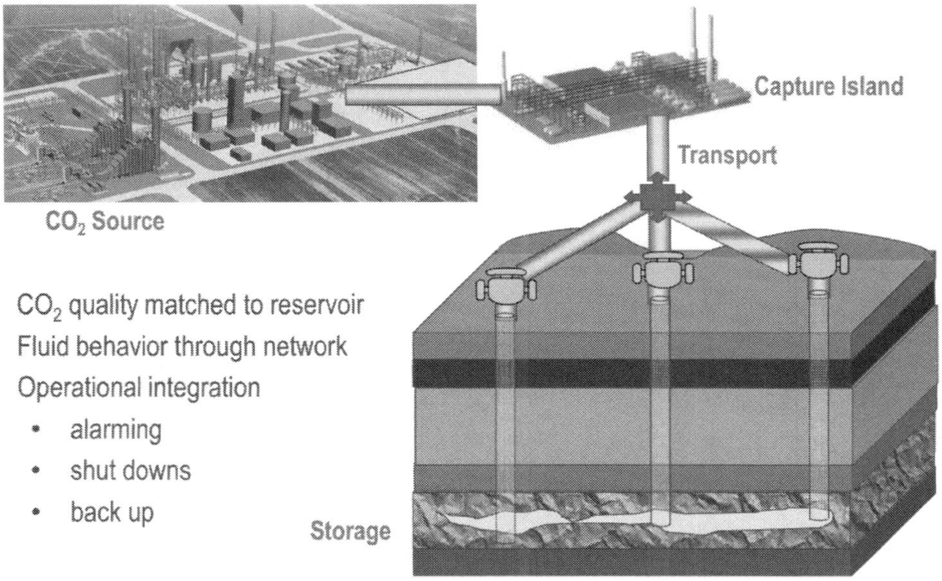

- CO_2 quality matched to reservoir
- Fluid behavior through network
- Operational integration
    - alarming
    - shut downs
    - back up

The Risk Management Matrix

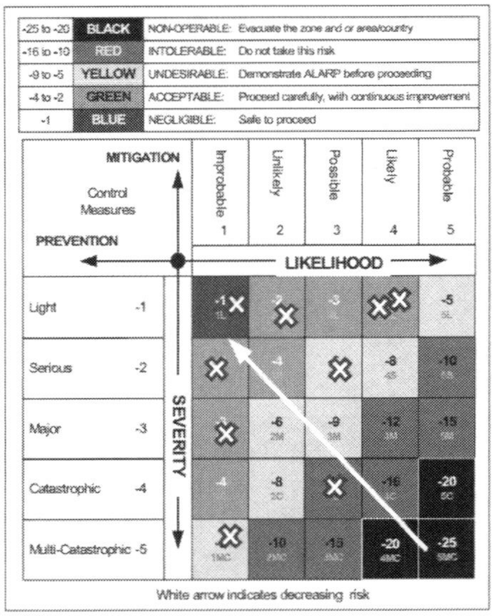

Responses

- reduce likelihood (PREVENT)
- reduce severity (MITIGATE).

Tasks:

- Intelligently construct "Scenarios" that can be modeled.
- Efficiently apply simulation resources.

Hazard Analysis and Risk Control Standard SLB-QHSE-S020

CO_2 Monitoring – 3 objectives

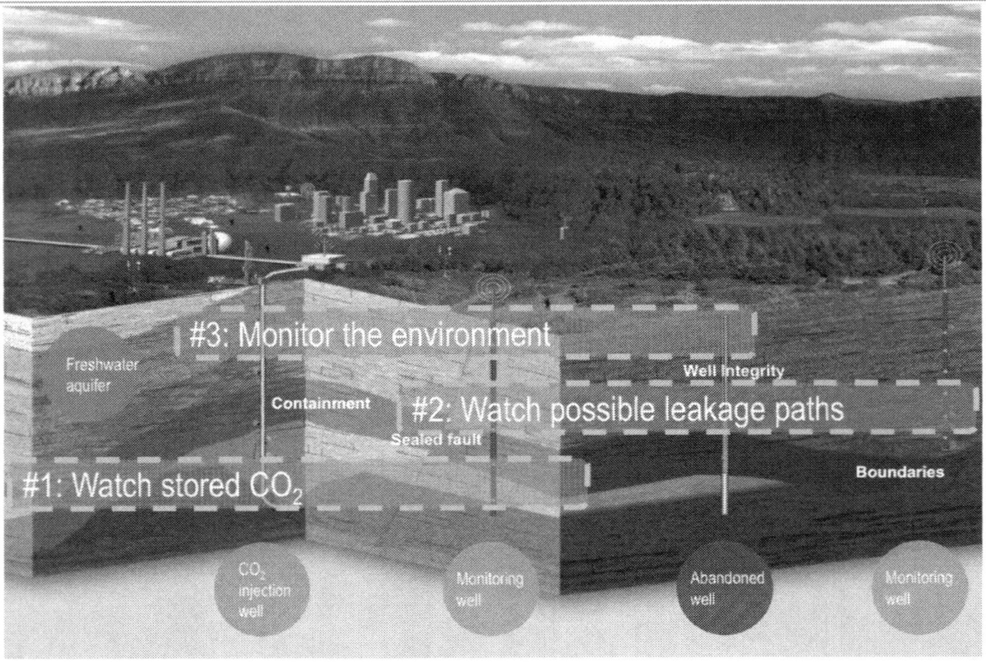

CO2 Injection Well

Schlumberger
Carbon Services

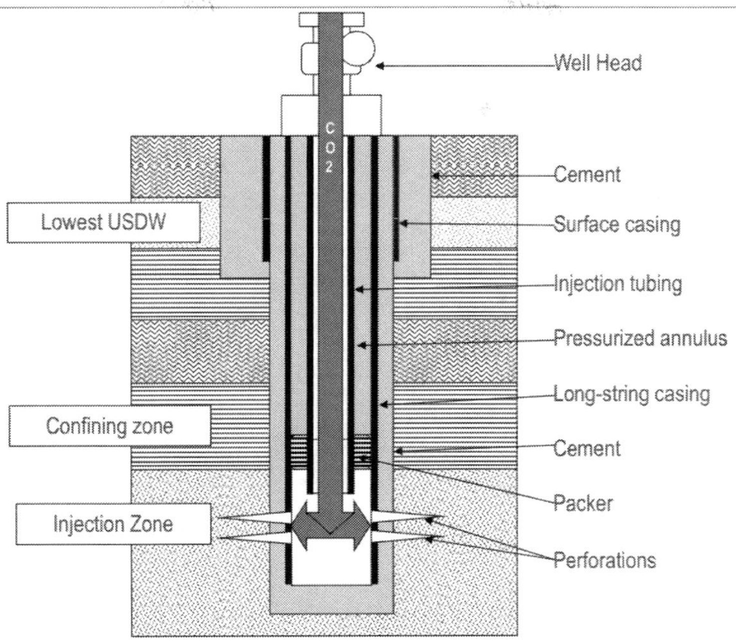

Executing a cement job

Schlumberger
Carbon Services

- Cement is pumped down inside the casing.
- Cement then passes in the gap (or annulus) between the casing and open hole.
- It is displaced into position with drilling fluid or water.
- Spacers are often pumped ahead and behind to clean the hole and protect the cement.

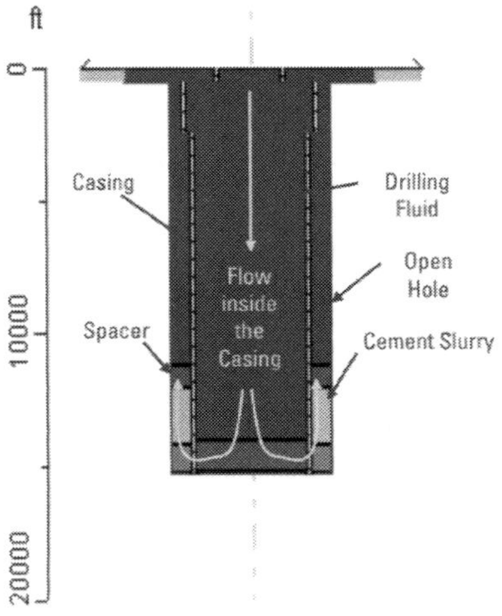

Fluid Sequence

Leading Causes of Cementing Failures

Schlumberger
Carbon Services

- Incomplete Mud Removal
- Breakdown of formation during cementing
- Well fluid flow during or after cement is placed
- Inappropriate cement slurry design
- Surface execution problems

UIC Injection Well Construction Requirements.

Schlumberger
Carbon Services

- "Casing and cement or other materials used in the construction of each Class VI well must have sufficient structural strength and be designed for the life of the geologic sequestration project. All well materials must be compatible with fluids with which the materials may be expected to come into contact"
- "Cement and cement additives must be compatible with the carbon dioxide stream and formation fluids and of sufficient quality and quantity to maintain integrity over the design life of the geologic sequestration project."
- "At least one long string casing, using a sufficient number of centralizers, must extend to the injection zone and must be cemented by circulating cement to the surface in one or more stages"
- "Tubing and packer materials used in the construction of each Class VI well must be compatible with fluids with which the materials may be expected to come into contact."

Materials: CO_2-Resistant Cement

Schlumberger
Carbon Services

Cements for remediation and construction of new wells can still be Portland based.

Field experience has shown that reactions in the ground are probably much slower than what has been seen in the lab [Crow et al., 2008 and Carey et al., 2006] but further research needs to be conducted. Field experience under much more severe (geothermal) conditions has shown complete degradation of well materials exposed to CO_2 [Milestone et al., 1986]

CO_2-resistant Portland cements have been used and are currently available.

- Some CO_2 resistant cements work by reducing the amount of calcium hydroxide (CH) and increasing the amount of calcium silica hydrate (C-S-H). C-S-H is more resistant to CO_2 attack than CH. Additives such as pozzolans provide additional silicon to create additional C-S-H.
- Additives such as bentonite which require a large increase in the water-to-cement (W/C) ratio should be avoided. High W/C can lead to accelerated cement degradation.

Construction: Cement sheath defects – liquid cement

Schlumberger
Carbon Services

Placement defects

- Poorly centralized casing can lead to channels by bypassing the narrow side of the annulus
- Mud films can also create pathways and reduce zonal isolation

Gas migration during cement hydration may cause channels

- Driven by a drop in cement pore pressure during hydration

Thermomechanical effects

- Cracks caused by cement failure in compression/tension, microannuli caused by debonding at the interfaces with casing and/or rock

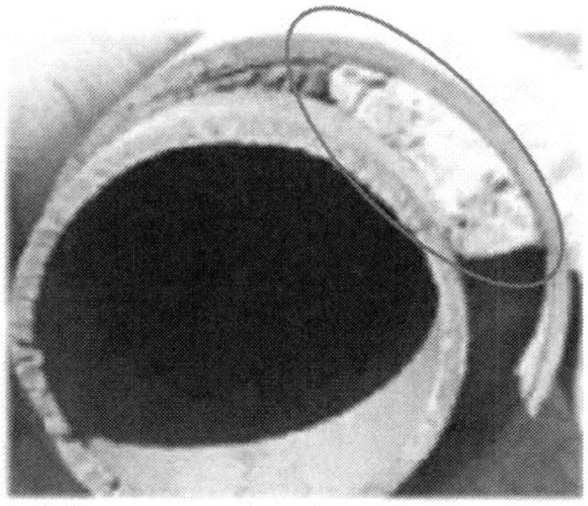

These can all lead to direct communication to the casing and may increase transport by establishing a communication path vertically

Construction practices should minimize the chances for defects in the well to be created and should include the use of centralizers and proper mix design.

Materials: Casing and Elastomers

Schlumberger
Carbon Services

Typical carbon steel casing is vulnerable to degradation by anodic dissolution, exposure to chloride laden waters, and exposure to carbonic acid [Anstice et al, 2005 and Stone et al., 1989]

- A passive layer is created between the cement and the casing that protects the casing from degradation
- Lack of cement or degradation of the cement can lead to a breakdown of the passive layer allowing for corrosion of the steel

More corrosion resistant alloy casing do exist (stainless steel etc.)

Elastomers can be degraded by CO_2. Degradation can include explosive decompression, blistering, or others.

UIC Injection well operating requirements– Annular Pressure

Schlumberger
Carbon Services

"The owner or operator must maintain on the annulus a pressure that exceeds the operating injection pressure, unless the Director determines that such requirement might harm the integrity of the well or endanger USDWs."

Concerns

- May cause a leak in the casing
- May damage the cement isolation capacity
- May damage the surrounding formation

EXAMPLE

- Injection pressure at the top of the well of 2000 psi
- Surface temperature of 50°f and a gradient of 1°f per 100ft
- Annular fluid 10% KCl with a gradient of 0.46 psi/ft
- Annular pressure 100 psi above injection pressure at the top of the well
- A fracture pressure gradient of 0.8 psi/ft
- Injection depth 7000 ft
- Variable CO_2 density and gradient based on constant densities over 100-foot intervals calculated using REFPROP [Lemmon et al., 2007]

Annular Pressure

Schlumberger
Carbon Services

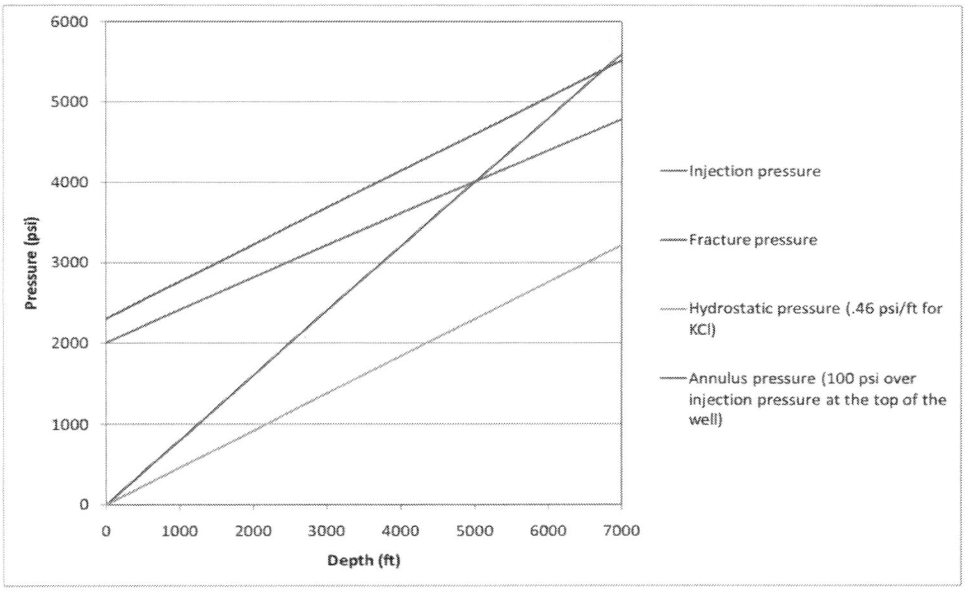

Summary

Schlumberger
Carbon Services

A successful CCS project development plan requires an holistic project approach encompassing CO_2 capture, transport, and GS aspects of the CCS chain.

Understanding the characteristics of the geological storage site is particularly important to reduce uncertainty and appropriately manage any potential risks.

Meeting the requirements of the rule can occur in a seamless fashion as part of a comprehensive workflow.

A workflow that treats the stages of a GS project as a continuous process will generate the information needed for permitting as part of the process of developing a project.

The CarbonWorkFlow process is iterative. The Design stage combines the needs of the project with safety and risk to create the most efficient project plan possible. As the design stage of the project progresses, solutions of challenges, such as those raised by the properties of CO_2 and well design requirements, will be met based on the needs of the project using the site-specific data collected during the Appraisal/Characterization phase.

Monitoring, Verification and Accounting (MVA) in Retention and EOR Settings

Michael H. Young
Susan D. Hovorka
Bureau of Economic Geology
Jackson School of Geoscience
The University of Texas at Austin

Presented at the AIChE Regional Process Technology Conference
October 6, 2011, Galveston, Texas

SECARB ED

BUREAU OF ECONOMIC GEOLOGY

Outline

- Definitions and Context
- Background on Subsurface Geology
- Monitoring Concepts and Approaches
- Verifying Conceptual Model
- The Iterative Loop of Monitoring and Modeling
- Accounting (or, Detecting) Change in CO_2
- Contingencies

SECARB ED

BUREAU OF ECONOMIC GEOLOGY

Definitions and Goals

- Monitoring Instruments – a device or sensor that collects information about the site environs
- Monitoring Systems – a system that collects the output of sensors
- Monitoring Strategy – a set of monitoring systems that emphasize and support a specific concept or philosophy
- Monitoring Program - a set of monitoring strategies, including data collection intervals, analytical methods and data analysis

---

- *Confirm predictions of containment based on site characterization*
- *Maintain confidence during injection - observations should be close to model predictions*
- *Confirm no unacceptable consequences result from injection*

SECARB ED

BUREAU OF ECONOMIC GEOLOGY

Regional Geologic Setting – Cross Section

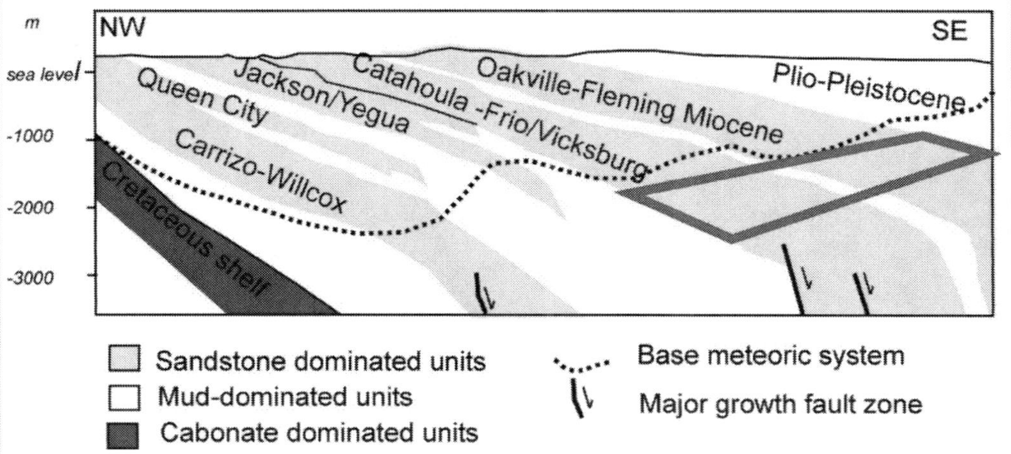

Modified from Galloway and others, 1982

SECARB ED

BUREAU OF ECONOMIC GEOLOGY

Gulf of Mexico Basin is composed of individual formations, creating a thick wedge

Miocene

- 20,000 feet thickness
- Sand prone
- favorable depths for sequestration (2,500-12,000) in TX State Waters

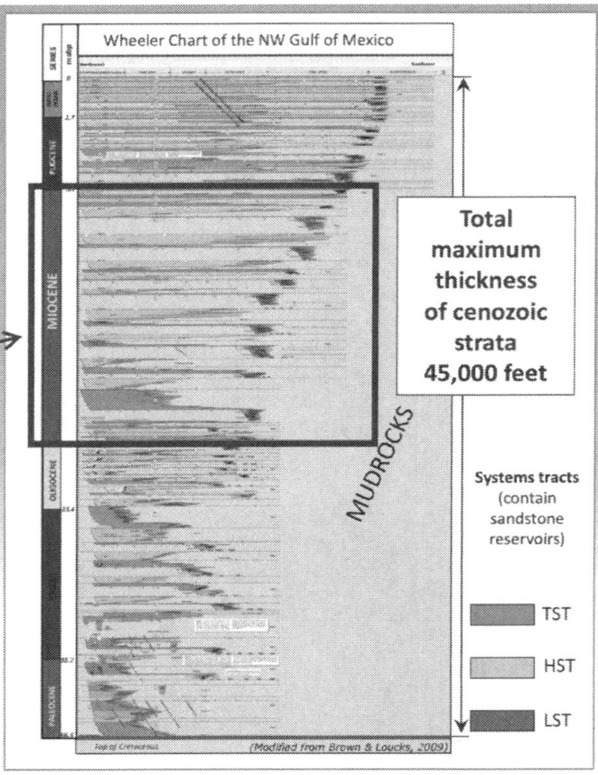

SECARB ED

Heterogeneity in logs and cores show complicated depositional history

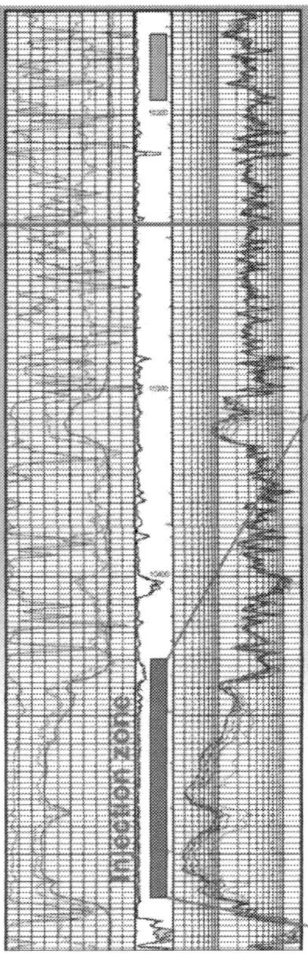

Bureau of Economic Geology

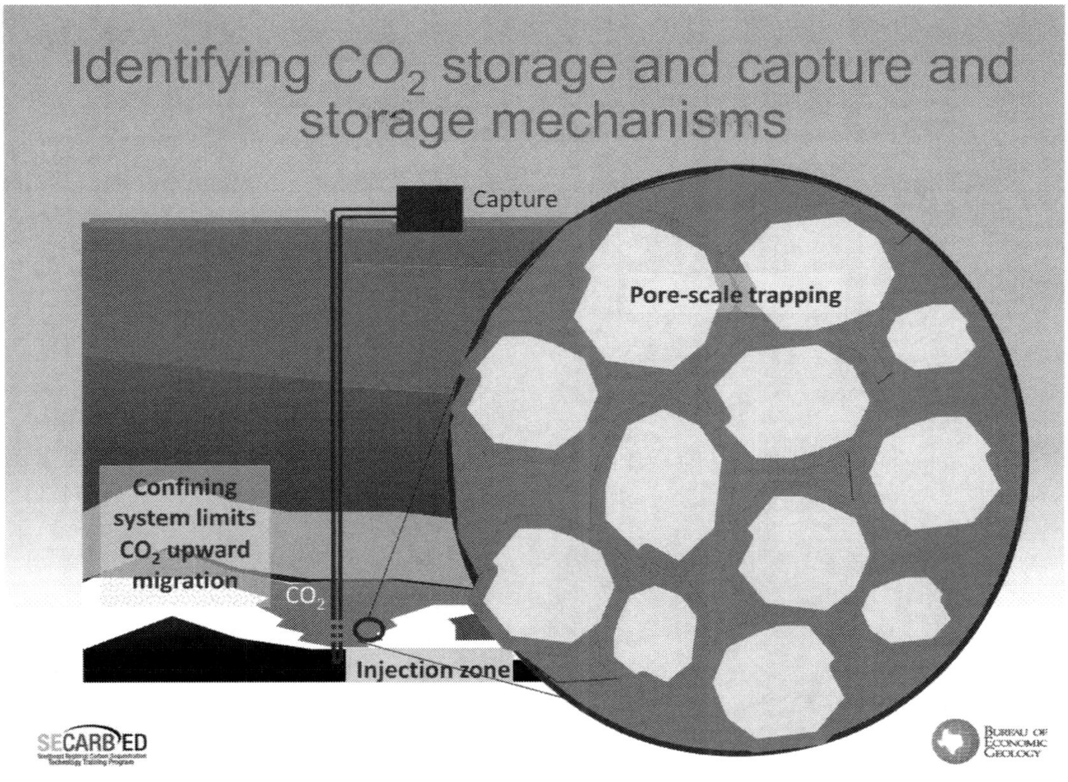
Identifying CO_2 storage and capture and storage mechanisms
Capture
Pore-scale trapping
Confining system limits CO_2 upward migration
CO_2
Injection zone
SECARB ED
BUREAU OF ECONOMIC GEOLOGY

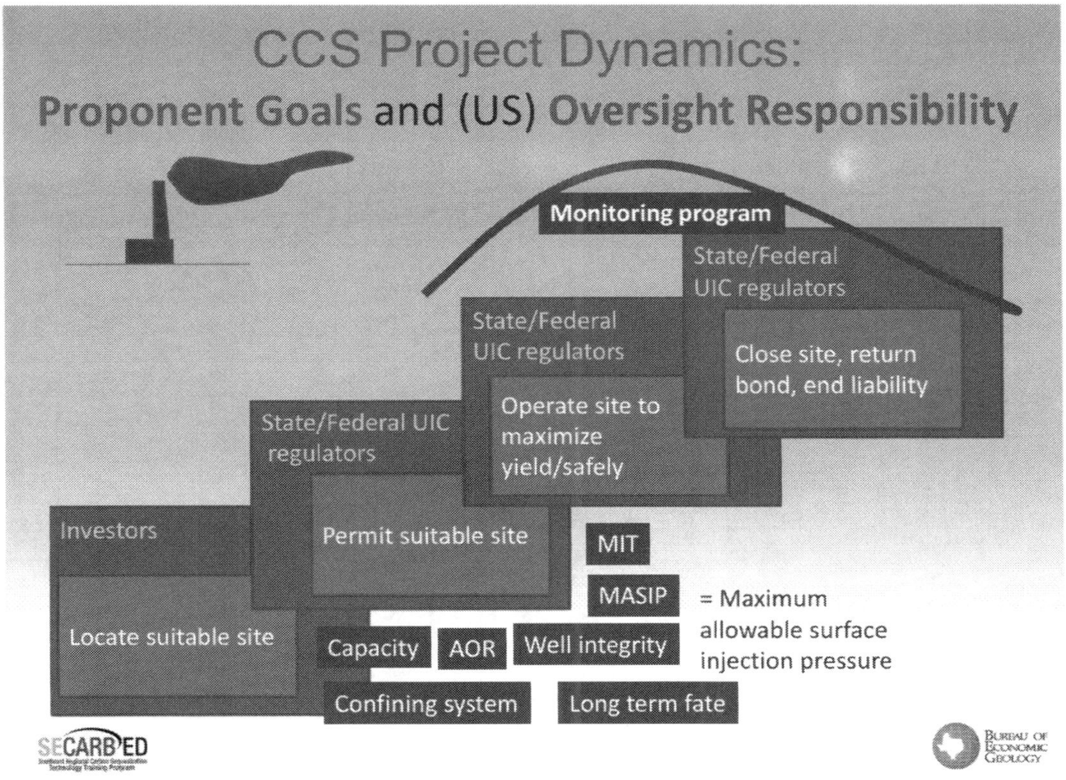
CCS Project Dynamics:
Proponent Goals and (US) Oversight Responsibility
Monitoring program
State/Federal UIC regulators
Close site, return bond, end liability
State/Federal UIC regulators
Operate site to maximize yield/safely
State/Federal UIC regulators
Permit suitable site
Investors
Locate suitable site
MIT
MASIP
= Maximum allowable surface injection pressure
Capacity
AOR
Well integrity
Confining system
Long term fate
SECARB ED
BUREAU OF ECONOMIC GEOLOGY

How EOR sequestration differs from storage-only (brine) sequestration

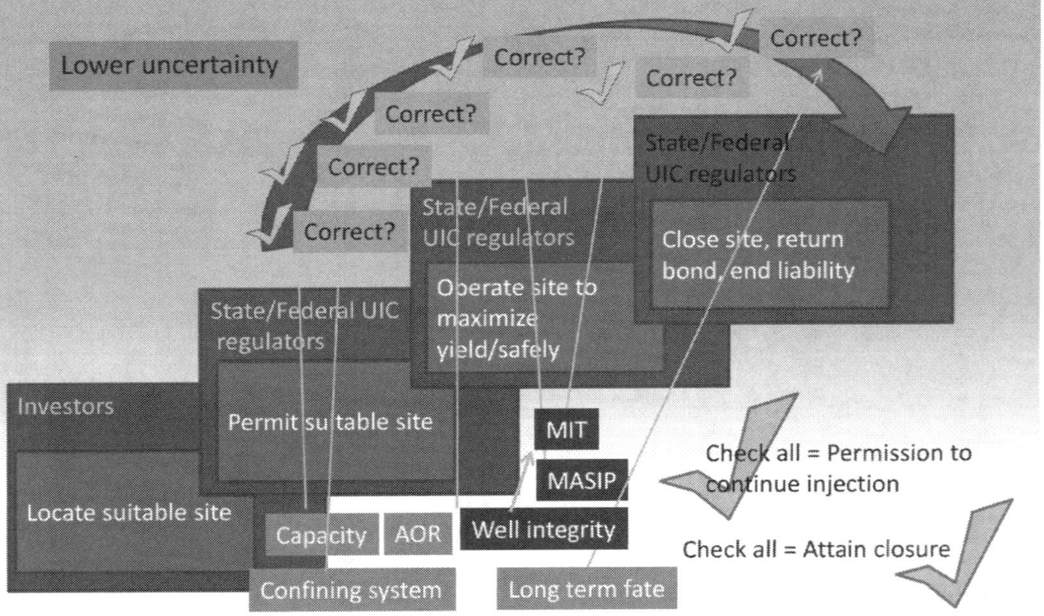

Monitoring For a Skeptic

- If CO_2 was not contained in the reservoir, where could it be?
- How would we know if it was there?
- How robust is the monitoring program to detect leaks?
- *Serious* consideration of questions that disturb stakeholders

SECARB ED Southeast Regional Carbon Sequestration Technology Training Program

Bureau of Economic Geology

Loch Ness Dilemma in a Monitoring Program: Proving a Negative

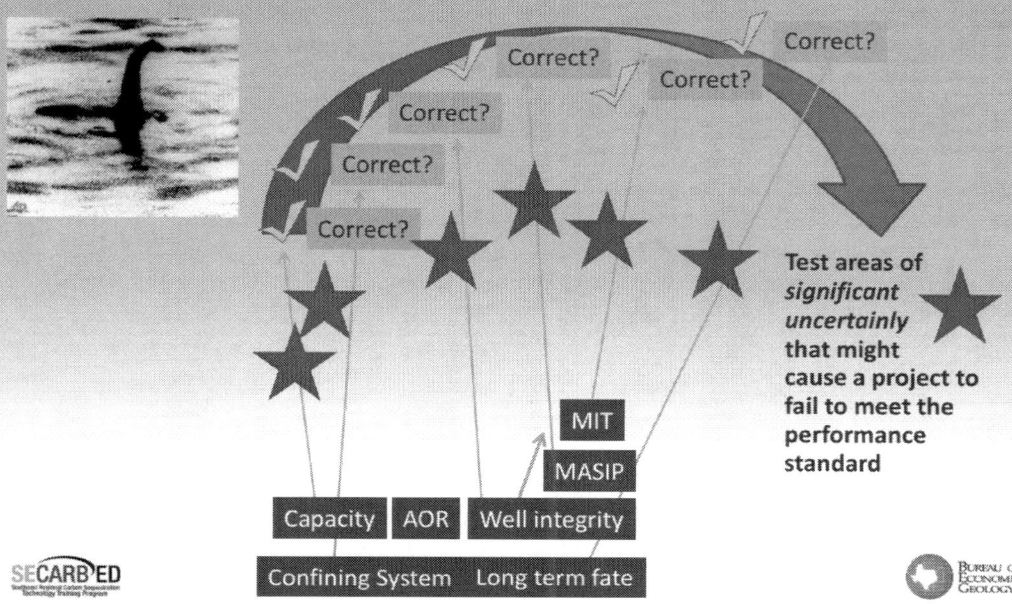

SECARB ED

BUREAU OF ECONOMIC GEOLOGY

Motivation for Monitoring

- **Historic Motivations**
  - Groundwater and surface water protection
  - Historic damages (e.g., salinization)
- **Current Motivations**
  - Benefit to the atmosphere
  - Follow the \$ - Who pays gap between cost of capture and purchase price of CO_2? Taxpayer, and ultimately electricity rate payer
  - Liability (is this a real issue?)
  - Public concerns/values/standards

SECARB ED

BUREAU OF ECONOMIC GEOLOGY

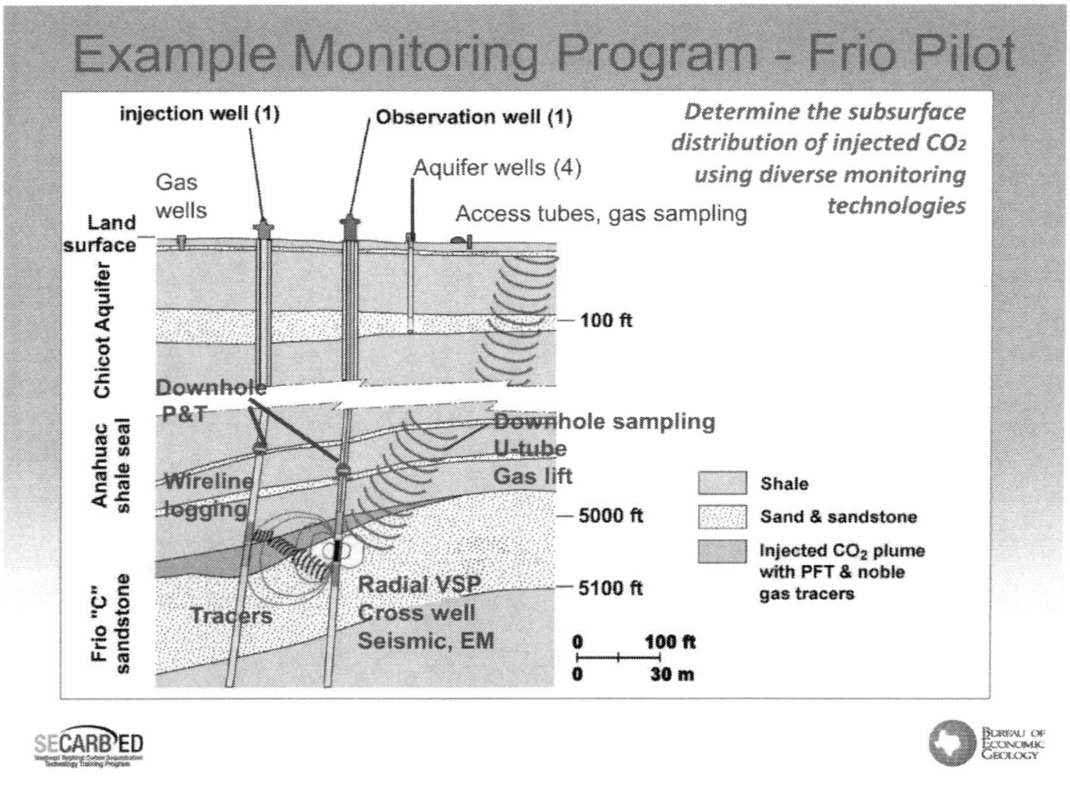
Example Monitoring Program - Frio Pilot
injection well (1)
Observation well (1)
Determine the subsurface distribution of injected CO_2 using diverse monitoring technologies
Aquifer wells (4)
Gas wells
Access tubes, gas sampling
Land surface
Chicot Aquifer
100 ft
Downhole P&T
Downhole sampling
U-tube
Gas lift
Anahuac shale seal
Wireline logging
5000 ft
Shale
Sand & sandstone
Injected CO_2 plume with PFT & noble gas tracers
Frio "C" sandstone
Radial VSP
Cross well
Seismic, EM
5100 ft
Tracers
0 100 ft
0 30 m
SECARB ED
Bureau of Economic Geology

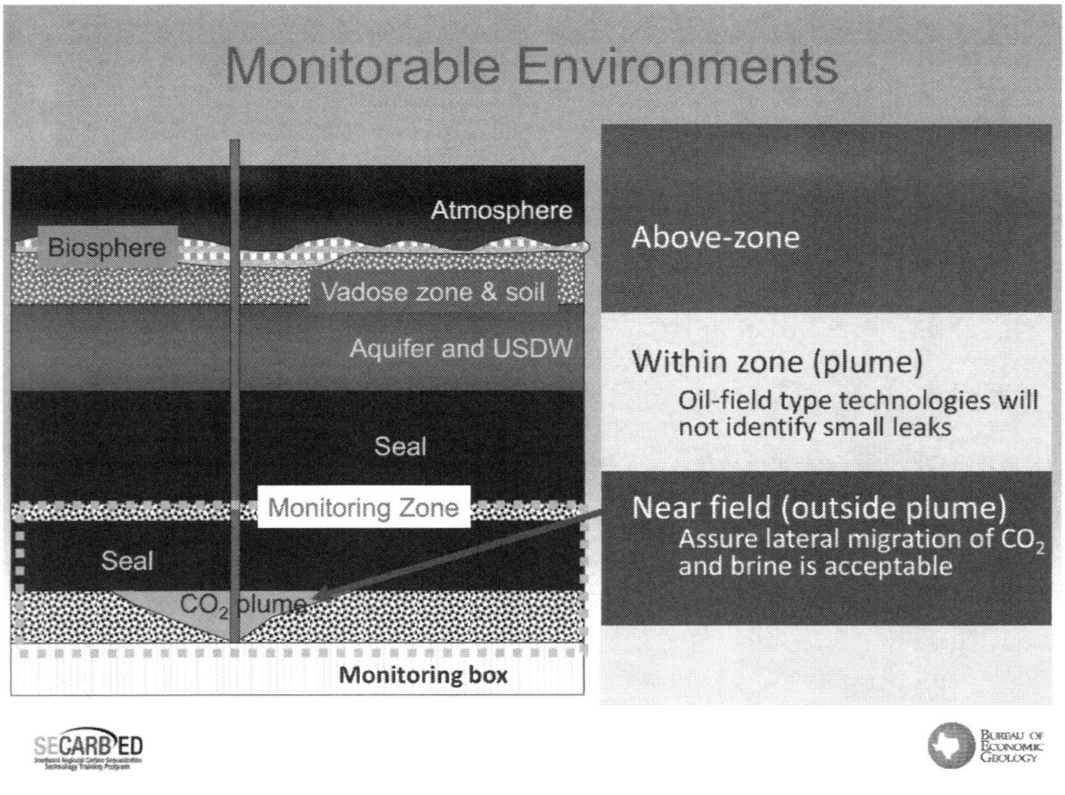
Monitorable Environments
Atmosphere
Biosphere
Vadose zone & soil
Aquifer and USDW
Seal
Monitoring Zone
Seal
CO_2 plume
Monitoring box
Above-zone
Within zone (plume)
Oil-field type technologies will not identify small leaks
Near field (outside plume)
Assure lateral migration of CO_2 and brine is acceptable
SECARB ED
Bureau of Economic Geology

Atmospheric Monitoring

- Direct detection of flux to atmosphere (key to climate impact)
- Background levels can fluctuate substantially with wind direction (need to establish upwind background sources of carbon
- Detection is complicated by systematic error, complexity of data processing

SECARB ED
Southeast Regional Carbon Sequestration Technology Training Program

Bureau of Economic Geology

Soil Gas Monitoring

- Widely using to monitor CO_2 from volcanoes
- Relatively low cost, time- integrated estimates of flux
- Possible interference from microbial respiration will bias estimates of CO_2 release
- Background levels of CO_2 change with soil temperature and water content
- Isotopes and gas ratios can reduce biases

SECARB ED
Southeast Regional Carbon Sequestration Technology Training Program

Bureau of Economic Geology

Groundwater Monitoring

- Monitors a major risk receptor
- Dispersion in aquifers can create larger targets for detection
- Detectability of leakage signal will vary depending on aquifer chemistry
- Could be combined with natural or introduced tracers

SECARB ED Southeast Regional Carbon Sequestration Technology Training Program

BUREAU OF ECONOMIC GEOLOGY

Monitoring Phases and Transitions

- Monitoring phases:
  - Pre-operational monitoring:
    - baseline/background characterization, establish ranges in parameters, higher frequency, model creation
  - Operational monitoring:
    - higher frequency, verify CO_2 volumes/balance, model testing/revising
  - Post-operational monitoring:
    - lower frequency, verify containment, conceptual/numerical model testing/updating (as needed)
- Design monitoring program with transition in mind

SECARB ED Southeast Regional Carbon Sequestration Technology Training Program

BUREAU OF ECONOMIC GEOLOGY

Likely Key Monitoring Activities In Gulf Coast EOR Field

- Well performance – long term stability
- Fault seal performance
- CO_2 Sweep – plume-edge tracking
- Surface facilities carbon balance

SECARB ED

BUREAU OF ECONOMIC GEOLOGY

Pay Attention to Data That Disturb Entrenched Beliefs

Traps and seals that held oil will hold CO_2

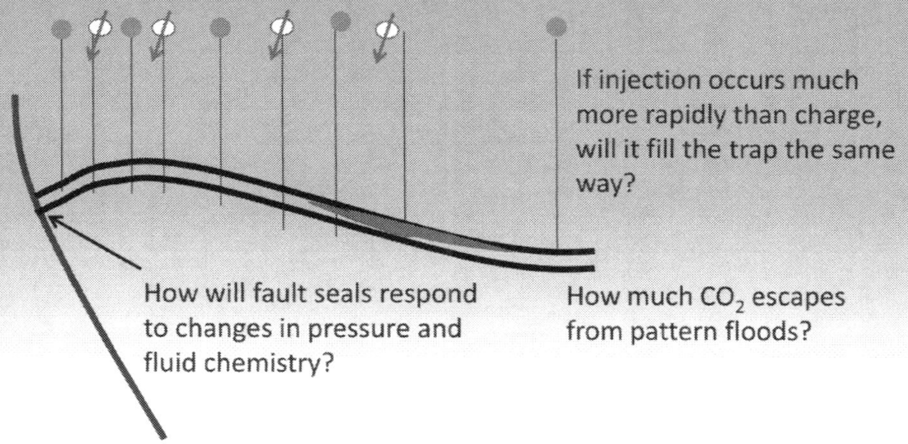

SECARB ED

BUREAU OF ECONOMIC GEOLOGY

Monitoring to Verify CO_2 Containment

- Imaging CO_2 in the subsurface
- Measuring pressure changes
- Assessing compositional changes in subsurface gas/water
- Match to predictive model

SECARB ED

Bureau of Economic Geology

Tools and Processes: Cross Well ERT

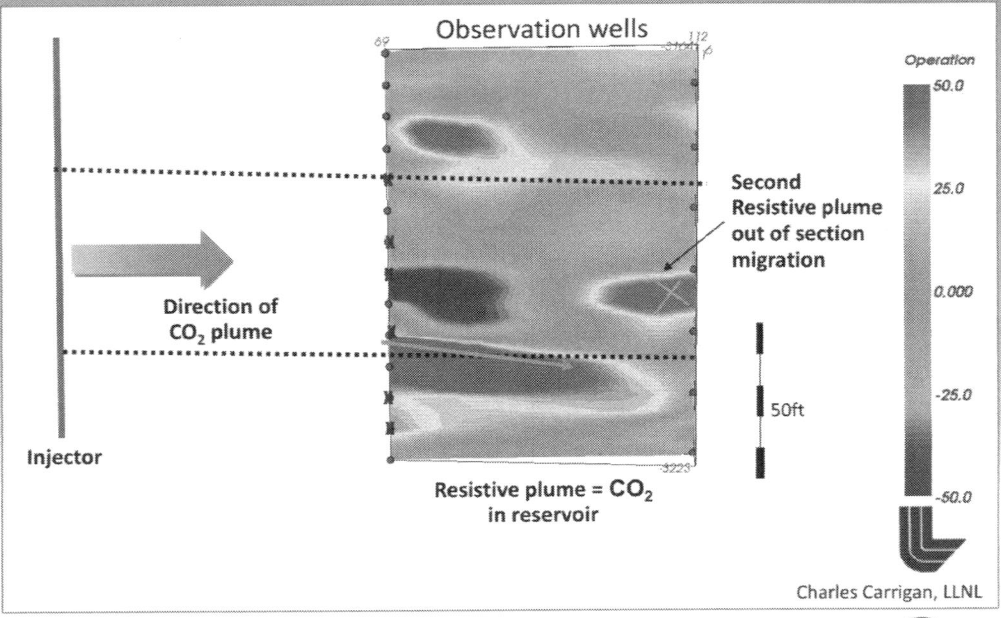

SECARB ED

Bureau of Economic Geology

Tools and Processes: Pressure Monitoring

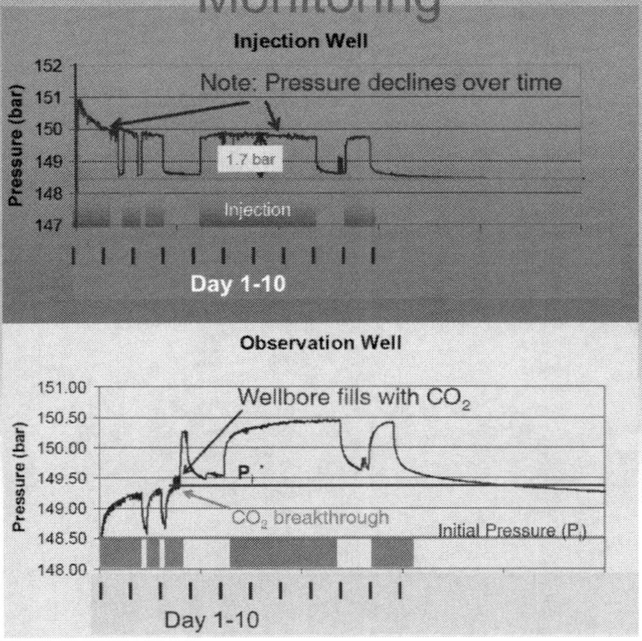

SECARB ED

Bureau of Economic Geology

Modeling / Monitoring – Iterative

- Modeling identifies variables that might control CO_2 post- injection migration
- Measurements made over a short time frame and small distance confirm correct value for these variables
- Better conceptualized and calibrated models used to extend injections

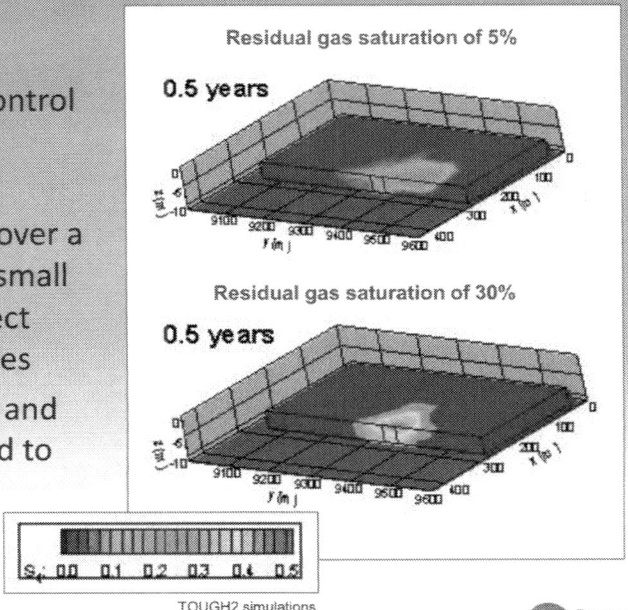

TOUGH2 simulations
C. Doughty LBNL

SECARB ED

Bureau of Economic Geology

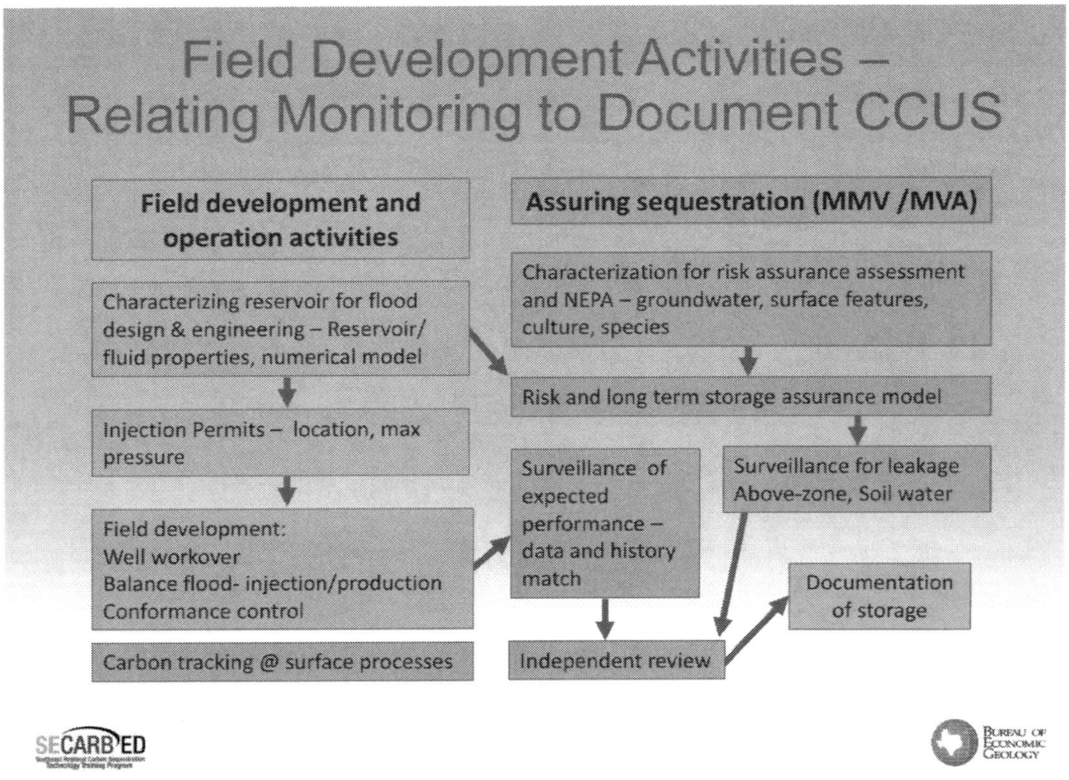
Field Development Activities –
Relating Monitoring to Document CCUS
Field development and operation activities
Assuring sequestration (MMV /MVA)
Characterizing reservoir for flood design & engineering – Reservoir/ fluid properties, numerical model
Characterization for risk assurance assessment and NEPA – groundwater, surface features, culture, species
Risk and long term storage assurance model
Injection Permits – location, max pressure
Surveillance of expected performance – data and history match
Surveillance for leakage Above-zone, Soil water
Field development:
Well workover
Balance flood- injection/production
Conformance control
Documentation of storage
Carbon tracking @ surface processes
Independent review
SECARB ED
BUREAU OF ECONOMIC GEOLOGY

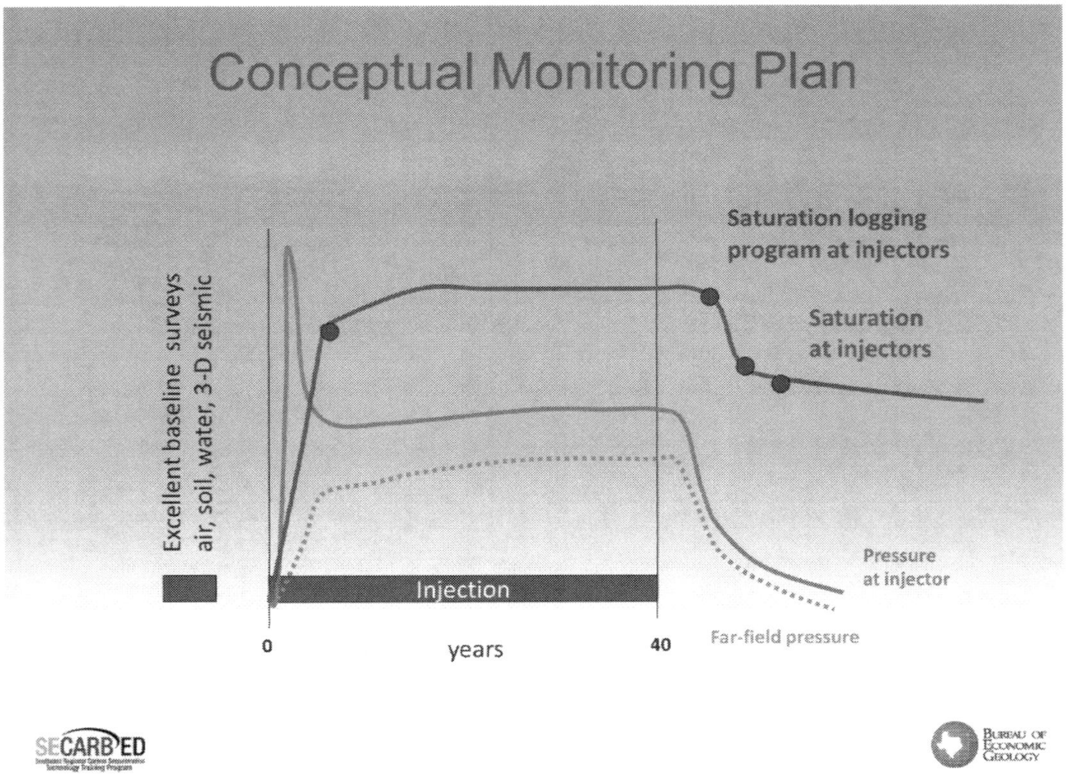
Conceptual Monitoring Plan
Saturation logging program at injectors
Saturation at injectors
Excellent baseline surveys air, soil, water, 3-D seismic
Injection
Pressure at injector
Far-field pressure
0
years
40
SECARB ED
BUREAU OF ECONOMIC GEOLOGY

Monitoring Finds Unacceptable Response

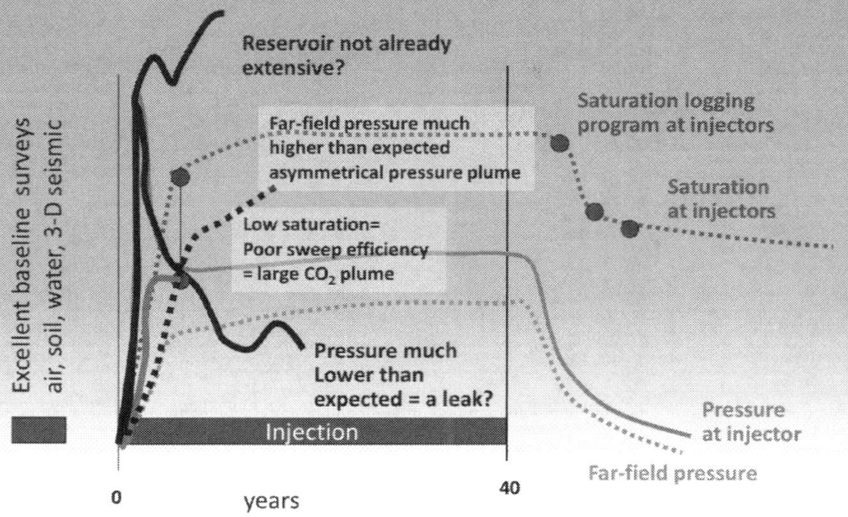

Monitoring Program with Contingencies

- Standardized, dependable, durable instrumentation, reportable measurements
- Frequent pressure measurements above-zone and in-zone – documents conformance
- Episodic saturation logging (at injectors?) syn-and post-injection documents sweep
- Trigger points:
  - an unexpected measurement initiates pre-planned monitoring program to assess origin of response

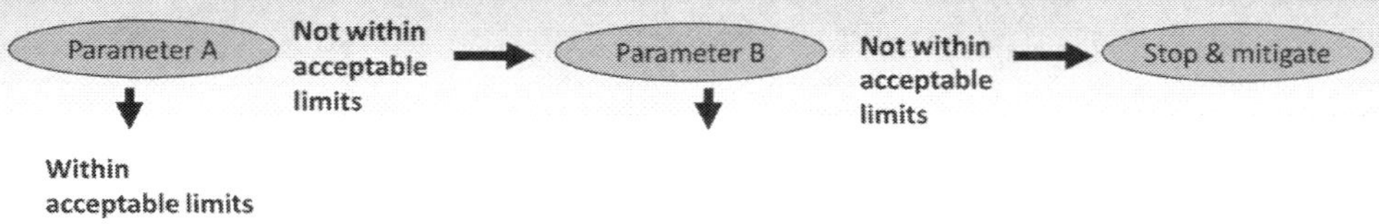

Plan for Contingencies

Unanticipated monitoring results would trigger possible contingencies:

- Enhanced monitoring
- Cessation of injection
- Mitigative action

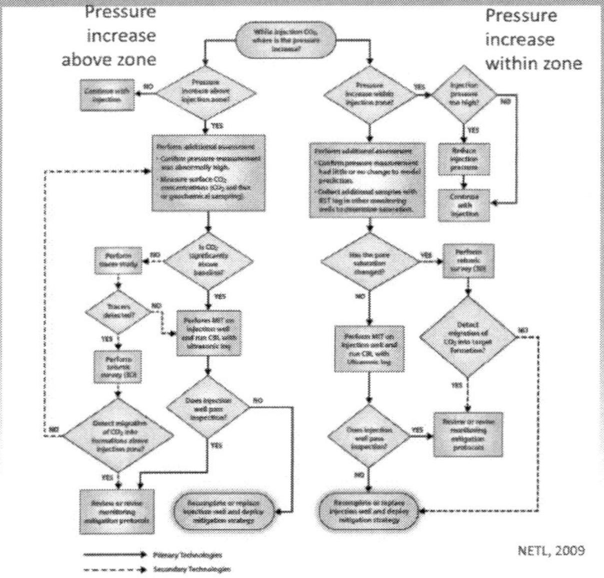

SECARB ED

BUREAU OF ECONOMIC GEOLOGY

Conclusions

- Attracting anthropogenic CO_2 is a promising way to expand EOR
- Attracting CCUS (CO_2 EOR) is a promising way to expand anthropogenic CO_2 containment
- Monitoring to ensure favorable storage conditions will attract supplies of anthropogenic CO_2
- Monitoring skills are practiced by EOR operators today
- Successful monitoring links different types of data and builds understanding of site environs and processes

SECARB ED

BUREAU OF ECONOMIC GEOLOGY

Additional Resources:

Gulf Coast Carbon Center, Bureau of Economic Geology, University of Texas at Austin
www.gulfcoastcarbon.org
http://www.beg.utexas.edu/gccc/forum/codexhome.php

World Resources Institute, 2008, CCS guidelines for carbon dioxide capture, transport, and storage, Washington DC
http://www.wri.org/publication/ccs-guidelines (specifically see p 53-105)

Cooper, Cal and CO_2 Capture Project, 2009, A Technical Basis for Carbon Dioxide Storage
http://www.co2captureproject.org/pubsearch.php (requires registration)

General Technical Support Document for Injection and geologic sequestration of carbon dioxide: subparts RR and UU (Clean Air Act) Greenhouse Gas reporting program
http://www.epa.gov/climatechange/emissions/downloads10/Subpart-RR-UU_TSD.pdf

EPA class 6 rules: US Environmental Protection Agency, Rules and Regulations
http://www.gpo.gov/fdsys/pkg/FR-2010-12-10/pdf/2010-29954.pdf

NETL, 2011, Best Practices Manual: Monitoring, Verification and Accounting of CO2 stored in deep geologic formations
http://www.netl.doe.gov/technologies/carbon_seq/refshelf/MVA_Document.pdf

IEAGHG monitoring tool developed by the British Geological Survey
http://www.co2captureandstorage.info/co2tool_v2.2.1/index.php.

SECARB ED

Bureau of Economic Geology

Gulf Coast Carbon Center (GCCC)

BEG Team

Susan Hovorka
Tip Meckel
J. P. Nicot
Rebecca Smyth
Ramon Trevino
Katherine Romanak
Changbing Yang
Dave Carr
Sigrid Clift
Jiemin Lu
Seyyed Hosseini
Jong Won Choi
Vanessa Nunez
Khandaker Zahid
Carey King
students
Scott Tinker
Ian Duncan

Collaborators

BEG- CEE
LBNL
LLNL
ORNL
USGS
New Mexico Tech
Mississippi State U
U of Mississippi
BP
ARI
SECARB
SWP
UT-PGE
UT- CCEP
UT- DGS
Univ Edinburgh

NETL: Bruce Brown
SSEB: Jerry Hill

IA sponsors (2011)

bp
Chevron
BG GROUP
KINDER MORGAN
Luminant
Saudi Aramco
Schlumberger Carbon Services
ExxonMobil
nrg
Entergy
EDF Environmental Defense Fund
ConocoPhillips
Jackson School of Geosciences

Gulf Coast Carbon Center

Bureau of Economic Geology

Acknowledgements

- SECARB-Ed is supported by the U.S. Department of Energy's (DOE) National Energy Technology Laboratory as part of the American Recovery and Reinvestment Act of 2009 under DE-FE0001930.
  - www.netl.doe.gov/technologies/carbon_seq/arra/training.html
- The Bureau of Economic Geology is a subcontractor to Southern States Energy Board, the Principal Contractor of SECARB-Ed to DOE.
  - http://www.secarb-ed.org/

Photo: Daniel Foster/Flickr

The Need for Constructive and Rational Dialogue on Risk

Mannan, M. S., Mary Kay O'Connor Process Safety Center

Originally presented on: 10/6/2011 0:00:00 - 0:45:00

The driving forces for change and sustainable development in a society are growing population, rising standard of living, perception of risk, and society's choices based on these driving forces. As the population is growing there is also at the same time a rise in standard of living. These changes are eminent not only in the United States, but particularly when you travel abroad to countries like China and India. And when the standard of living increases, the perception of risk or how willing you are to tolerate risk also changes. When you don't have a square meal to eat, risk means something quite different. But when you have three good meals, a nice house to live in, and a car to drive, your perception of risk is completely different. And then there comes the issue of society's choices, and as people's perspectives change, the choices made also change. And that's where a constructive and rational public dialogue on risk is essential.

China currently has an annual growth rate of 11% while the Indian economy is growing at about 8%. On a per capita basis, the United States uses 25 times more energy than China and on a similar comparative bases, the United States uses 62 times more energy per capita than India. In addition, when one takes into account the population of the two countries combined (about 2 and a half billion), the math is very simple but scary. So, if the growth rates of the world economies keep on increasing at the rate they are, the question is where are the resources going to come to feed the growth.

Given the back drop of the facts above, we do have to think about what our options are on a day-to-day basis. Clearly, "not in my back yard" is not an approach that will give us sustainable options. This paper presents a case history that puts this dialogue on risk "front and center." A facility engaged in the production of pesticide, had an accident in 2008, it was inspected by OSHA after that, and citations were issued. It was also inspected and investigated by the EPA and a host of other agencies, including the Chemical Safety and Hazard Investigation Board (CSB). The facility used methyl isocyanate in the production of the pesticide, but the 2008 incident did not involve methyl isocyanate, neither was any part of the plant handling methyl isocyanate involved in the incident. Following the incident, however; the plant went through extensive redesign that included reduction of inventory of methyl isocyanate and major reconfiguration. However, prior to startup of the redesigned plant, a lawsuit was filed in federal court claiming that the plant posed an "unreasonable risk" to the neighbors. This paper describes the events following the lawsuit and ultimate outcome of the chain of events.

CSB U.S. Chemical Safety and Hazard Investigation Board

Lessons Learned at the U.S. Chemical Safety Board

AIChE Regional Process Technology Conference

John Bresland
Board Member
U.S. Chemical Safety Board

Galveston, Texas
October 6, 2011

1

www.csb.gov

CSB U.S. Chemical Safety and Hazard Investigation Board

Disclaimer

The PowerPoint presentation given by John Bresland, Board Member, the United States Chemical Safety and Hazard Investigation Board to the AIChE Regional Process Technology Conference is for general informational purposes only. The presentation represents the individual views of the Board Member and all references, conclusions or other statements regarding current ongoing CSB investigations are preliminary in nature and do not represent a formal adopted product of the Board. Users of this presentation should also note that the contents were compiled solely for this presentation. For specific and accurate information on completed investigations, please refer to the final printed version by going to the CSB website at www.csb.gov. and clicking on the specific report desired under completed investigations

2

www.csb.gov

CSB U.S. Chemical Safety and Hazard Investigation Board

What I'm Going to Talk About

- What's new at the Chemical Safety Board?
- Lessons I've learned in my 9 years at the CSB
- Is the industry having fewer incidents?

3

www.csb.gov

CSB U.S. Chemical Safety and Hazard Investigation Board

The U.S. Chemical Safety Board

- Authorized by the 1990 U.S. Clean Air Act
- Offices in Washington, DC and Denver, CO
- Budget $10.8M in FY 2011
- Three Board members – two vacancies
- Staff – 41

4

www.csb.gov

CSB U.S. Chemical Safety and Hazard Investigation Board

CSB BOARD MEMBERS

John
Bresland

Rafael
MOURE-ERASO

Chairman

Mark
GRIFFON

5

www.csb.gov

CSB U.S. Chemical Safety and Hazard Investigation Board

Recently completed investigations

- DuPont, Belle, West Virginia
- Public Safety at Oil and Gas Storage Facilities
- Texas Tech Chemical Laboratory Incident
- Bayer CropScience, Institute, West Virginia

6

www.csb.gov

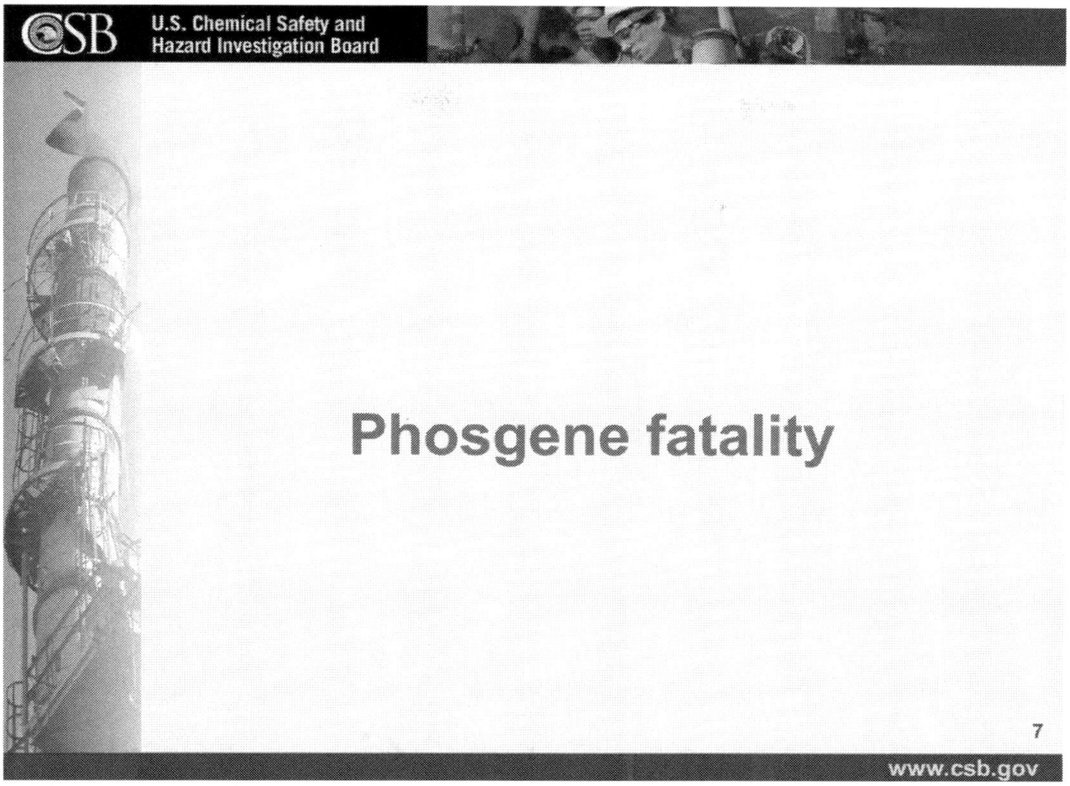
CSB
U.S. Chemical Safety and
Hazard Investigation Board
Phosgene fatality
7
www.csb.gov

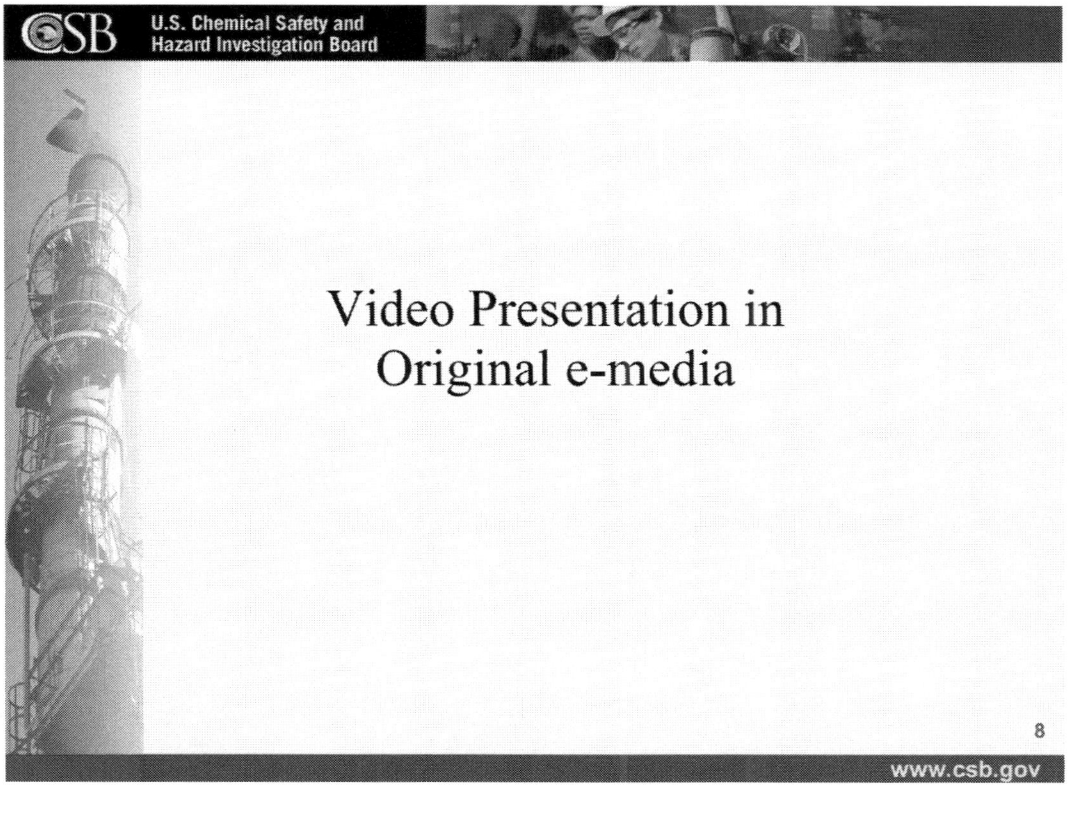
CSB
U.S. Chemical Safety and
Hazard Investigation Board
Video Presentation in
Original e-media
8
www.csb.gov

U.S. Chemical Safety and
Hazard Investigation Board
Delphi Oil gas production site
Carnes, Mississippi
www.csb.gov

U.S. Chemical Safety and
Hazard Investigation Board
Texas Tech University
www.csb.gov

U.S. Chemical Safety and Hazard Investigation Board

Current Major Investigations

- Gulf of Mexico
- Tesoro, Anacortes, Washington
- Citgo hydrogen fluoride release and fire
- Hoeganaes metal dust fatalities
- Caribbean Refining explosion and fire
- Silver Eagle refinery explosion

11

www.csb.gov

CSB U.S. Chemical Safety and Hazard Investigation Board

Tesoro refinery – Anacortes, Washington

www.csb.gov

CSB U.S. Chemical Safety and Hazard Investigation Board

What I've learned at the Chemical Safety Board

14

www.csb.gov

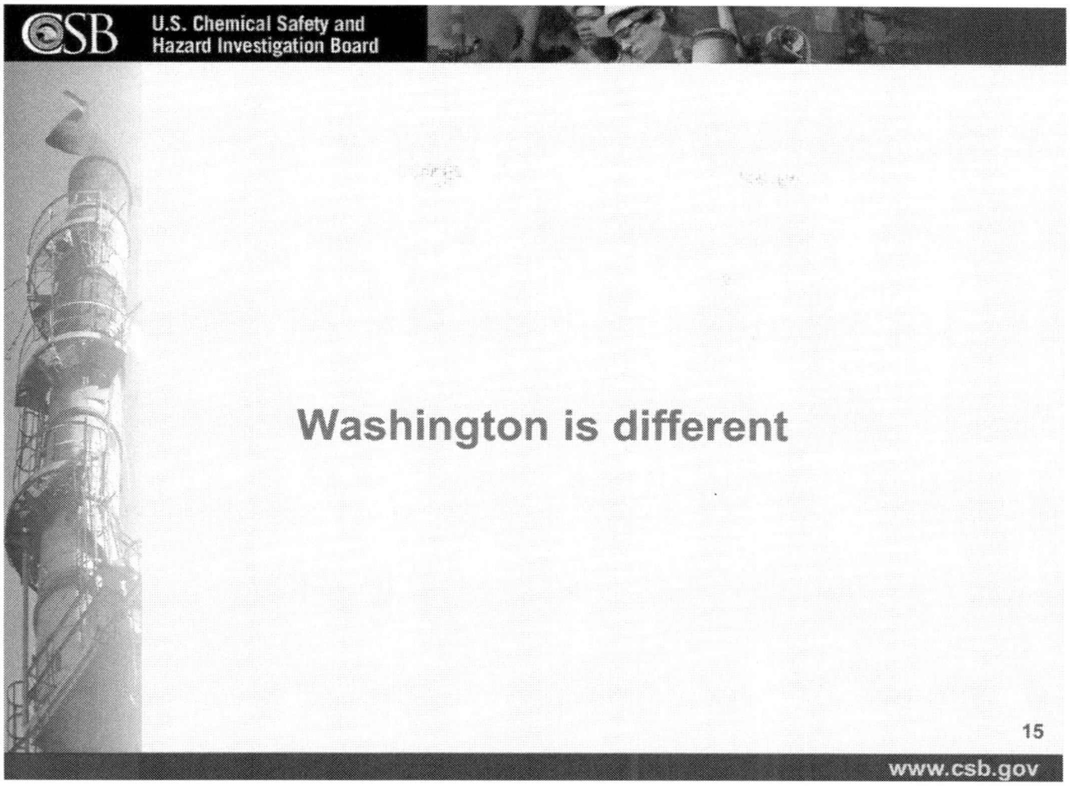
U.S. Chemical Safety and Hazard Investigation Board
Washington is different
15
www.csb.gov

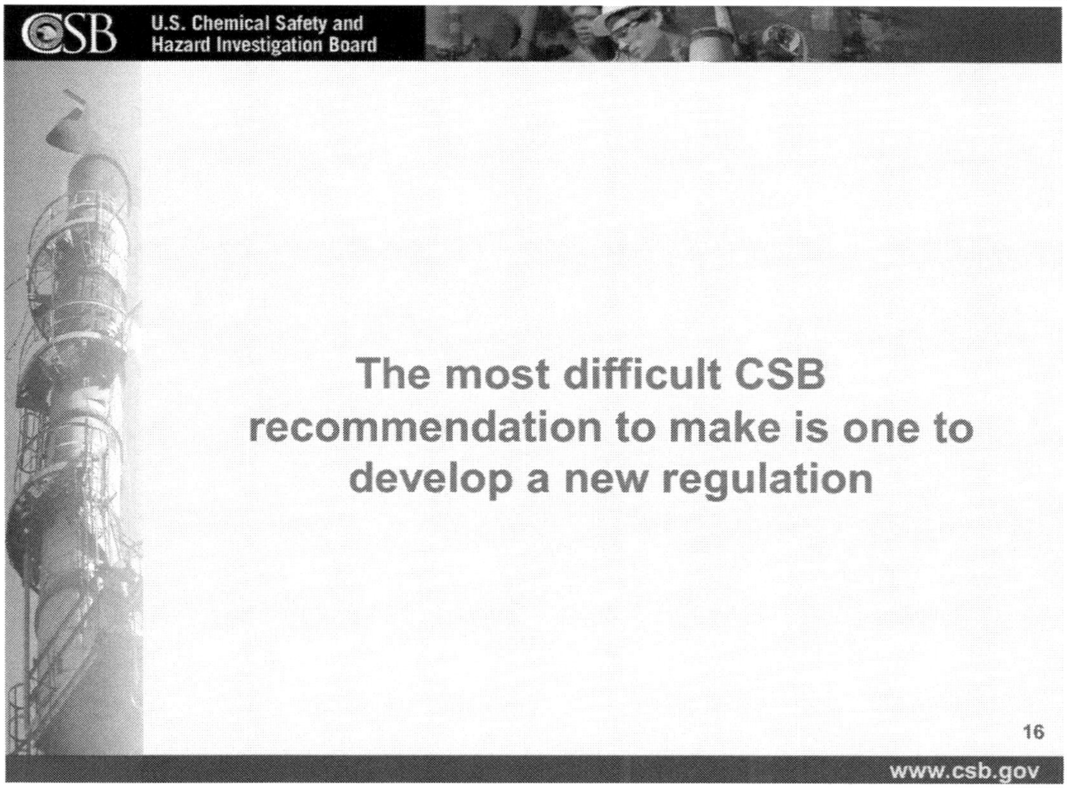
U.S. Chemical Safety and Hazard Investigation Board
The most difficult CSB recommendation to make is one to develop a new regulation
16
www.csb.gov

CSB U.S. Chemical Safety and Hazard Investigation Board

The OSHA PSM Standard

- It's an excellent basis for a strong process safety program
- It needs continuing resources and support from management
- It's critical to focus on mechanical integrity

17

www.csb.gov

CSB U.S. Chemical Safety and Hazard Investigation Board

www.csb.gov

CSB U.S. Chemical Safety and Hazard Investigation Board

One incident can ruin years of building a reputation

19

www.csb.gov

CSB U.S. Chemical Safety and Hazard Investigation Board

Keep your community informed

20

www.csb.gov

Sometimes I get perplexed by the actions of insurance companies

- In a complex chemical plant, details matter
- Everyday diligence is essential

U.S. Chemical Safety and
Hazard Investigation Board
Bad things can happen to good people
23
www.csb.gov

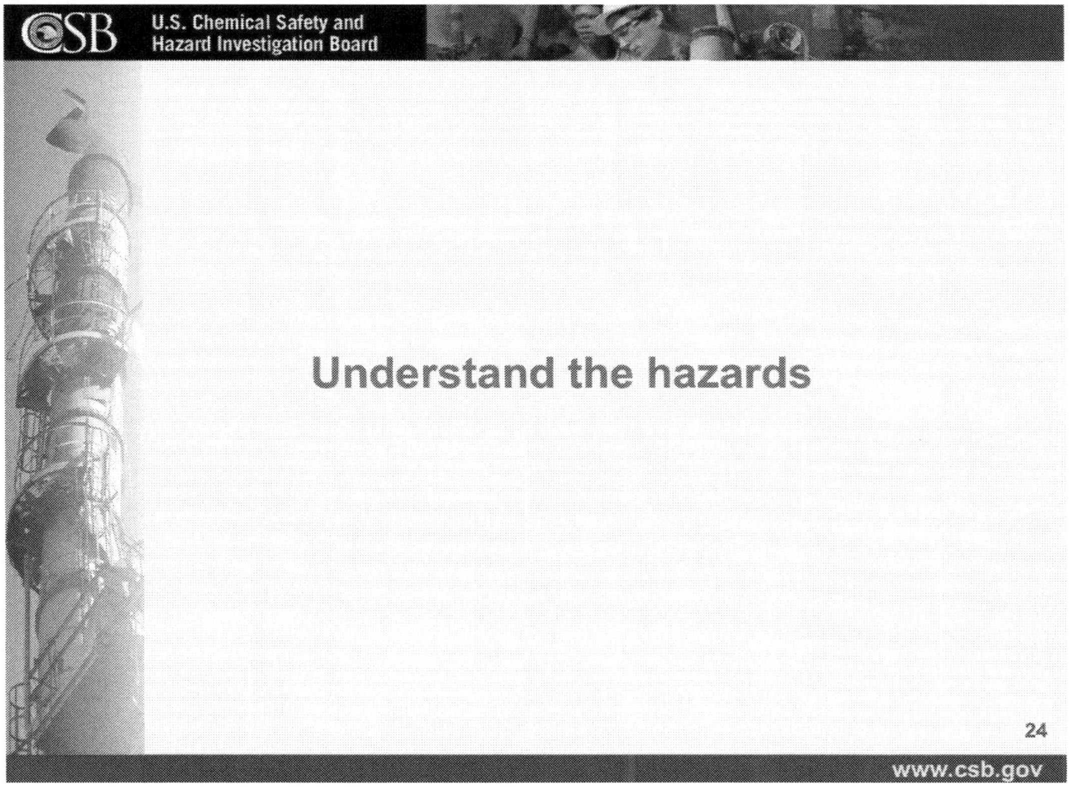
U.S. Chemical Safety and
Hazard Investigation Board
Understand the hazards
24
www.csb.gov

Imperial Sugar Company Refinery, Port Wentworth, Georgia February 7, 2008

If you want to get into the chemical manufacturing business, make sure you know what you are doing.

CSB
U.S. Chemical Safety and
Hazard Investigation Board
T2 Laboratories Jacksonville, Florida
December 19, 2007

U.S. Chemical Safety and Hazard Investigation Board

Individual failure vs. management failure

29

www.csb.gov

CSB U.S. Chemical Safety and Hazard Investigation Board

BP Texas City

- Evacuate trailers
- Relocate trailers
- Eliminate ignition sources
- Eliminate blowdown drums
- Malfunctioning instrumentation
- Inadequate control board display
- **Tower bottoms valve closed**
- Operator fatigue
- No supervisor in control room

31

www.csb.gov

CSB U.S. Chemical Safety and Hazard Investigation Board

Are we having fewer chemical plant/refinery incidents?

32

www.csb.gov

U.S. Chemical Safety and Hazard Investigation Board

Questions?

www.csb.gov

john.bresland@csb.gov

33

www.csb.gov

Protect Health, Safety and Environment in Engineering Projects

Ralph Ellis, Myra Ray, Alan English, Jack Chosnek & Vic Edwards
AIChE Regional Process Technology Conference, October 6, 2011

KVÆRNER

Ralph Ellis, Project Director

- Born Perth, Western Australia, 1969
- Bachelor of Engineering, Mechanical (with honors) 1990.
- 21 Years of Project Experience Across Australia & America Covering
  - Petrochemical Refining & Hydrocarbons
  - Chemical Refining
  - LNG Regasification
  - Materials Handling & Materials Processing
- Experience in Engineering, Procurement & Construction of Greenfield Facilities, co-located Additions to Existing Facilities & Area Turnarounds

KVÆRNER

Make Safety Second Nature

- **The Foundation of Health, Safety and Environmental Success is a Commitment by All to their Personal & Collective Outcomes.**
- **Management Must Provide the Example and the Leadership.**
- **Establish Forums where the Entire Team can Contribute & Seek Open Participation.**
- **Everyone in the Organization must Value and "Own" HSE**

3

KVÆRNER

Early Planning for Future Success

- **Build around a Corporate HSE Management System & Support from Corporate Leadership.**
- **Conduct an HSE Analysis to Select the Risk Identification Methods to be Applied.**
- **Develop a Project HSE Management Plan to Document the Principles and Process.**
- **Organize the Project Team for HSE Success.**
- **Schedule HSE Critical Activities & Ensure that the Necessary Inputs are Available.**

4

KVÆRNER

Integrate HSE in Design Early in a Project

- Early Integration of HSE in Design is Essential.
- Create the HSE Basis during FEED
- Early Integration in FEED Prevents Costly Changes in EPC
- Complete HSE in Design during Detail Engineering.

5 KVÆRNER

FEED HSE in Design

- Create a Strong Commitment to HSE by All Team Members.
- Incorporate HSE Principles and Practices in the Project Design Basis.
- Practice HSE in Design Throughout the Project.
- Conduct Reviews which Utilize Increasingly Defined Data as the Project is Developed.
- Manage Change during the Design Process.
- Avoid Drilling Down in Detail before the Base Data can support this.

6 KVÆRNER

FEED HSE in Design – Risk Engineering

- Conduct a Preliminary Hazards Analysis (HAZID), including Inherently Safer Technology, when PFDs & Plot Plan are Available.
- Model the Consequences of Hazardous Scenarios & Feedback the Results to the Design Team.
- Hold a P&ID Design Review Prior to the PHR (HAZOP) to Verify that the Design is Ready to be HAZOPed.
- Conduct a Preliminary Process & Utility PHR (HAZOP)
- Complete Preliminary LOPA and SIL Assessments.
- Conduct a Facility Siting Study.
- Track PHR Recommendations to Successful Resolution.

KVÆRNER

FEED HSE in Design - Environment

- Create a Project Environmental Management Plan.
- Identify Potential Waste Streams and Design to Eliminate or Minimize.
- Assess Baseline Noise Levels and Model Anticipated Noise Sources.
- Plan in order to Minimize Adverse Impacts on the Environment during Construction.
- Ensure that Construction Teams are Skilled in Sensitive Treatment of the Environment.
- Design to Minimize Adverse Impacts on the Environment during Operation.
- Ensure that the Facility Design meets all Regulatory Requirements.
- Pay Attention to the Federal, State & Local Environmental Protocols & Permitting Requirements.

KVÆRNER

FEED HSE in Design – Fire Protection

- Incorporate Fire Protection Strategy in Project Design Basis Consistent with Owner Standards & Relevant Codes.
- Identify Fire Hazardous Areas and Fire Water Demand.
- Develop Fire Protection Water System Design.
- Specify and Locate Fire and Gas Detection and Alarm System.
- Specify Fire Proofing for Key Structures and Vessels.
- Specify Miscellaneous Fire Protection and Safety Equipment.

KVÆRNER

FEED HSE in Design - Safety

- Minimize Exposure to Occupational Health and Safety Hazards.
- Review the Plot Plan and Equipment Arrangement Drawings.
- Employ Human Factors Engineering in Facility Design.
- Conduct Electrical Hazardous Area Classifications.
- Design for Constructability and Construction Safety.
- Develop the Commissioning & Start-up Sequence During FEED.

KVÆRNER

Questions?

11 © Kvaerner 2011 | 14.10.2011 KVÆRNER

Copyright and disclaimer

Copyright
Copyright of all published material including photographs, drawings and images in this document remains vested in Kvaerner and third party contributors as appropriate. Accordingly, neither the whole nor any part of this document shall be reproduced in any form nor used in any manner without express prior permission and applicable acknowledgements. No trademark, copyright or other notice shall be altered or removed from any reproduction.

Disclaimer
This Presentation includes and is based, inter alia, on forward-looking information and statements that are subject to risks and uncertainties that could cause actual results to differ. These statements and this Presentation are based on current expectations, estimates and projections about global economic conditions, the economic conditions of the regions and industries that are major markets for Kværner ASA and Kværner ASA's (including subsidiaries and affiliates) lines of business. These expectations, estimates and projections are generally identifiable by statements containing words such as "expects", "believes", "estimates" or similar expressions. Important factors that could cause actual results to differ materially from those expectations include, among others, economic and market conditions in the geographic areas and industries that are or will be major markets for Kvaerner's businesses, oil prices, market acceptance of new products and services, changes in governmental regulations, interest rates, fluctuations in currency exchange rates and such other factors as may be discussed from time to time in the Presentation. Although Kværner ASA believes that its expectations and the Presentation are based upon reasonable assumptions, it can give no assurance that those expectations will be achieved or that the actual results will be as set out in the Presentation. Kværner ASA is making no representation or warranty, expressed or implied, as to the accuracy, reliability or completeness of the Presentation, and neither Kværner ASA nor any of its directors, officers or employees will have any liability to you or any other persons resulting from your use.

12 © Kvaerner 2011 | 14.10.2011 KVÆRNER

Auoth Bio

- Dr. Sanjeev Saraf is a senior associate in Exponent Failure Analysis Associates. Dr. Saraf's primary focus is on evaluating processes/products for increased safety, reliability, and economic feasibility.

Exponent

1

How to Ensure HAZOP Quality?

Sanjeev Saraf, Ph.D.
Subodh Medhekar, Ph.D., P.E., C.R.E.

Exponent Inc.

6th Oct. 2011

Exponent

ARE THERE GOOD AND BAD HAZOPS?

Exponent

3

Hazop Quality

- Commonly observed practices
  - Preparation
  - Experienced facilitator
  - Right team
- Audit
- Observer

Exponent

4

OBJECTIFY HAZOP QUALITY COMPONENTS

Exponent

5

BEFORE HAZOP

Exponent

6

Team Selection

- HAZOP is affected by
  - Knowledgebase
  - Group interaction
- Recommended Practice
  - Selection of team members should be based on professional background as well as personalities
  - Try bringing in new team members during HAZOP revalidation
  - Bring in professionals with requisite professional background who have not been involved in the process being HAZOPed

Exponent 7

Facilitator Selection

- Common selection criteria
  - Experience
  - Knowledge of the process
  - Leadership and management skills
- Drawbacks
  - Years of experience and intimate knowledge of a process may make the facilitator numb to certain risks and worse influence the HAZOP team's opinions.
  - Leadership skills in work environment do not necessarily transfer to managing relevant conversation during HAZOP session.
  - Dominant personality, often mistaken for leadership skills, of the facilitator may bias the HAZOP
- Recommended Practice
  - Neutral personality and ability to engage team
  - Interview the facilitator
  - Grasp of risk

Exponent 8

DURING HAZOP

Exponent

9

HAZOP Proceeding

- Use 3D views of the plant if available during a HAZOP. Most people can relate to 3D pictures better. Thus visualization keeps the team engaged and helps stimulate discussions.
- Dual projectors for pulling up supporting documentation
- 6-hour of HAZOP per day is recommended. Spending additional time may help you meet the schedule but the team may be going through the motions of HAZOP and sacrificing quality
- Sufficient breaks should be provided. We recommend at least a 10-minute break every 1 to 1.5 hours of HAZOP
- For HAZOPs that go beyond four weeks, we suggest a break between longer HAZOP

Exponent

10

HAZOP Proceeding

- Relevant case studies should be used to energize the HAZOP team and keep discussions lively.
- Provide a mechanism for HAZOP members to provide feedback anonymously during the study period
- We suggest a HAZOP team of 7 ± 2 people
- Managing group dynamics
  - Group thinking
  - False-consensus

Exponent

11

AFTER HAZOP

Exponent

12

Feedback

- Feedback from HAZOP members
  - Facilitator
  - Documentation
  - Duration
  - Suggestions

Exponent

13

Closing Out Recommendations

- Wording of HAZOP recommendations is extremely important. Recommendations must be clear, concise, unambiguous, and relevant
- A HAZOP recommendation may have multiple "action items."
  - Treat each HAZOP recommendation as a project – each project has the outcome and actions assigned to individuals along with deadlines. This way you will know why the HAZOP actions are not being closed out.
  - Instead of using Excel, we recommend a simple project management applications.
    - Example, Deskaway: $25/month (http://www.deskaway.com)

Exponent

14

Troubleshooting and Improving Operation of a Main Fractionating Tower

Pless, L., Tracerco
LaCasse, S., CHS Refining
Originally presented on: 10/6/2011 0:00:00 - 0:25:00

Following a refinery turnaround the Main Fractionator in the Ultra-low Sulphur Diesel unit was field tested and studied for its root cause of flooding. Gamma scans indentified two sources of flooding originating from a pair of draw downcomers. Neutron backscatter technique was used to investigate and confirm downcomer hydraulics to identify the root-cause of flooding in the Fractionator. The downcomer design was studied and determined to be conducive for plugging and flooding. A simpler downcomer design was proposed and implemented renovating a complicated piece of hardware that was conducive to plugging.

Distill-Max™

2011 AIChE Regional Process Technology Conference
Galveston, October 6, 2011

Dividing Wall Column – Reliable Solution for Efficient Refinery Separations

Bhari Bhujan Singh
Bill Townsend

KBR

Agenda

Distill-Max™

- Introduction to DWC
- DWC – Past & Present
- DWC – Drivers, Considerations, and Benefits
- Case Studies
- Conclusions

Dividing Wall Column (DWC): Concept

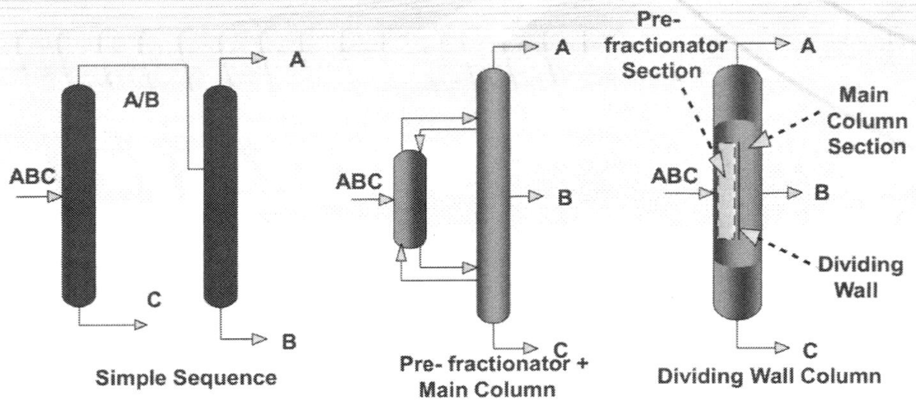

The DWC produces three purified products by effectively combining two distillation columns into one.

DWC: Working Principle

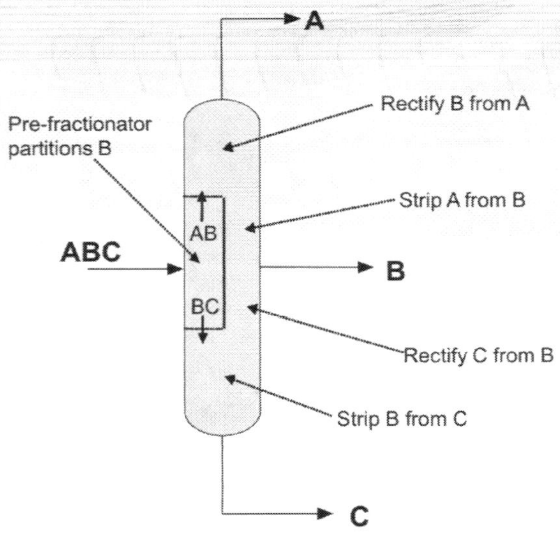

Dividing Wall Column (Three Product)

- **Minimum re-mixing of B**
  - **Re-mixing always present in conventional columns**
- **Prefrac and Main frac**
  - **Fully heat integrated via vapour links**

DWC: Thermodynamically Efficient

Simple Sequence | **Dividing Wall Column**

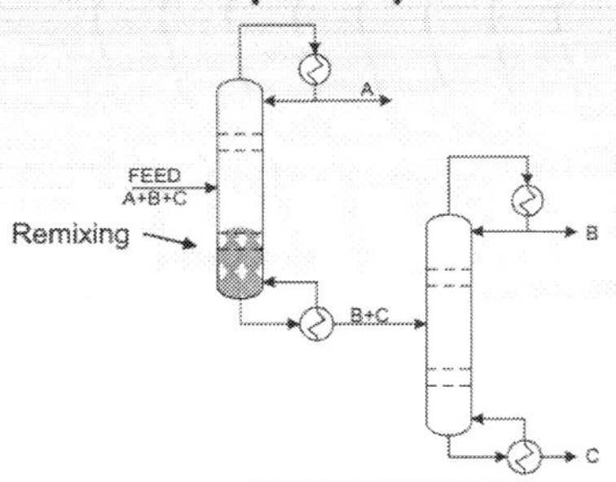

Remixing Effect in Direct Sequence

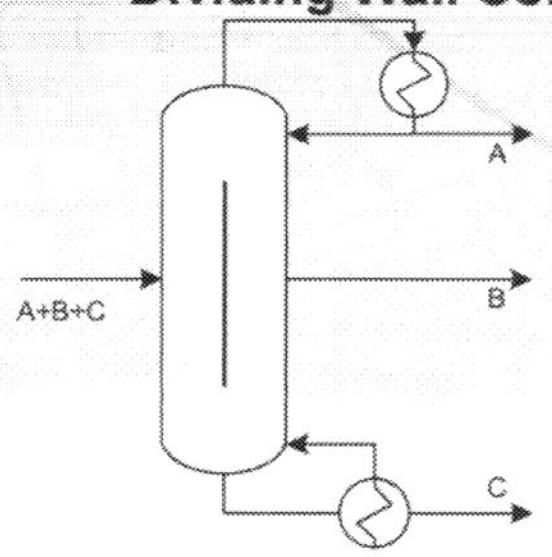

Thermally Coupled Prefractionator Arrangement (No Remixing)

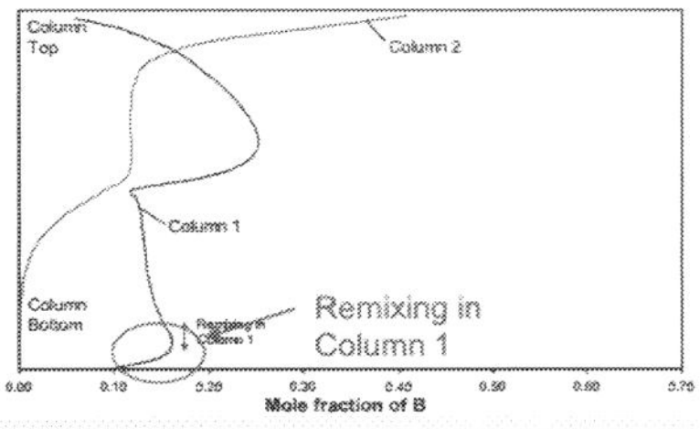

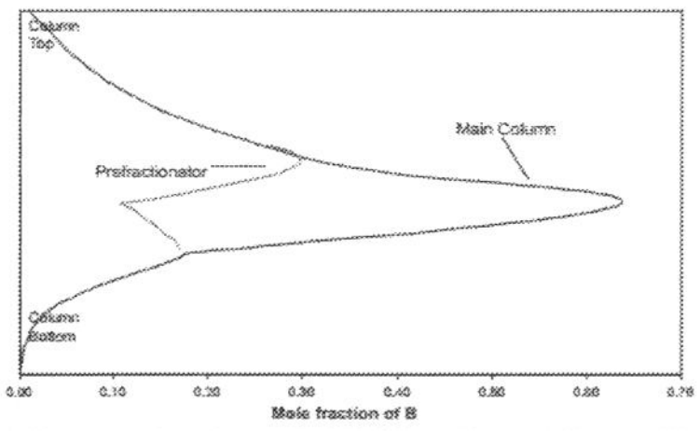

DWC Technology: The Past & Present

- Conceptually available since 1930s
- First commercialized in 1985
- Commercialization Delayed….
  - Inexpensive availability of energy
  - Lack of computational power
  - Concern about operability and control
- >100 DWCs now in operation
  - Chemicals & Refining

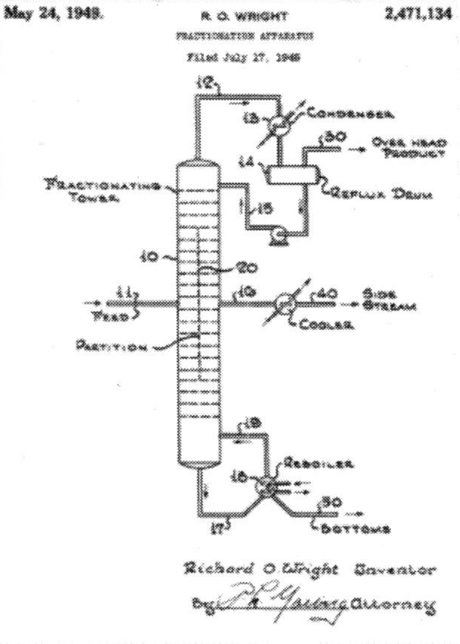

DWC: Drivers

Distill-Max™

- Changes in regulations
- Need to make additional products
- Plot space limitations
- Limited investment capital
- Maximizing return on investment
- Energy improvement initiatives
- Alternate use of existing assets

DWC: Benefits

Distill-Max™

- **Additional mass balance options**
  - Sharper separation
  - More stream cuts / more products
  - Better overall utilisation of process capacity and energy
- **Combines 2 columns in one shell (single reboiler and condenser)**
  - Enables one tower to make 3 'sharp-split' products
    - Middle product can be much purer than achievable in a conventional side-draw column
  - Heat integration via internal vapour and liquid links
  - Can be done as re-fit of an existing tower
- **Up to 25% capital savings for a grassroots application**
- **A useful tool for separation revamps**
- **Energy-efficient**
  - Up to 30% less energy than conventional 2-column sequence

©2011 KBR, Inc. All Rights Reserved.

DWC: Considerations

- Fully proven technology
  - Revamps / Refit
  - Grass-roots
- Must be targeted for use in correct process location
  - Applied to process flow-scheme in most economically advantageous location
  - Appropriate mass balance
  - Appropriate relative volatilities
  - Appropriate steam level

DWC Technology Development and Experience at KBR

- Development at UMIST (1988-1993)
  - Kellogg member of DWC Consortium
  - Pilot Scale Plant (Packed Column)
- Co-Operation Agreement with BP (1995)
- First DWC Commissioned
  - BP Coryton, 1998
  - Aviation Gasoline
  - 30 tray column revamped
- Grassroots DWC, Confidential Client, Japan
  - Start-up 2004
  - De-isopentaniser + Naphtha fractionation
  - KBR commercial reference unit
- 4 Grassroots DWC, Valero, USA
  - Start-up Q4-2010
  - Reformate Splitter

DWC: Questions to Ask

- **What are the economic drivers? Where else are savings realized or margins improved?**
- **Where in the process scheme is the best location for implementation of DWC?**
  - Effective scoping study is vital
- **How will the unit be operated, what are the alternate conditions, and how will feed change over time?**
  - Stability and robustness
    - Inherent to the design
    - Effective control scheme
  - Flexible
    - Feed changes
    - Spec changes

KBR project experience… from concept to full EPC implementation

Case Studies

Case A - Debutaniser to DWC

Case B - Naphtha Splitter to DWC

Introduction: Case A & Case B

Distill-Max™

Motivation

✓ To explore the energy saving potential in a system when "appropriate" column is revamped into Dividing Wall Column (DWC)

Debutaniser Process Study (Case A)

✓ Debutaniser converted into DWC

Naphtha Splitter Process Study (Case B)

✓ Naphtha Splitter converted into DWC

Case A: Introduction - Debutaniser Process Study

Distill-Max™

Purpose

✓ To explore the energy saving potential when Debutainser Column is revamped into Dividing Wall Column (DWC)

Original Work

✓ Base Case Debutaniser is modified into DWC to assess its impact on the network

- ✓ Main Fractionator PA duties are the same

Case A: FCC VRU Section - Base Case

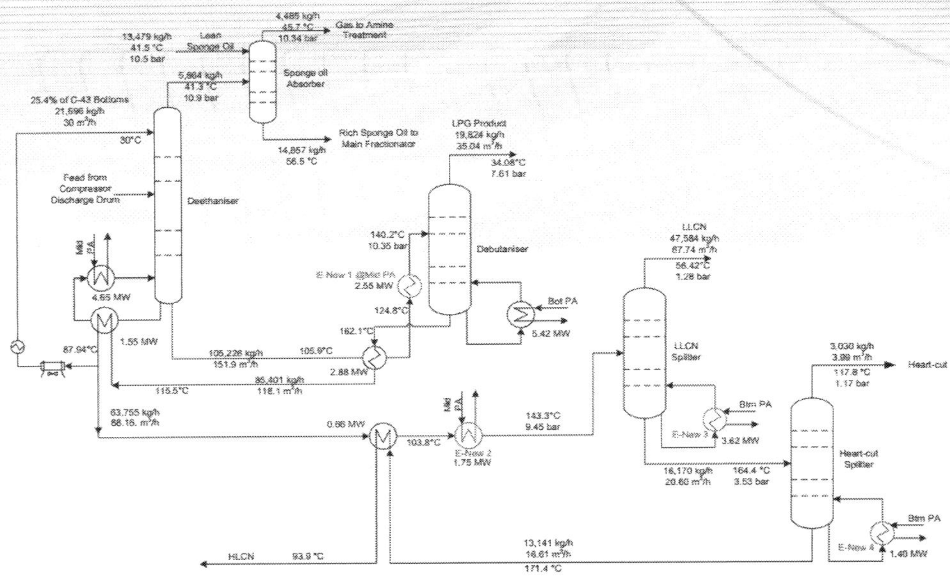

Case A: FCC VRU Section - Base Case + DWC

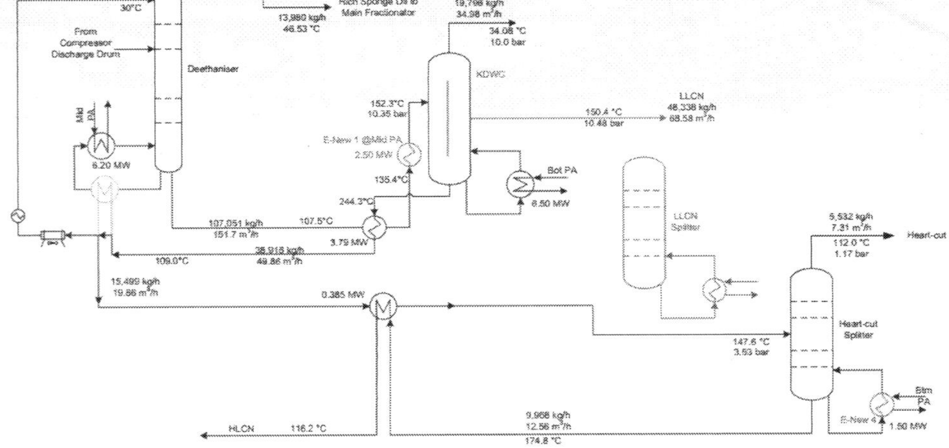

Case A: DWC Features

Distill-Max™

	Debutaniser	LLCN Splitter	Debutanizer DWC
	Base Case	Base Case	DWC
Reboiler duty (MW)	5.42	3.62	**6.50**
Condenser duty (MW)	8.64	6.18	**8.67**
LLCN Reid VP at 37.8°C BY API 5B1.1 (Naphtha) Method, bar	-	0.6918 (10.03 psi)	0.6923 (10.04 psi)

- DWC column can replace Debutaniser and LLCN Splitter operational performance
- Reboiler energy saving as compared to base case is 28.1%
- LLCN Splitter is taken out of operation
- The existing reboiler (design duty 8.5 MW) and condenser may be used
- Bottom PA can provide reboiler duty
- LLCN is taken as side-draw
- DWC bottom stream recycle helps Deethaniser performance
- Product ASTMs are close to Base Case

ASTM D-86* (LiqV%)	LLCN		Heart-Cut		HLCN	
	Base Case	DWC Case	Base Case	DWC Case	Base Case	DWC Case
10	44.5	44.5	106.0	101.4	133.9	137.2
90	103.9	110.1	127.1	129.9	159.9	162.8
Diff	59.4	65.6	21.1	28.5	26.0	25.6

*HYSIS API 1994

Case A: Conclusions - Debutaniser Process Study

Distill-Max™

DWC potential

- ✓ Able to meet product specifications
- ✓ Significant energy saving potential when Debutainser Column is revamped into Dividing Wall Column
  - Reboiler energy saving as compare to base case is 28.1%
- ✓ Payback is <1.5 years

Case B: Introduction - Naphtha Splitter Process Study

Distill-Max™

✓ Scope of Work
- To explore the energy saving potential when Naphtha Splitter is revamped into Dividing Wall Column

✓ Feed from FCC Debutanizer is processed in a three product naphtha splitter

✓ Naphtha Splitter
- Produces light gasoline as overhead product
- Produces heavy gasoline as bottom product
- Produces intermediate gasoline from the side stripper

Case B: Naphtha Splitter - Base Case

Distill-Max™

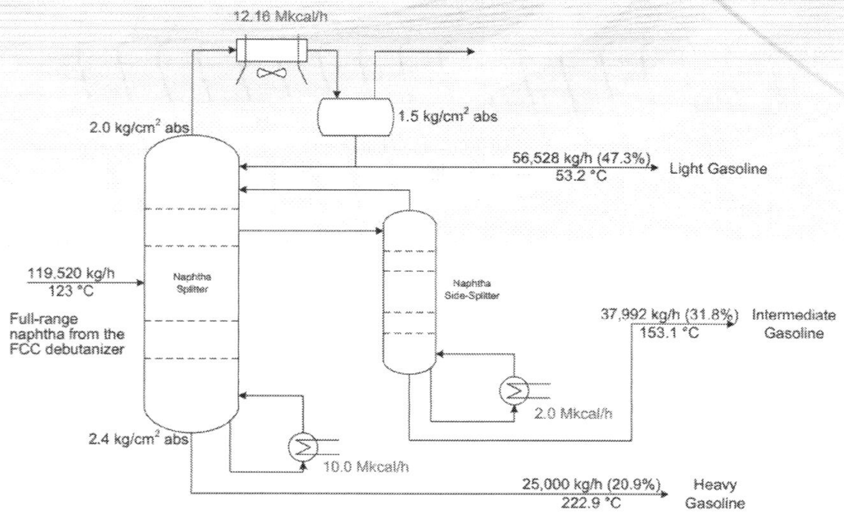

Case B: Naphtha Splitter – DWC Case

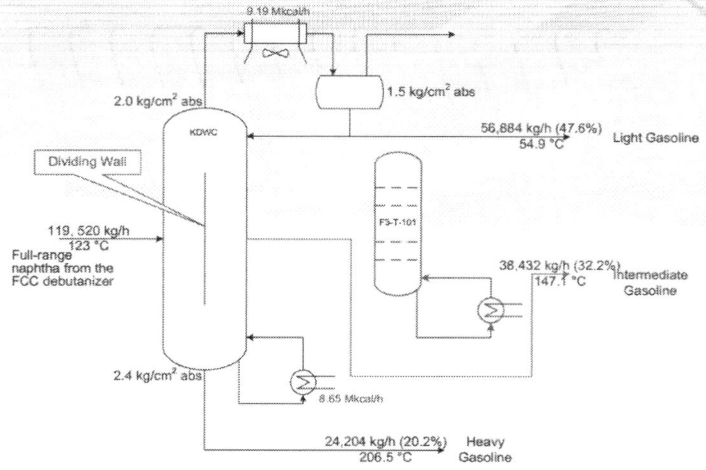

Reboiler Energy Saving ~28%

Case B: DWC Features

	Naphtha Splitter Case	DWC Case
Reboiler Duty, Mkcal/h		8.65
Naphtha Splitter	10.00	
Side Stripper	2.00	
Condenser Duty, Mkcal/h	12.16	9.19
Reboiler Energy Saving		~28%

- DWC can replace Naphtha Splitter and Naphtha Side-Stripper operational performance
- Meeting product specifications
- Reboiler energy saving is about 28.0%
- Naphtha Side-Stripper is taken out of operation
- Naphtha Side-Stripper Reboiler is out of operation
- Naphtha Splitter existing reboiler and condenser may be used
- Intermediate Gasoline is taken as side-draw

Case B: Conclusions - Naphtha Splitter Process Study

DWC potential

- ✓ Able to meet product specifications
- ✓ Significant energy saving potential when Naphtha Splitter Column is revamped into Dividing Wall Column
  - Reboiler energy saving as compare to base case is 28.0%
- ✓ Payback is <1.5 years

Summary: Case A & Case B

DWC potential

- ✓ Significant energy saving potential when "appropriate" column is revamped into Dividing Wall Column
  - @ Case A - energy saving is about 28.1% as compared to base case
  - @ Case B - energy saving is about 28.0% as compared to base case
- ✓ Attractive payback
  - Typically <2 years

Points to Remember

Distill-Max™

- ✓ DWC is a fully proven technology
  - Applicable to grassroots & revamp
  - It offers capital & energy saving potential
  - Can be applied to both tray & packed towers
- ✓ Scoping study is important to justify techno-economic potential of DWC project
- ✓ DWC design process is far more complex than conventional distillation column
  - Rigorous design procedure enables to guarantee its performance
- ✓ Mainly used to separation ternary mixtures. However, significant potential yet to be explored in
  - Separation of more than three component
  - Reactive DWC
  - Azeotropic DWC
  - Extractive DWC
- ✓ DWC will be a standard distillation tool in future

Extending the Operating Range of Cross-flow Trays

Izak Nieuwoudt
R&D Director
Koch-Glitsch

October 2011

3rd AIChE Regional Process Technology Conference, Galveston, TX

KOCH
KOCH-GLITSCH

The Problem

- Cross-flow tray capacity drops off in spray regime
- Efficiency in spray regime not as good as in froth regime
- A significant number of towers operate in the spray regime
- Conversion to packing is not always an option
- *What are the options for capacity increase and maintaining efficiency?*

KOCH
KOCH-GLITSCH

Cross-flow tray at 85% jet flood

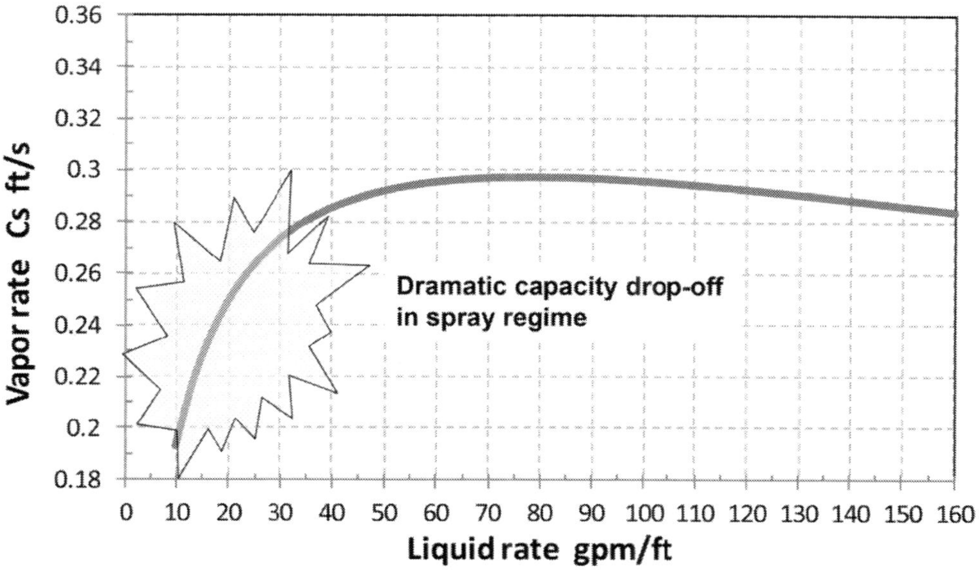

KOCH
KOCH-GLITSCH

Good bi-phase quality in froth regime

KOCH
KOCH-GLITSCH

Poor bi-phase quality in spray regime

KOCH
KOCH-GLITSCH

Potential Solution

- **Packing?**
- **Tray with reduced entrainment and decent efficiency**
  - Installation in towers equipped with cross-flow trays
  - Can accommodate reduced tray spacing
  - Simplicity
  - Keep cost under control
- **Centrifugal action needed to overcome gravity limitation**
- **Cross-flow needed for good efficiency**
- **Avoid the spray regime where possible**
- **Cross-flow ULTRA-FRAC® Tray**

KOCH
KOCH-GLITSCH

Cross-flow ULTRA-FRAC® Tray

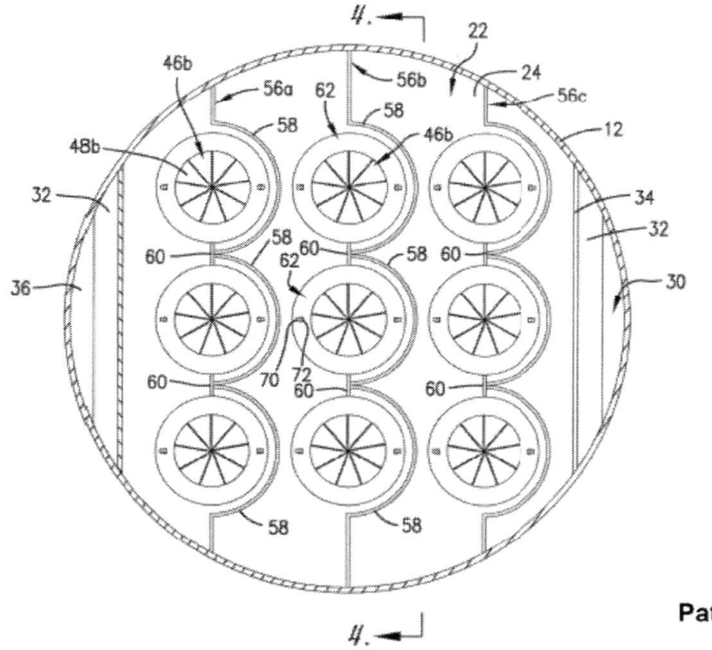

Patent pending

KOCH
KOCH-GLITSCH

Cross-flow ULTRA-FRAC® Tray

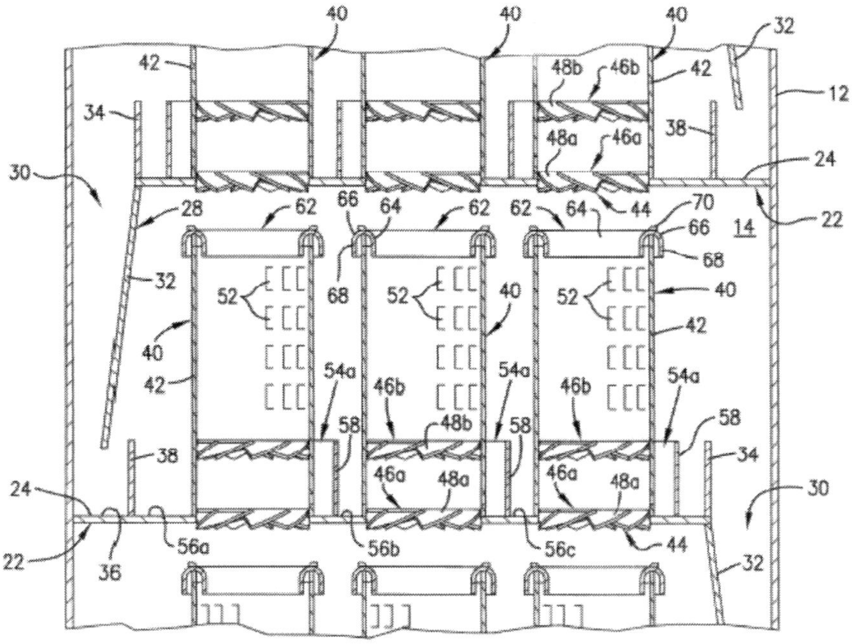

Patent pending

KOCH
KOCH-GLITSCH

Koch-Glitsch:
Piloting The Way to Improved Technology

KOCH
KOCH-GLITSCH

Tray efficiency - 16" Tray spacing
Low pressure operating case

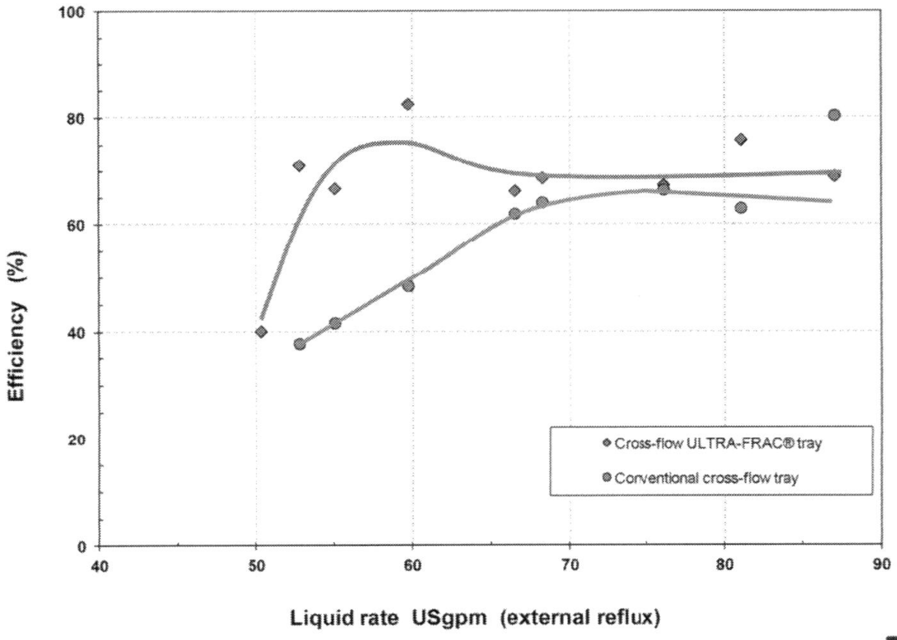

KOCH
KOCH-GLITSCH

Cross-flow ULTRA-FRAC® tray capacity

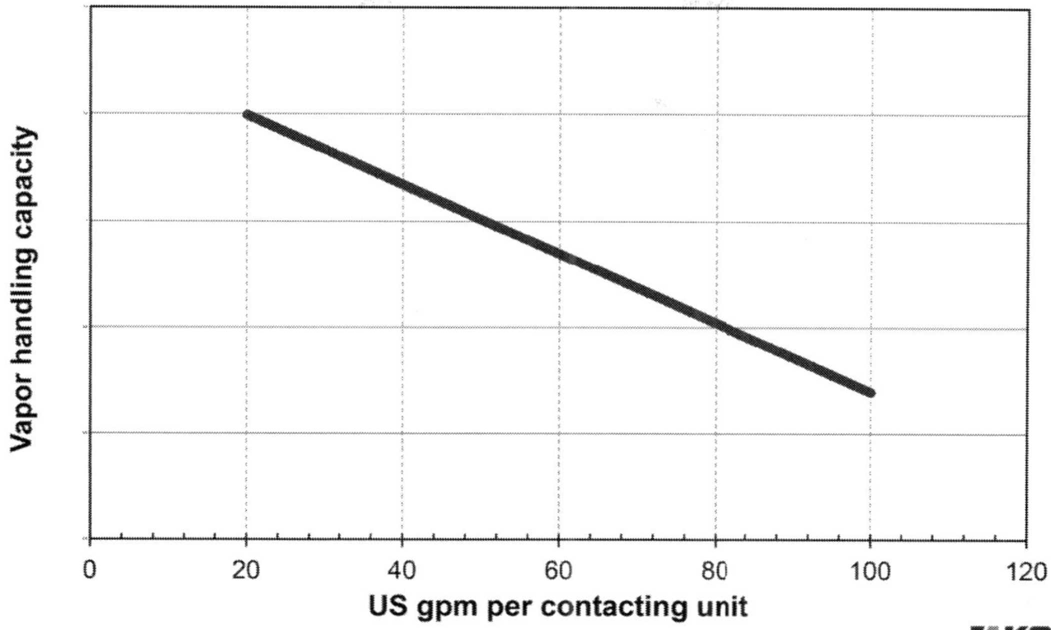

KOCH
KOCH-GLITSCH

Capacity increase - 4 ft tower & 16" TS

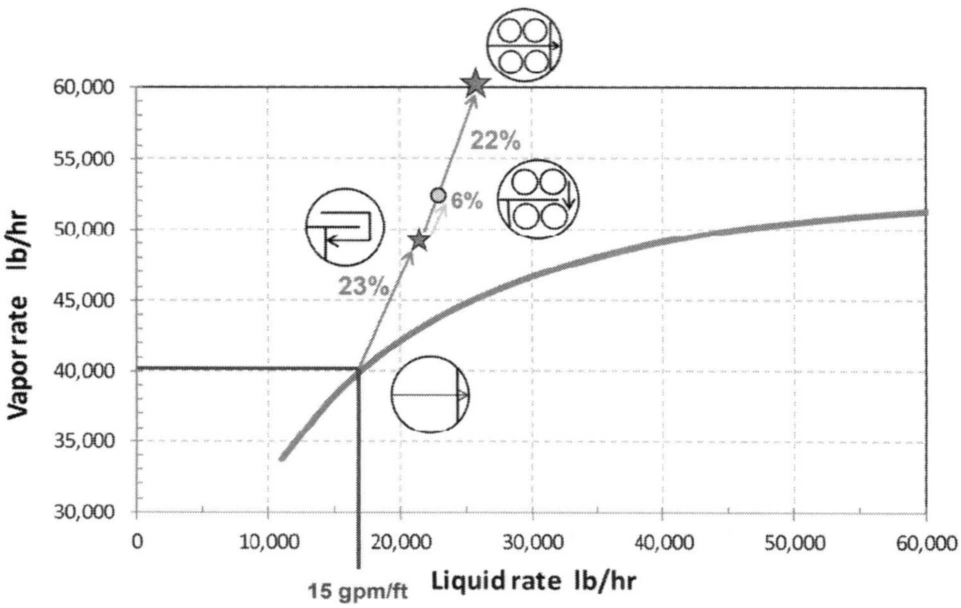

KOCH
KOCH-GLITSCH

Design example 2: 80" tower & 18" TS

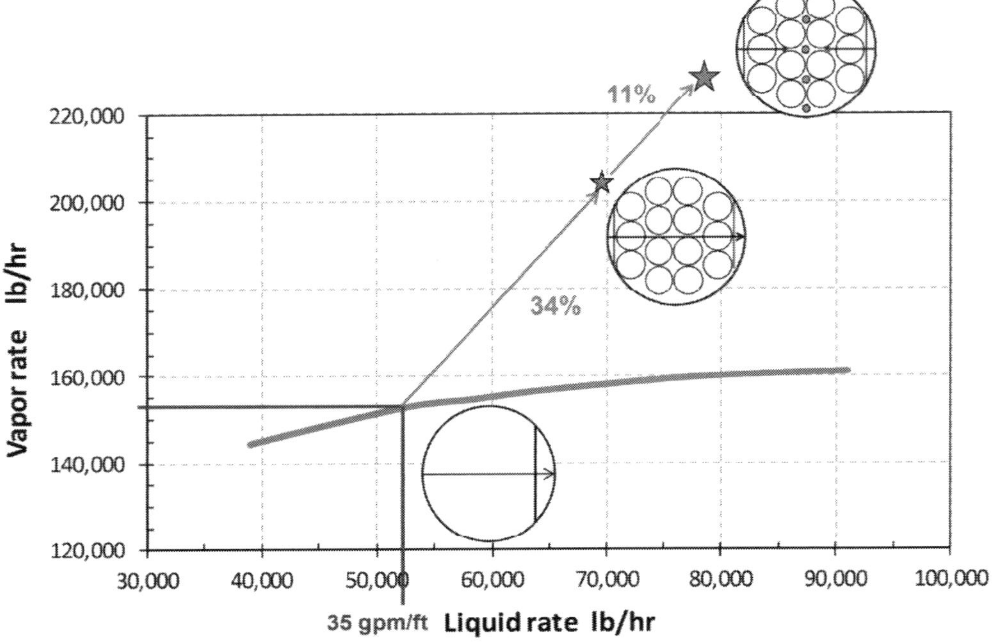

KOCH-GLITSCH

Conclusions

- **Tower capacity can be extended by using the Cross-flow ULTRA-FRAC® tray to replace conventional cross-flow trays operating in the spray regime**
- **The efficiency of the Cross-flow ULTRA-FRAC® tray is comparable with the efficiency of cross-flow trays**
- **The performance of the Cross-flow ULTRA-FRAC® tray has been confirmed in a large diameter hydrocarbon test tower**
- **The Cross-flow ULTRA-FRAC® tray exhibits desirable properties:**
  - High capacity
  - Good turndown
  - Good efficiency
  - Can replace conventional trays on 1-for-1 basis
  - Range of tray spacings, flow passes and flow arrangements

www.koch-glitsch.com

KOCH-GLITSCH

Thank You!

- **Comments?**
- **Questions?**

KOCH-GLITSCH

Thoughts on Troubleshooting: Tips from Those Who Do It Best

Ruth R. Sands
Mass Transfer Consultant
DuPont Engineering Research & Technology
Heat, Mass & Momentum Transfer

3rd AIChE Regional Process Technology Conference
October 6-7, 2011
Galveston, TX

2

Agenda

Motivation

- Training for new or inexperienced process engineers
- How do we develop the knack in ourselves and others?

Tips on Troubleshooting

- Interviews and advice from expert troubleshooters
  - Highlighted some of the most interesting advice
  - Not complete by any means!

The Case of the Inexplicable Inefficiency

- Example of difficult troubleshooting problem
- Loss of column efficiency following extensive project work

06 Oct 2011

References

Books

- Kister, Henry Z., Distillation Troubleshooting, Hoboken, New Jersey: John Wiley & Sons, Inc., 2006.
- Kister, Henry Z., Distillation Operation, New York: McGraw Hill, 1990.
- Mostia, Jr., William L., Troubleshooting: A Technician's Guide, 2nd edition, NC: ISA, 2006.

Interviews

- Jose Bravo, Chief Scientist, Shell Global Solutions
- Henry Kister, Senior Fellow, Fluor Corporation
- Ed Longwell, Principal Engineer, DuPont
- Jeff Loomis, Senior Engineer, DuPont
- Nick Sands, Technology Fellow, DuPont
- Teresa Thomas, Senior Engineer, DuPont
- Jean Trotter, Consultant, DuPont
- Bob Trotter, Engineering Fellow (retired), DuPont

The Case of the Inexplicable Inefficiency

Background

Tower originally designed for different service

- Taller and larger diameter than required

Fatigue cracking and corrosion leads to frequent repairs

Project approved to replace and repair trays in 2007

Goals of project are:

- Reduce / eliminate weeping where possible
  - Eliminate fatigue cracking
  - Save on future maintenance costs
- Design for anticipated expansion rates

Valve Tray Tower

25 Trays, Caged Multi-Weight Valves

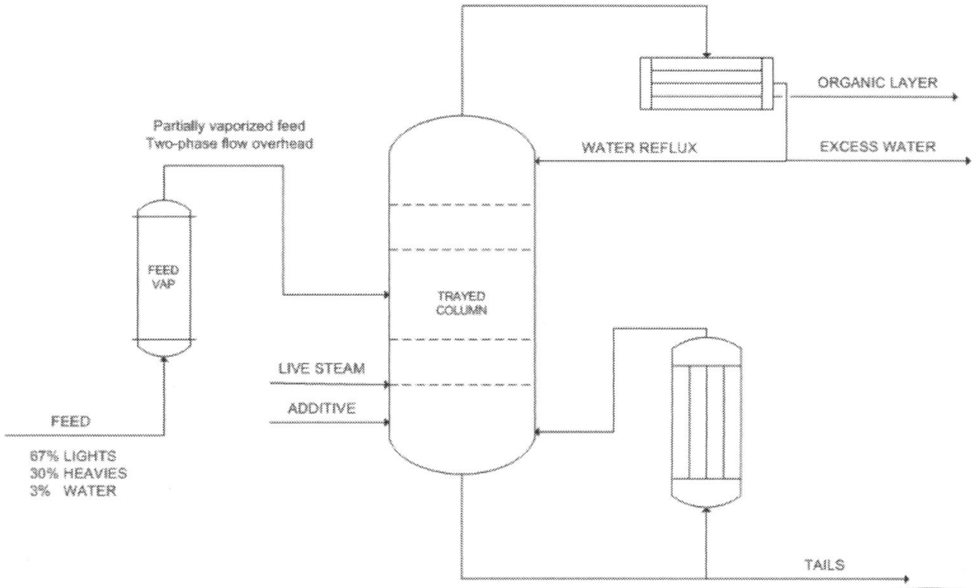

7

Column Work During September 2007 Shutdown

Replaced upper trays

- Lower valve count and blocked exit weir to reduce weeping and increase efficiency

Replaced trays below feed for expansion rates

- Increased valve count

Repaired bottom trays

- Replaced damaged panels and valves
- Pitting found on tray support rings
  - Unexpected and new issue

06 Oct 2011

DUPONT

8

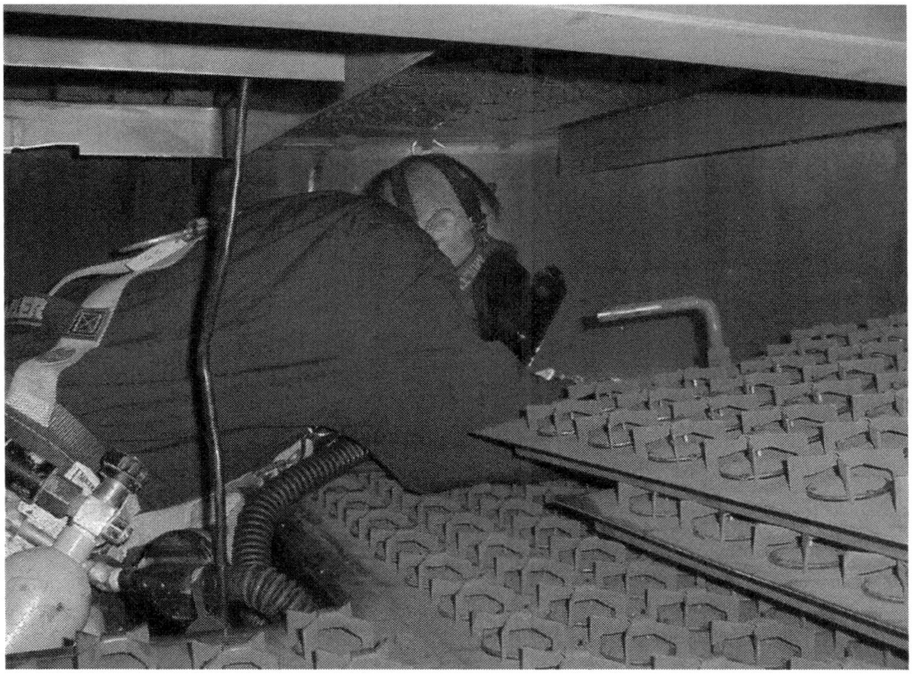

06 Oct 2011

DUPONT

Problem after Start-up

2-3 ppmw lights in tails

Prior to this shutdown, lights are non-detectible in tails

Depressed temperature profile just below feed

Effective loss of 2 theoretical stages or 3-5 real trays

Immediate Steps Taken to Mitigate Problem

Attempt to regain column temperature profile

- Increased feed vaporizer exit temperature 2 C
- Increased sparge steam
- Increased bottoms temperature control point 5 C

Avoid downstream problems

- Redirect stream with lights to appropriate place
  - Easily done with existing equipment
- Slight increase in operating costs
- No issue with capacity or product quality
- Gives time!

11

Immediate Steps Taken to Mitigate Problem

Attempt to regain column temperature profile

- Increased feed vaporizer exit temperature 2 C
- Increased sparge steam
- Increased bottoms temperature control point 5 C

Avoid downstream problems

- Redirect stream with lights to appropriate pl...
  - Easily done with existing equipme...
- ... in operating cos...
- ...apacity or pro...

08 Oc...

DUPONT

Tray 13 good indicator for lights leakage

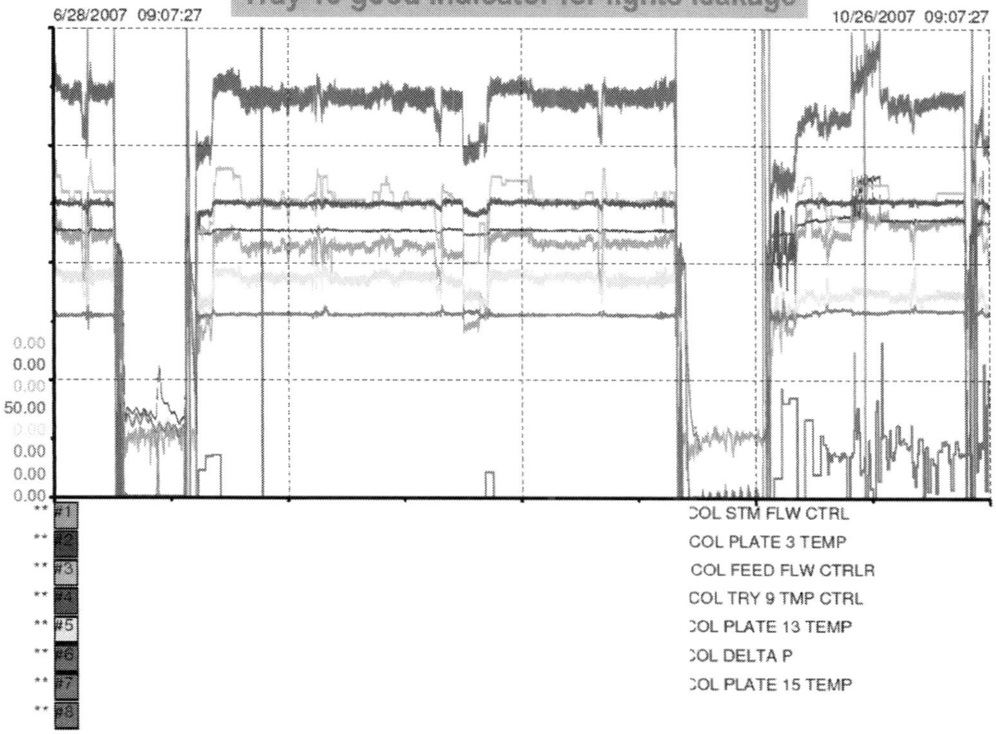

13

Column Temperature Profile

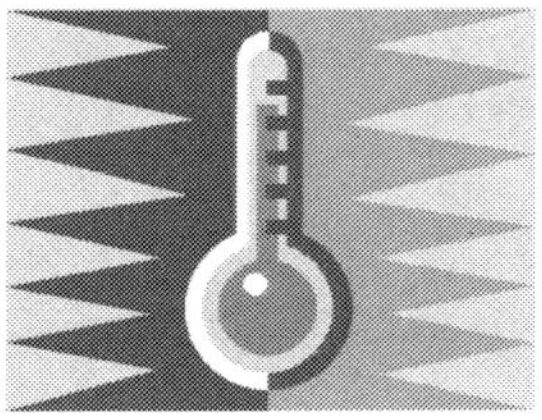

06 Oct 2011

14

Potential Causes of Inefficiency

Tray damage

- Or loss of tray manways

Feed composition change

- Feed vaporizer tube leak causing increased water in feed
- Calandria tube leak
- Upstream issue causes increased lights in feed
- Larger volume of liquid feed

Vapor Cross Flow Channeling

- By-passing

06 Oct 2011

Potential Causes of Inefficiency

Tray damage

- Or loss of tray manways

Feed composition ch

- Feed vaporizer tube le
- Calandria tube leak
- Upstream issue caus

Good troubleshooters understand the basics well.

Use calculations, models, experiments, etc. to check theories.

Know what's important and what's not!

Tray Damage

Pressure drop profile

- dP lightly lower than expected

Gamma scan results

- Indicates that all trays are present and functioning
  - No indication of weeping, entrainment, damage
  - Liquid level on all trays
- Completed 5 chords to look for inconsistencies or evidence of vapor cross flow channeling
- Low end of the expected dP, so still could have missing or damaged trays

Manways were closed properly following work

- QC completed

Tray Damage

Pressure drop profile

- dP lightly lower than exp

Gamma scan results

- Indicates that all tray
  - No indication of w
  - Liquid level on all
- Completed 5 chord
  nelin
  exp

Know what to expect before you start.

What are the lights, heavies?

What temperature, pressure profile should expect?

What quality of separation should expect?

Do simplified material and energy balance for tower.

Have a feel for what is reasonable based on a variety of experiences.

Gamma Scan

Feed Composition Changes

Increased water in column feed

- Feed sample confirms normal amount of water in feed
  - Top temps would be higher if feed water concentration was high
- Feed Vaporizer tube leak suspected due to high steam consumption
  - Energy balance around Feed Vaporizer sub-cooler indicates that increased steam consumption caused by steam blowing through vaporizer
- Calandria tube leak?
  - Steam valve % open normal compared to historical values
- Does not explain increased leakage in tails

Increased lights in column feed

- Decreased liquid feed flowrate due to higher feed vaporization
- Does not explain depressed temperature profile around feed section
- % valve opening for lights is normal compared to historical values

Feed Composition Changes

Increased water in colu

- Feed sample confirms n
  - Top temps would
- Feed Vaporizer tub
  - Energy balance a ased
    steam consumptio
- Calandria tube leak?
  - Steam valve % open normal compared to values
- Does not explain incre age in tails

Do not overlook the obvious.

However, correcting obvious problems doesn't necessarily solve the whole issue.

Patience to stick with it.

ts in column feed

feed flowrate due to higher feed vaporization

depressed temperature profile around feed section

for lights is normal compared to historical values

Larger Amount of Sub-Cooled Liquid Feed

Much greater portion of feed entering column as cold liquid

- Temperature profile supports theory

Would result in loss of tray efficiency

- Trays below feed become heat transfer trays

Calandria duty would have to be much higher

- Vaporizer is inefficient, but is providing some vaporization
- Does not support this theory

Larger Amount of Sub-Cooled Liquid Feed

Much greater portion of f... uid

- Temperature profile suppo...

Would result in loss ...

- Trays below feed be...

Calandria duty w...

- Vaporizer is inefficien...
- Does not support this ...

"Good troubleshooters have a willingness to accept the data rather than their own theories."

The mental model both helps and hurts the troubleshooting process!

"Don't let the data get in the way of a good theory"

Vapor Cross Flow Channeling or By-Passing

Trays just below feed meet two of the three criteria for vapor cross flow channeling

Bad tray design?

- Trays only operate well at higher rates?
- Tray design per industry correlations looks okay
  - Re-confirmed with vendor
- Current conditions are ~70% of design rates

Gamma scans give no evidence of weeping or entrainment

- Took 5 chords including parallel to inlet and exit weirs

Troubleshooting Team Pathforward

Tower Scan Results

- No obvious problems were identified
- Requested opinions from vendor and other DuET Consultants
- Internal inspection in mid-November
- Requested quote for original tray-design replacement parts

Testing Plan

- Increase feed vaporizer exit temperature 2-4 C
- Vary sparge steam
  - Sparge steam at low condition results in 6-10 ppmw lights in tails
  - Sparge steam at high condition results in 2-3 ppmw lights in tails
- Decrease bottoms control temperature by 2 C
- Increased rate test
  - If increased rates improves performance, indicates by-passing is cause
- Take sample of tails

Troubleshooting Team Pathforward

Tower Scan Results

- No obvious problems we
- Requested opinions fro
- Internal inspection in n
- Requested quote fo

Testing Plan

- Increase feed va

proves ... ing is cause

s

Involve others; LISTEN.

Testing strategy advice varies considerably! Depends on your situation!

- *Testing should begin with the easiest to prove or disprove and not based on how likely the theory.*
- *Do all practical tests before making a permanent change*
- *Apply carpet-bombing methods.*

"Here's what happened."

- Internal inspection revealed 5 open manways
- Manways had unusual locking handle design
- QC procedures updated to include detailed instructions for close-up
- Replaced 5 trays below feed with old design
  - Carpet-bombing methods prevailed.

27

Other Good Tips from the Experts

06 Oct 2011

28

Ways to Challenge the Mental Model

Is the process at steady state?

Visualize what is happening

- Imagine yourself as a pocket of liquid or vapor looking for the easiest path

Think of everyday analogies

- Processes are the same that occur in the kitchen, bathroom, or yard

06 Oct 2011

Believe Your Instruments

Believe your instruments …unless you have a good reason not to

- Do not start by questioning your instruments
- What scenario could cause this data to be true?
- "Instruments report information as they see it"
- Err on the fail safe side

People as Sources of Information

Operators can be powerful resources.

- They know how to make their own equipment work.
- What do you see?
- What do you do in response? What do you to fix it?

LISTEN to all sources.

- See value from different viewpoints (level, experience, function, personalities)
- Seek out unsolicited opinions & contrary views.

LISTEN more than talk.

Use a learning attitude.

- Implying that someone screwed up is a sure way to get no cooperation.
- Having an attitude of wanting to learn more about the system and what can be improved will gain cooperation

31

Learn to Troubleshoot

Incident investigations

- Gives ideas for modes of failure
- Good investigations cover people, equipment, and system

Participate in start-ups

- New plants especially
- Compresses the learning!
- Start-ups from shut-downs are also opportunities

Learn through examples

- There is a reason that most troubleshooting books and courses are a series of examples.
- Failure surveys show that failures are repetitive.
- A person that learns from others' failures is wise!

06 Oct 2011

DU PONT

32

Where to Start For Distillation Systems

Know what to expect

- Look at the VLE – what are the lights, heavies
- Do a simple material and energy balance
- What temperature, pressure profile should you expect?
- What quality of separation should expect?

Is the column at steady-state?

- Is the top pressure stable?
- Is the reflux stable?
- Is the dP in the expected range for these conditions?
- Reboilers and condensers are often the culprit

06 Oct 2011

DU PONT

Get Help!

Some people have a knack for troubleshooting

Everyone can improve their skills through practice

Partner with a good troubleshooter!

Final Thoughts

Good troubleshooters have solid technical skills AND good interpersonal skills.

- Have solid grasp of fundamentals
- Use material and energy balances
- Know what's important
- Use calculations and models to support understanding of basics
- Patient
- Good listeners
- Open

Takes experience!

Possible to learn and improve!

The miracles of science™

06 Oct 2011

SULZER

Sulzer Chemtech

Divided Wall Columns – Tray Technology

Daniel R. Summers | October 6, 2011

SULZER

Sulzer Chemtech

Divided Wall Column (DWC) Background

- **Energy Efficiency in today's Refinery and Chemical Plant is Already very Important**
- **Fractionation uses 40 to 50% of the Energy in today's Refinery or Chemical Plant**
- **After one applies Heat Integration, Vapor Recompression, Feed PreHeat and Pinch technology – there's not much left.**
- **For those applications with a substantial "Middle Boiler", a DWC may be an attractive way to reduce Energy Costs.**

SULZER
Sulzer Chemtech

Divided Wall Column Concept

- Energy savings incentives up to 30%
- Efficiency improvement due to better fractionation
- Capital saving incentives up to 30%

Applicable when:

- Middle boiling component is in excess
- When the middle boiling component represents a significant portion of the feed (DWC is attractive as long as the split between light and middle component is at least difficult as it is between middle and heavy)

Disadvantages:

- Different operating pressure is required than direct sequence
- DWC will be taller and larger in diameter than any of the columns in sequence

SULZER
Sulzer Chemtech

Divided Wall Column History

- Divided Wall Columns (DWC) are not new
- Sulzer has been using them since 1984
- Really have not been accepted in the industry until the early 2000's
- Most applications are packed – not trayed
- Sulzer has 34 Towers in operation and 2 more being readied
- Only Six of these applications employ trays
- Tower diameters range up to 6 meters (19.5 feet)

SULZER
Sulzer Chemtech

Divided Wall Column 2-Column Simulation

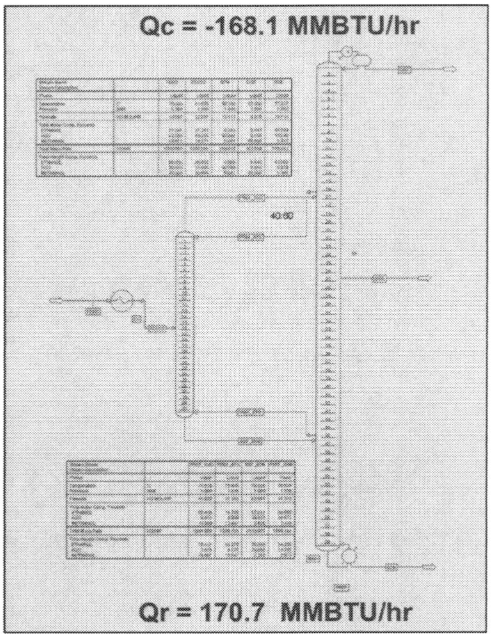

THERMAL DUTIES ARE APPROXIMATELY 25% LOWER THAN TWO COLUMNS IN SERIES

Methanol, Ethanol, Water Example

SULZER
Sulzer Chemtech

Divided Wall Column 4-Column Simulation

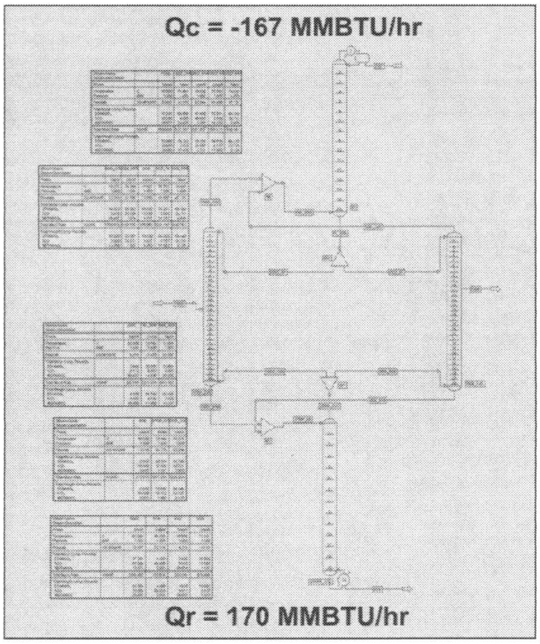

SAME SIMULATION AS PREVIOUS SLIDE BUT WITH DIFFERENT ARRANGEMENT OF THE UNIT OPERATIONS

Methanol, Ethanol, Water Example

Divided Wall Column Simulations

SULZER
Sulzer Chemtech

	2-column system	4-column system
Advantages	-all specs are in one column -more clear concentration profile -most of the system is in one column -easy initial setup	-clear structure -easy changing of NTS -faster converging -accurate split definition -easy changing of split -flows between column parts are easy to read
Disadvantages	-inaccurate definition of split -changing of NTS is more difficult -longer convergence time -needs first estimation of streams	-separate concentration profiles for each column -needs first estimation of streams

Simulation

SULZER
Sulzer Chemtech

- Personally have only simulated the 2 Column version
- Be sure to use same # theoretical stages on both sides of the dividing wall because of similar tray efficiencies – However may not always be true
- Setting top and bottom composition spec and setting the side draw rate forces a material balance
- To minimize energy consumption, one must vary;
  - feed tray location
  - draw tray location
  - overall number of trays
  - the liquid split between feed & draw side
  - the vapor split between feed & draw side

SULZER

Simulation Arrangement - Aromatics

Sulzer Chemtech

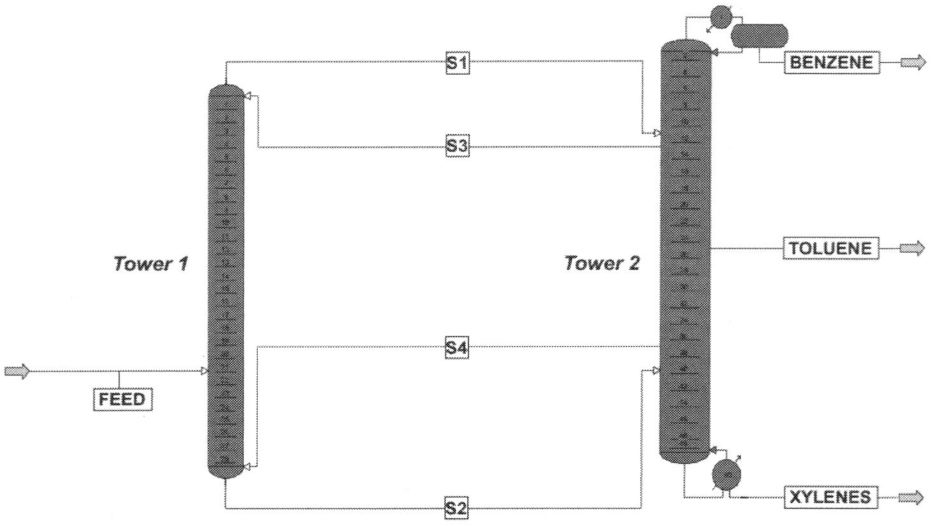

DWC - Galveston/ October 6, 2011 - 9

SULZER

DWC Heat and Material Balance

Sulzer Chemtech

(Units = Wgt%)	Feed	Benzene	Toluene	Xylenes
Hexane and lighter	8.87	42.29	0	0
Benzene	7.82	37.08	0.15	0
Heptane	3.47	14.6	1.4	0
Methyl CycloHexane	0.15	0.24	0.34	0
Toluene	30.28	5.66	94.56	4.16
2-Methyl Heptane	1	0.13	3.34	0.05
EthylBenzene	5.38	0	0.1	10.63
P-Xylene	5.67	0	0.04	11.11
M-Xylene	12	0	[illegible]	[illegible]
O-Xylene	7.44	0	[illegible]	[illegible]
Trimethyl Benzene & Heavier	17.92	0	[illegible]	[illegible]

This is a 28.5% savings in energy in comparison to the Base case!

Reboiler Duty = 80.2 MMBTU/hr

DWC - Galveston/ October 6, 2011 - 10

Simulation Arrangement – Aromatics Example

SULZER
Sulzer Chemtech

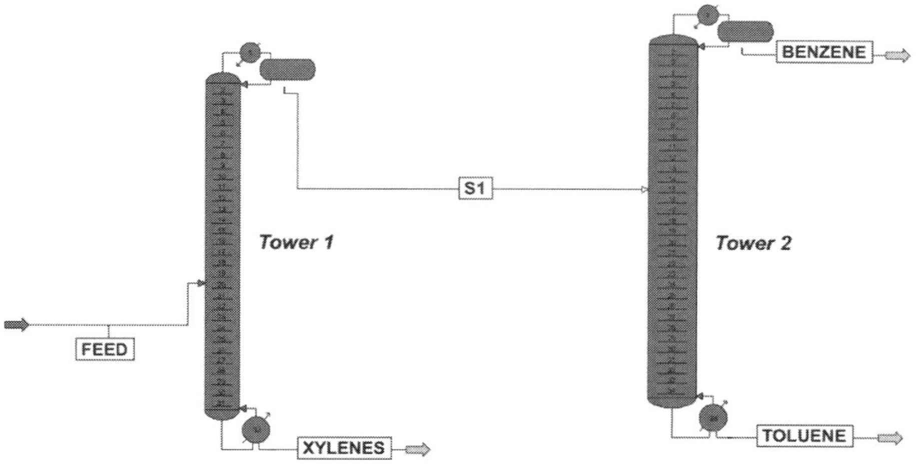

Base Design Heat and Material Balance

SULZER
Sulzer Chemtech

(Units = Wgt%)	Feed	Benzene	Toluene	Xylenes
Hexane and lighter	8.87	42.30	0	0
Benzene	7.82	37.25	0.02	0
Heptane	3.47	14.45	1.54	0
Methyl CycloHexane	0.15	0.22	0.35	0
Toluene	30.28	5.66	94.56	4.16
2-Methyl Heptane	1	0.12	3.32	0.06
EthylBenzene	5.38	0	0.11	10.62
P-Xylene	5.67	0	0.04	11.21
M-Xylene	12	0	0.06	23.76
O-Xylene	7.44	0	0	14.74
Trimethyl Benzene & Heavier	17.92	0	0	35.46

Total Reboiler Duty = 112.1 MMBTU/hr

Tray Loads

SULZER
Sulzer Chemtech

6 Primary Hydraulic Sections of a Divided Wall tower

	Top	Above Feed	Below Feed	Above Draw	Below Draw	Bottom
Vapor Load, lb/hr	485400	321500	370000	166300	166100	568100
Liquid Load, lb/hr	411900	178300	576800	236000	136100	744900

Tray Area Controlling Loads

Tray Dimensions

SULZER
Sulzer Chemtech

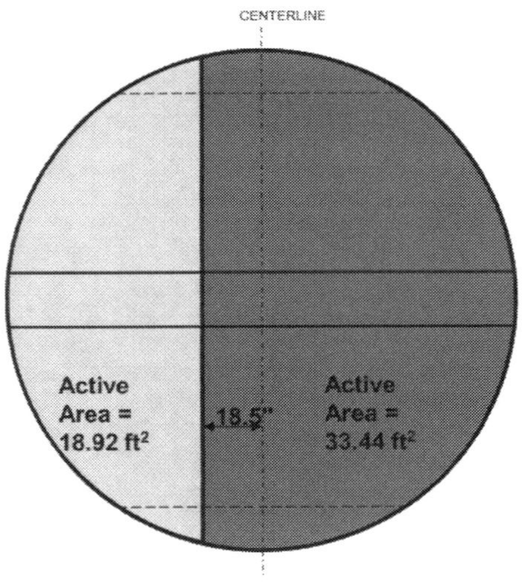

Center Downcomer Design

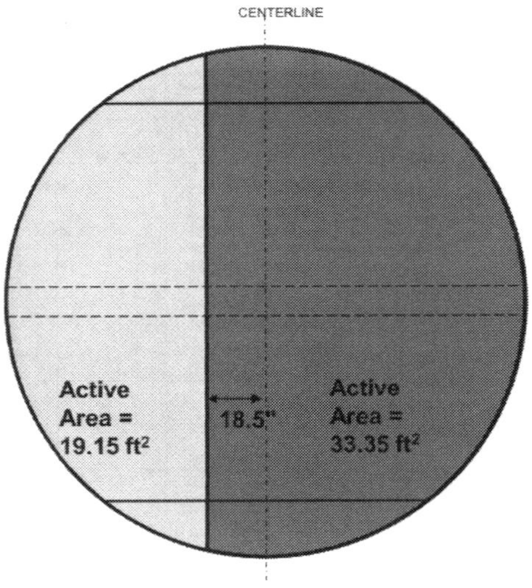

Side Downcomer Design

Tray Layout

SULZER

Sulzer Chemtech

2-Pass Trays because of the relatively High Weir Loading

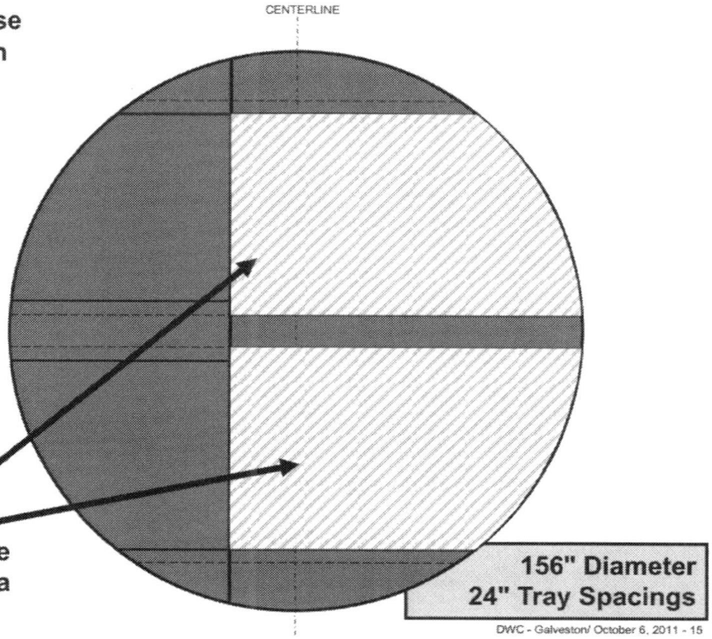

Below Feed Vapor Loads determines the Wide Side Active area

Tray Layout

SULZER

Sulzer Chemtech

Above Draw Liquid determines the Narrow Side DC Width

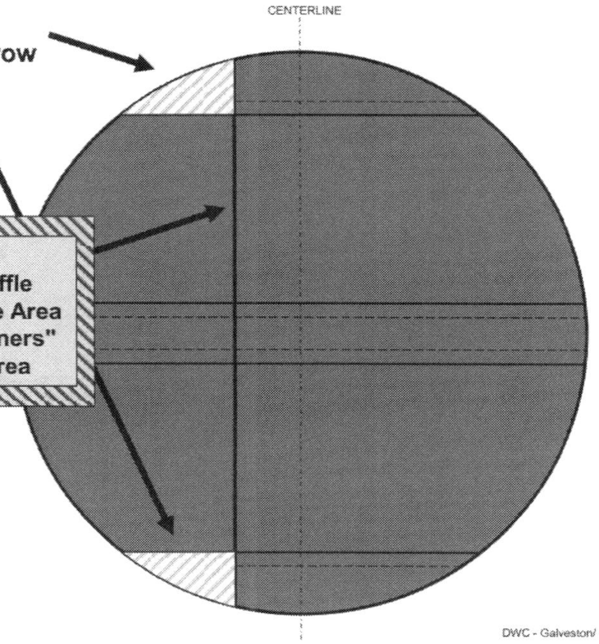

Caution:
Placement of the Baffle based simply on Active Area can lead to these "corners" being too small in area

Design Hydraulics

SULZER
Sulzer Chemtech

Name:	Top	Above Draw	Below Draw	Above Feed	Below Feed	Bottom
Trays	#1-12	#13-32	#33-51	#13-37	#38-51	#52-65
Vapor:						
%Flood	61	61	57	60	76	72
Dry Tray Pressure Drop, in H2O	1.45	1.37	1.37	1.42	1.85	1.57
Spray Regime Factor	4.26	5.3	4.15	3.76	5.22	5.31
% MVG Open Area	13	12.4	12.4	14.2	14.2	14.5
Liquid:						
Downcomer Velocity, ft/sec	0.159	0.37	0.209	0.103	0.349	0.283
Weir Loading, gpm/in.	5.86	10.44	5.92	3.43	11.6	10.47
Downcomer Froth Backup, %	36	44	34	30	51	49
Tray Pressure Drop, mmHg	4.49	4.93	4.42	4.08	5.74	5.18

Pressure Drop Across the 39 trays on the Feed Side =182.4 mmHg
Pressure Drop Across the 39 trays on the Draw Side =182.6 mmHg

Pressure Drop

SULZER
Sulzer Chemtech

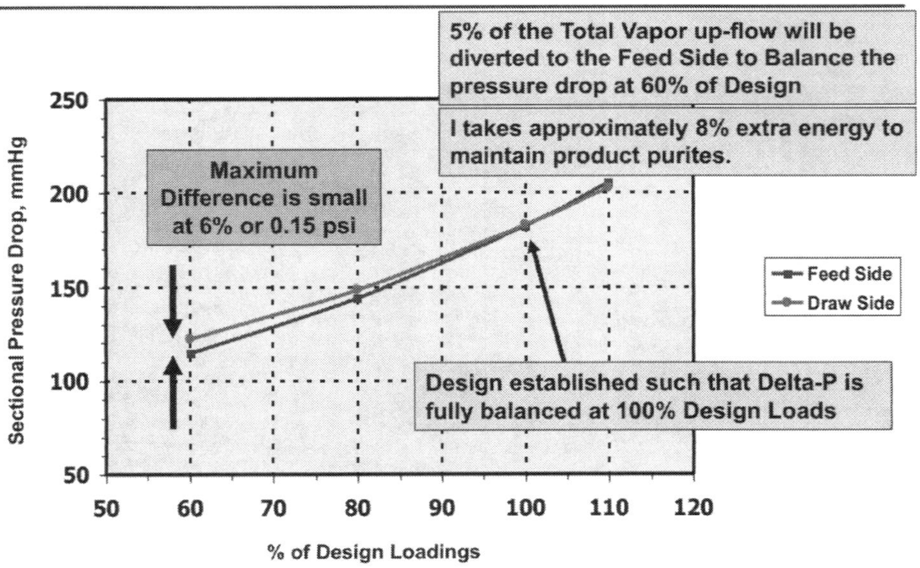

Tower Layout

SULZER
Sulzer Chemtech

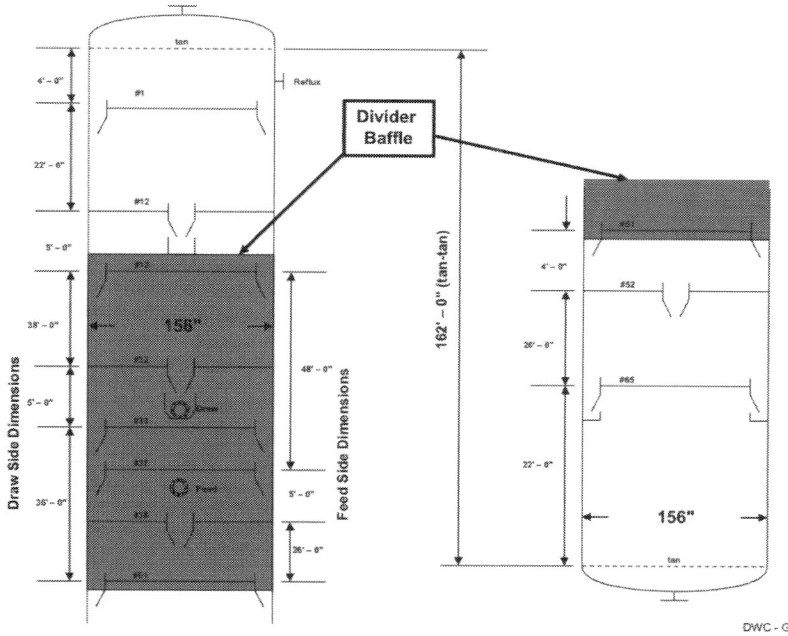

Tray Design Notes

SULZER
Sulzer Chemtech

- The dividing wall is used for support and welded to the vessel wall
- Dividing Wall does NOT need to be placed down the tower Centerline
- Vapor distributes by pressure drop
- Liquid can be split internally or externally

SULZER

Tray Design Notes

Sulzer Chemtech

- Can use any type tray deck
- Downcomers can be full length or truncated
- Pressure drop per tray should be similar on both sides
- Keep weir loading and open area proportional
- Trays have the advantage of easily allowing multiple draw points
- Trays do have higher pressure drop than packing

DWC - Galveston/ October 6, 2011 - 21

SULZER

2-Pass Tray Design

Sulzer Chemtech

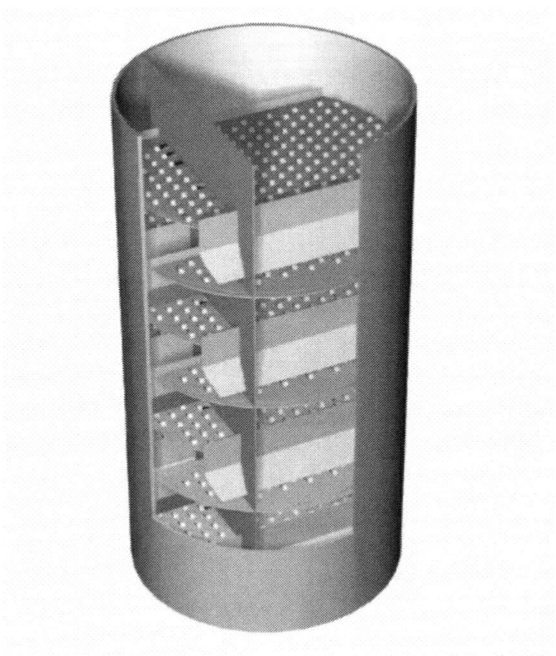

DWC - Galveston/ October 6, 2011 - 22

2-Pass Tray Design

SULZER
Sulzer Chemtech

DWC - Galveston/ October 6, 2011 - 23

Conclusions

SULZER
Sulzer Chemtech

- DWC are becoming more Common
- Simulations of DWC are Tedious to minimize Energy Consumption
- Tray Designs are Tricky but this Presentation should Help
  - Watch the Area and Weir Length of the smaller side Downcomer
  - Maintain Flow Path length to maximize Tray Efficiency
  - Watch turndown – will need extra reboil/reflux because of imbalanced pressure drop. Build in extra pressure drop at design but prefer to limit turndown
- An External Liquid Split is preferred - to better control the liquid to each side of the baffle – but is not necessary.

Thank You

DWC - Galveston/ October 6, 2011 - 24

Texas Board of Professional Engineers

Professional Practice
Update / Ethics

C. W. Clark, P.E.
Director of Compliance & Enforcement
October 6, 2011

What does it mean to be a

Professional Engineer ?

2

Texas Engineering Practice Act

"The purpose of this chapter is to protect the public health, safety, and welfare..."

3

1937

4

5

Competency

PE can only practice in areas where competent

6

Competency

Education

Experience

Exams

Competency

How would you demonstrate YOUR engineering competency?

Pr es al E ne ?

What does an to you?

Professionalism – NSPE Example

- Engineers shall:
  - Hold paramount the safety, health, and welfare of the public.
  - Perform services only in areas of their competence.
  - **Issue public statements only in an objective and truthful manner.**
  - **Act for each employer or client as faithful agents or trustees.**
  - **Avoid deceptive acts.**
  - **Conduct themselves honorably, responsibly, ethically, and lawfully** so as to enhance the honor, reputation, and usefulness of the profession.

Professionalism – ASME Example

- Engineers uphold and advance the integrity, honor and dignity of the engineering profession by:
  - Using their knowledge and skill for the enhancement of human welfare;
  - ***Being honest and impartial, and serving with fidelity their clients (including their employers) and the public; and***
  - Striving to increase the competence and prestige of the engineering profession.

Professionalism – ASCE Example

- Engineers uphold and advance the integrity, honor and dignity of the engineering profession by:
  - using their knowledge and skill for the enhancement of human welfare;
  - ***being honest and impartial and serving with fidelity the public, their employers and clients;***
  - striving to increase the competence and prestige of the engineering profession; and
  - supporting the professional and technical societies of their disciplines.

Professionalism – TBPE

- Chapter 137 Subchapter C – Professional Conduct and Ethics
  - §137.53 - Engineer Standards of Compliance with Professional Services Procurement Act
  - §137.55 - Engineers Shall Protect the Public
  - ***§137.57 - Engineers Shall be Objective and Truthful***
  - §137.59 - Engineers' Actions Shall Be Competent
  - ***§137.61 - Engineers Shall Maintain Confidentiality of Clients***
  - ***§137.63 - Engineers' Responsibility to the Profession***

Texas Board of Professional Engineers 2011 13

Professionalism – TBPE

- *§137.57 - Engineers Shall be Objective and Truthful*
  - *Misleading, Fraud, Deceit*
  - *Conflict of Interest*
- *137.61 - Engineers Shall Maintain Confidentiality of Clients*
- *§137.63 - Engineers' Responsibility to the Profession*
  - *Faithful Agent for Employers and Clients*
  - *Actions of Others who report to you*
  - *Conduct Business in a Respectful Manner (billing, language, etc.)*
  - *Injure Professional Reputation of Others*
  - *Retaliation*

Texas Board of Professional Engineers 2011 14

Latest News From The Board

Texas Board of Professional Engineers 2011 15

New Board Members

New Board Member Appointments

Lamberto J. Balli, P.E.
Houston

Carry A. Baker, Esq.
Amarillo

Board Member Re-Appointment

James A. Greer, P.E.
Keller

Texas Board of Professional Engineers 2011 16

Agency Organization & Responsibilities

- Executive Director: Lance Kinney, P.E.
- 30 Staff / Office in Austin
- Review and process all licensing applications
- Coordinate FE and PE Exams (~6900 / yr)
- Investigate all enforcement cases (~800 / yr)
- Administer Licenses for 54,000 PE's
- Administer Registrations for 8,800 Firms

Texas Board of Professional Engineers 2011 17

2011 Legislative session

- Bills filed – Regular Session
  - HB 2067 – Petroleum Engineering
  - HB 2284 – Building Design
  - SB 652 – Sunset

- Bills filed – Special Session
  - HB 3 – Windstorm: TWIA, TDI, and TBPE

Texas Board of Professional Engineers 2011 18

2011 Legislative session

HB 2067 – Petroleum Engineering

- Allows specific petroleum engineering practice in Texas by P.E.'s from other states that allow reciprocal practice by Texas P.E.'s.
- Effective Date – 5/28/2011

2011 Legislative Session

HB 2284 – Engineer / Architect Overlap

- Addresses the overlap issues between architects and engineers.
- Clarifies activities that are engineering and architecture.
- Creates a permitting process for engineers, that meet certain requirements and have specified experience, to practice architecture.

2011 Legislative Session

HB 2284 – Engineer / Architect Overlap

- Effective Date – September 1, 2011
- Texas Board of Professional Engineers (TBPE) & Texas Board of Architectural Examiners (TBAE) staff are working together on implementation
- FAQ on TBAE / TBPE websites
- Frequent Communication with Licensees and Registrants

2011 Legislative session

HB 2284 – Engineer / Architect Overlap

- Modifies Both Acts
- Specifies:
  - What is engineering in building design
  - What is architecture in building design
  - What can be designed by either

2011 Legislative Session

HB 2284 – Engineer / Architect Overlap

- Requires TBAE to create a list of engineers eligible to do architecture
- List will be available on TBAE site
- Applicant for list must be:
  - PE as of Jan 1, 2011
  - In good standing with TBPE

2011 Legislative Session

HB 2284 – Engineer / Architect Overlap

- Apply by January 1, 2012
- Minimum of 3 projects built before Jan 1, 2011
- Review process being determined by TBAE in consultation with TBPE
- Appeal Process

2011 Legislative session

HB 2284 – Engineer / Architect Overlap

- Statutory Joint Task Force to recommend engineering / architecture practice
- JAC Repealed

2011 Legislative session

SB 652 – Sunset Review

- Moves Sunset Review for TBPE and TBAE to 2013 from 2015
- SDSI is also under review in 2013
- Process started September 1

2011 Legislative Session

HB 3 – Texas Windstorm

- Windstorm Inspector Program
- PEs must be qualified to do design in high-wind areas
- TBPE will create and maintain a list of qualified PEs

2011 Legislative Session

HB 3 – Texas Windstorm

- Statutes effective September 28, 2011
- Rules must be adopted by December 1, 2011
- TBPE / TDI Working on details

New Board Rules

- Exam duration
  - 8 Hour exams
- Reference Statements
  - Electronic Submittal
- International Temporary License
  - Clarifying intent of 3 year limit
- Firm registration surrender

New Board Rules

- November
  - Applications for Engineering Educators
  - Board Committees
- February
  - Canadian Temp License

Proposed Board Rules

- Intn'l Temp License References
- Exam Registration Fees
- Structural Exam schedule
- Exam Analysis options
- Name changes
- Student Loans

Texas Board of Professional Engineers 2011

Proposed in August

- Legislative Implementation
  - JAC rule deletion
  - Windstorm rules addition
  - Sanction Table changes

Texas Board of Professional Engineers 2011

Online Applications

- Currently Online
- Application Form
- Online Payment
- Future
  - SER's
  - References
  - Other Documents

Texas Board of Professional Engineers 2011 33

Reminders

- **Continuing Education**
  - Random Audits Ongoing
  - Keep Documentation for 3 years
  - Be prepared to substantiate your 15 hours
  - Activity Log / Removed from TBPE
  - Enforcement Cases (up to $2,500+)
  - Over 65 is Not Exempt from CE

Texas Board of Professional Engineers 2011 34

Engineering Practice Act - Reminders

- Exemptions
  - From licensure
    - What is "Public"
    - Public Works
  - From Professional Fee
  - From Continuing Education
    - No Over 65 Exemption

Texas Board of Professional Engineers 2011 35

Reminders

- **Criminal Convictions**
  - All New Licensees
  - Required to Report within 30 days
  - Board will Review Incidents
    - Severity
    - Recidivism
    - Relevant to Engineering
    - Other Factors

Texas Board of Professional Engineers 2011 36

Board Rules - Reminders

- Engineer's Seal and Signature
  - What to Seal
    - Texas Seal for Texas work
  - Preliminary Documents
  - Electronic Seals and Signatures

Texas Board of Professional Engineers 2011 37

More Board Updates

- Video of Board Meetings on Website
- ECHO System
  - Online Payments
  - Renewals (PE / Firm)
  - Personal Info Updates

Texas Board of Professional Engineers 2011 38

Update on NCEES

- 2011 Annual Meeting
  - Industrial Exemption – Discussed Model Law Change
  - BS+30 Alternatives
    - 6 Years Exp + Mentorship + Add'l Courses
  - CBT – Work on Converting PE Exam
  - Building Commissioning – Passed Model Law Change

Texas Board of Professional Engineers 2011 39

Computer Based Testing

- Computer Based Testing Task Force
- Approved at National NCEES last year
- Start with the FE and FS exams
- Expected national implementation by 2014

Texas Board of Professional Engineers 2011 40

Outreach

FY	Attendees	Presentations
2008	5,739	100
2009	6,928	120
2010	9,145	157
2011	8,954	153

- Webinars
- Includes K-12
- E-Week

Texas Board of Professional Engineers 2011 41

C & E Statistics

- **FYs 2010-2011**
  - 1478 Cases Opened
    - Public, Internal, Anonymous
  - 1464 Cases Closed
    - 977 Voluntary Compliance (67%)
    - 82 Disciplinary Actions
    - 192 Dismissed / No Violation

Texas Board of Professional Engineers 2011 42

C&E Statistics

- **Case Types**
  - Cases against PE's
  - Unlicensed Practice
  - Illegal Use of Title
  - Corporate Name
  - Firm Registration

43

Engineering Ethics

Case Files

CONFIDENTIAL

44

From the Case Files

- Based on TBPE Enforcement Cases from 2009-2011
- Names and Situations Changed
- All in Newsletter / on Website
- Focus on Ethical / Practice Issues

45

From the Case Files

DFW Metroplex

- Pedestrian Bridge Project
- Out-of-State Licensed PE
- Prepared Engineering plans & performed calculations for the bridge project
- Not licensed in Texas

- *Only engineers licensed in Texas can offer or provide engineering for Texas projects.*

46

From the Case Files

Out of State Licensee

- Building collapsed in Wisconsin
- One death and $100M in damages
- Investigation and Civil litigation revealed that the engineer's calculations and design of building support system was deficient
- Wisconsin Board Revoked P.E.'s licensee

- *Texas P.E. licensed in another state is subject to sanction if the violation in that state would be a violation in Texas*

47

From the Case Files

Dallas, TX

- Individual used "P.E." and a PE number after his name in an email correspondence
- Electrical engineering designs were included
- Board records determined he was not a licensed P.E.

- *Unlicensed Practice of engineering in Texas is a violation subject to sanction*

48

From the Case Files

San Antonio, TX

- Individual offered engineering services.
- License was delinquent
- Only PE in firm
- *Unlicensed Practice of engineering in Texas is a violation subject to sanction; firm registration issue*

49

From the Case Files

Corpus Christi, TX

- PE sealed a signed a TDI WPI-2 for windstorm inspection certifying that a room addition met code.
- Later found that the roof covering was not installed properly and resulted in leakage and damage to the interior. Also found that the window and the foundation was not code compliant
- P.E.'s employee had visited once and relied on the contractor's word that all was built properly
- *P.E. must not give misleading certifications, must exercise reasonable care and diligence to ensure employees perform required inspections, and be a faithful agent to their client.*

50

From the Case Files

Austin, TX

- PE submitted an Architectural Barriers Proof of Submission form to TDLR more than 20 days after plans were issued.
- *PE must submit plans in a timely manner*
- *TBPE can enforce for violations of other code requirements*

51

From the Case Files

Houston, TX

- PE working for engineering firm did consulting work 'on the side' for another engineering firm.
- Did work for firm #2 while at work for firm #1.
- Signed agreement and did not disclose.
- *PE must engage in professional business activities; faithful agent to employer; possible conflict of interest issues.*

52

From the Case Files

Houston, TX

- Other Variations:
  - 'Moonlighting' for Self on Company Time
  - Starting Spin-Off Company on Work Time

53

From the Case Files

Waco, TX

- PE sealed a hand written foundation design plan.
- Certified that repairs were performed correctly.
- Sealed design of contractor and did not do inspection.
- *PE shall only seal work done by them or performed under their direct supervision.*
- *"Plan Stamping"*

54

From the Case Files

El Paso, TX

- PE Contracted with an 'engineering firm' to do building design.
- 'Firm' used his seal on other letters and designs.

- *Keep control of your seal – physical and electronic versions*

Texas Board of Professional Engineers 2011 55

From the Case Files

Tyler, TX

- Engineer hired to perform engineering services for dog kennel business.
- Sealed designs with deceased PE seal (and signature).

- *Verify the name and PE # of plans you review / accept.*

Texas Board of Professional Engineers 2011 56

From the Case Files

Austin, TX

- Engineer hired to perform engineering services and design.
- Sealed designs with fake PE seal using his name and PE number of deceased PE.

- *Verify the name and PE # of plans you review / accept.*

Texas Board of Professional Engineers 2011 57

From the Case Files

San Antonio, TX

- Professional Engineer issued subdivision plans to his client that were unsealed.
- Plans were not marked preliminary indicating why they were being issued and lacked a limitation of their use.
- Plans also contained numerous errors

- *Texas P.E.'s must execute due care and diligence in performance of their engineering business and follow Board Rules on sealing, plans and identifying preliminary documents with proper caveats*

Texas Board of Professional Engineers 2011 58

From the Case Files

Dallas, TX

- Engineering firm was contracted to perform testing and site observation of a retaining wall construction project
- Responsible P.E. issued a document which stated that construction was performed in accordance with the design
- Photos revealed that construction was not in compliance.

- *Board Rule requires P.E. to exercise care and diligence to prevent partners, associates, and employees from engaging in conduct that if done by the P.E. would be a violation*

Texas Board of Professional Engineers 2011 59

Professionalism

- Complaint was filed on engineer by client regarding competence and ethics.
- Denial and Cover-up
- A professional would/should have "owned" the issue and disclosed it to the client in a timely fashion.
- A repair plan could have easily addressed the issues.
- Case would have never been filed or possibly dismissed.

Texas Board of Professional Engineers 2011 60

Professionalism

- Communication
- Contract
- Calculations
- Helps Avoid Complaints

Texas Board of Professional Engineers 2011 61

From the Case Files

The Valley, South TX

- PE claimed performance of engineering services for water and sewer, as well as residential improvements related to ORCA.
- County and Grant $ involved.
- Payments approved for businesses owned by relatives.
- Work not done; conflict of interest not disclosed.

- *Multitude of violations*
- *$30K / 5 Year Suspension / TTU Ethics*

Texas Board of Professional Engineers 2011 62

From the Case Files

Laredo, TX

- PE Submitted 16 re-roof and windstorm certifications for a high school (10 roofs / 6 whole buildings).
- Photos showed long term roof wear, cacti growing on roofs, exposed decking, etc.)
- No substantiating documentation.
- *Same person as previous violation*

- *Multitude of violations*
- *Recidivism*
- *Revocation*

Texas Board of Professional Engineers 2011 63

From the Case Files

- Enforcement Is Critical to Protecting the Health, Safety, and Welfare of the Public
- We need your help to report potential violations!

Texas Board of Professional Engineers 2011 64

Questions?

Texas Board of Professional Engineers 2011 65

Contact TBPE

- 1917 IH-35 South
- Austin, TX 78741
- Phone: 512-440-7723
- FAX: 512-440-5715
- Web: http://www.tbpe.state.tx.us

Texas Board of Professional Engineers 2011 66

The First Professor of Chemical Engineering in Space
Originally presented on: 10/6/2011 10:31:00 - 11:00:00

Albert Sacco, Jr. (born May 3, 1949) is an American chemical engineer who flew as a Payload Specialist on the Space Shuttle Columbia on shuttle mission STS-73 in 1995.

Born in Boston, Massachusetts, Sacco completed a bachelor's degree in chemical engineering from Northeastern University in Boston in 1973, and then a Ph.D. in chemical engineering from the Massachusetts Institute of Technology in 1977. He then joined the faculty of the Worcester Polytechnic Institute, becoming a full professor and rising to department head in 1989.

Sacco accepted the position of Dean of the Edward E. Whitacre Jr. College of Engineering at Texas Tech University, effective January 1, 2011.[1] He currently serves in this role.

He formerly was the George A. Snell Distinguished Professor of Engineering at Northeastern University. He is was Director of the Center for Advanced Microgravity Materials Processing.

Sacco flew as a payload specialist on STS-73, which launched on October 20, 1995, and landed at the Kennedy Space Center on November 5, 1995. The 16-day mission aboard Columbia focused on materials science, biotechnology, combustion science, and fluid mechanics contained within the pressurized Spacelab module. He has over 70 publications (including book chapters) in the areas of carbon filament initiation and growth, catalyst deactivation, and zeolite synthesis.

He married the former Teran Lee Gardner, of Massachusetts, in November 1971. They have four children. His hobbies include scuba diving; he is a certified scuba diving instructor.

Algenol's Direct to Ethanol Process

McCool, B., Algenol Biofuels
Sharma, P., Algenol Biofuels

Originally presented on: 10/7/2011 0:00:00 - 0:25:00

Algenol's Direct to Ethanol process has been demonstrated to be technically feasible with large scale photobioreactors, but in order to assess economic viability and environmental impact, the entire system must be analyzed. Integrating technoeconomic and lifecycle analyses with the design of upstream, downstream and photobioreactor components, and evaluating a range of rational alternatives/uncertainties, represent essential steps in making the transition from the lab to commercial production. This talk will describe the key questions as they pertain to analysis of highly distributed systems such as Algenol's, and highlight the major modeling and analysis challenges.

JOHN ZINK
CLEAN COMBUSTION. CLEAR SOLUTIONS.™
Ultra-Low NOx Burner Arrangements in Furnace Revamps:
Utilization of CFD to Prevent and Mitigate Adverse Furnace Flow Patterns
Shirley Chen, Jaime A. Erazo, Jr., Bailey Hendrix and Kirk Wendel
©2011 by John Zink Company, LLC

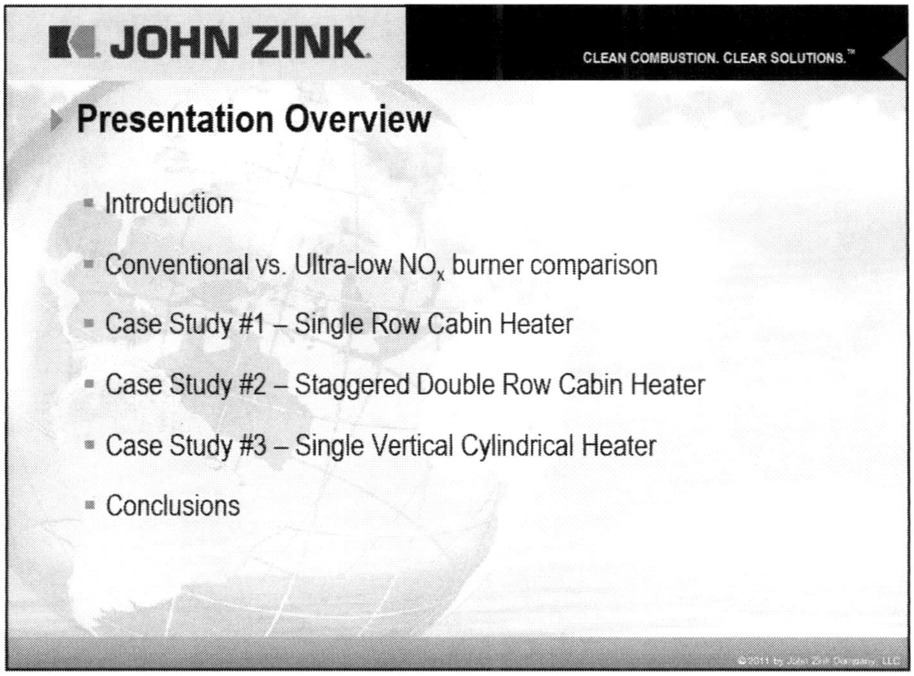
JOHN ZINK
CLEAN COMBUSTION. CLEAR SOLUTIONS.™
Presentation Overview
Introduction
Conventional vs. Ultra-low NOx burner comparison
Case Study #1 – Single Row Cabin Heater
Case Study #2 – Staggered Double Row Cabin Heater
Case Study #3 – Single Vertical Cylindrical Heater
Conclusions
©2011 by John Zink Company, LLC

JOHN ZINK — CLEAN COMBUSTION. CLEAR SOLUTIONS.™

Introduction

- As NO_x emission regulation continues to become more stringent, the need to retrofit existing heaters with Ultra-low NO_x burner technology will continue to rise.
- Ultra-low NO_x burners have certain characteristics which make them more susceptible to heater flow patterns than conventional burners.
- In order to mitigate any adverse heater effects, Computational Fluid Dynamics (CFD) can be used to customize the burner design to counteract adverse flow patterns.

JOHN ZINK — CLEAN COMBUSTION. CLEAR SOLUTIONS.™

Conventional Burners

- Short, intense flames (1-1.5 ft per MMBtu/hr flame lengths)
- No gas or air staging
- Small flame envelope
- High peak flame temperatures
- High oxides of nitrogen (NO_x) emissions

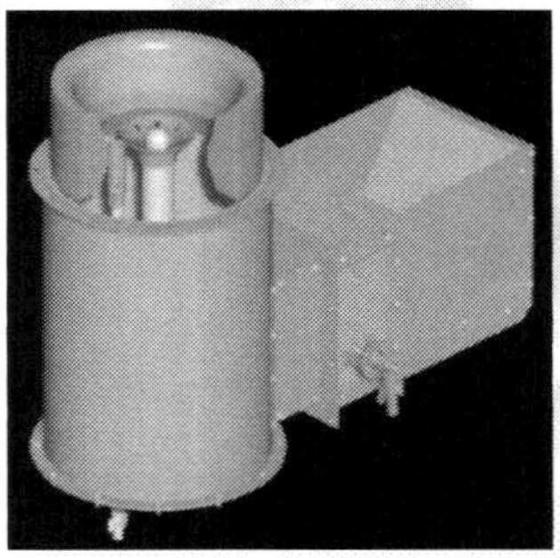

Figure 1: 3D Rendering of Conventional Burner

Process Heaters designed to utilize conventional burners were therefore optimized for process performance based on this small flame volume.

JOHN ZINK

CLEAN COMBUSTION. CLEAR SOLUTIONS.™

Ultra-Low NO_x Burners

- Staging of fuel or air
  - Staging results in longer flames
- Internal flue gas recirculation
  - Flue gas recirculation requires adequate burner to burner spacing
- Furnace flow patterns have greater effect on performance

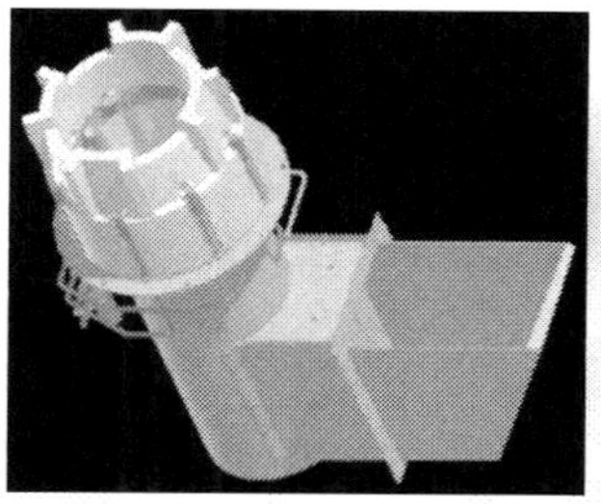

Figure 2: 3D Rendering of Ultra-low NO_x Burner (ULNB)

JOHN ZINK

CLEAN COMBUSTION. CLEAR SOLUTIONS.™

Case Study #1 – Single Row Cabin Heater

© 2011 by John Zink Company, LLC

JOHN ZINK

CLEAN COMBUSTION. CLEAR SOLUTIONS.™

Single Row Cabin

- Four cells connected to a common convection section
- Ducting into the convection section from the top of the radiant cell is located to one side of the radiant cell
- Tubes run horizontally along the length of the cell
- Conventional burners installed in the floor (firing vertically upward) in a single row centered between the tubes

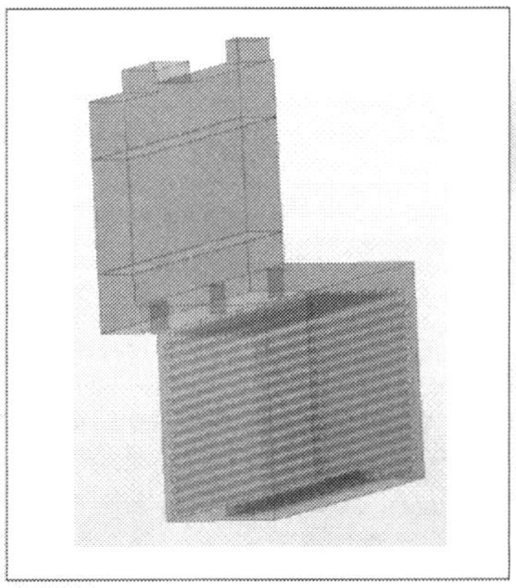

Figure 3: CFD Geometry of Single Row Cabin (one cell of the four cell crude heater).

JOHN ZINK

CLEAN COMBUSTION. CLEAR SOLUTIONS.™

- Ultra-low NO_x burners were modeled in the heater
- Longer than expected flame lengths (greater than 2/3 of the radiant cell height)
- Significant leaning of the flames at the top of the radiant cell towards the convection section inlet
- Flame impingement on process tubes

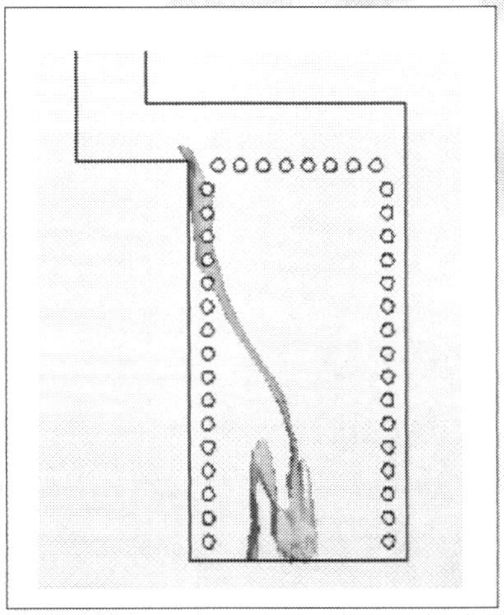

Figure 4: Standard ULNB flame impinging on heater tubes.

JOHN ZINK CLEAN COMBUSTION. CLEAR SOLUTIONS.™

- A pocket of stagnant flue gases containing relatively high concentration of CO was next to the burner. This side of the flame was oxygen deficient. This caused the flame to elongate

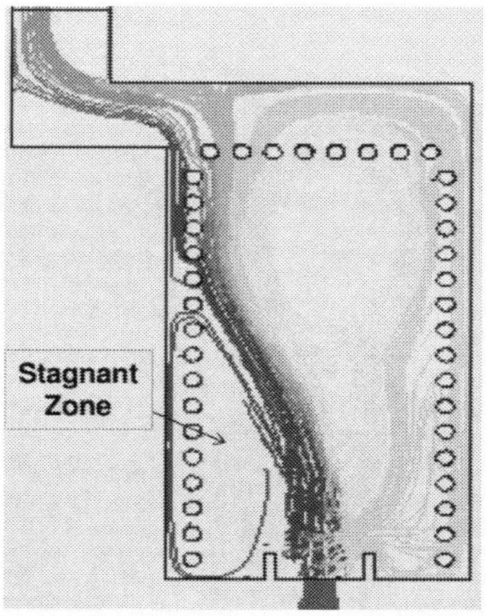

Figure 5: Flow pathline view of heater currents.

JOHN ZINK CLEAN COMBUSTION. CLEAR SOLUTIONS.™

- Inserted a baffle to determine if problem was heater or burner related
- Baffle extended from the inlet of the convection section
  - Caused the flue gas outlet to be directly above the burners
- End user decided against modifying the furnace
  - Burner modifications were investigated to counter the flue gas currents and asymmetric flue gas exit

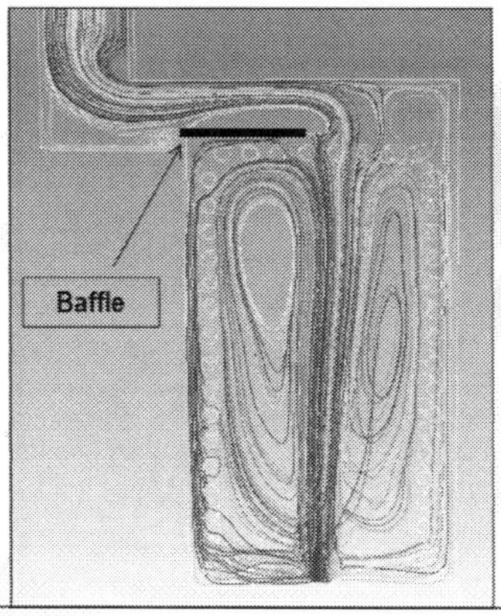

Figure 6: Flow pathlines used to show recirculation caused by heater configuration with baffle present.

JOHN ZINK

CLEAN COMBUSTION. CLEAR SOLUTIONS.™

Burner Modifications

- Reduced burner size
  - Increased the momentum of the air flowing through the burner
  - Provided more space between the burners for more efficient internal flue gas recirculation
- Fuel Biasing
  - Provided equal consumption of oxygen in the heater minimizing the flue gas that was oxygen deficient
- Changed drilling pattern to shorten flame

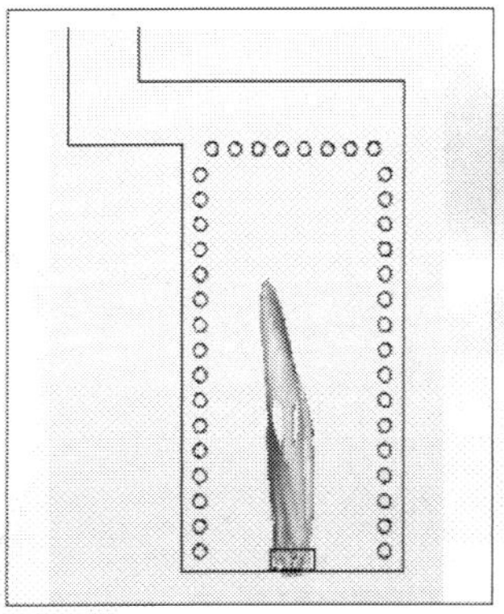

Figure 7: Flame envelope of modified ULNB.

JOHN ZINK

CLEAN COMBUSTION. CLEAR SOLUTIONS.™

Results

Positive: Straightened and significantly shortened flame lengths. Heater modifications were avoided

Trade-Offs: Reduced capacity and slightly higher NO_x emissions

JOHN ZINK
JOHN ZINK COMPANY LLC

JOHN ZINK

CLEAN COMBUSTION. CLEAR SOLUTIONS.™

Case Study #2 – Staggered Double Row Cabin Heater

JOHN ZINK

CLEAN COMBUSTION. CLEAR SOLUTIONS.™

Staggered Double Row Cabin

- Three separate cells with vertical tubes around the perimeter of each cell
- Convection section runs perpendicular across all three cells
- Burners staggered in two rows to meet API burner to tube spacing

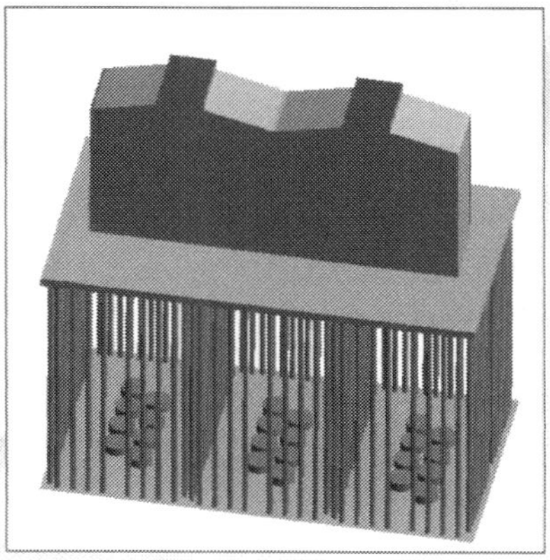

Figure 8: CFD Geometry of Staggered Double Row Cabin.

JOHN ZINK CLEAN COMBUSTION. CLEAR SOLUTIONS.™

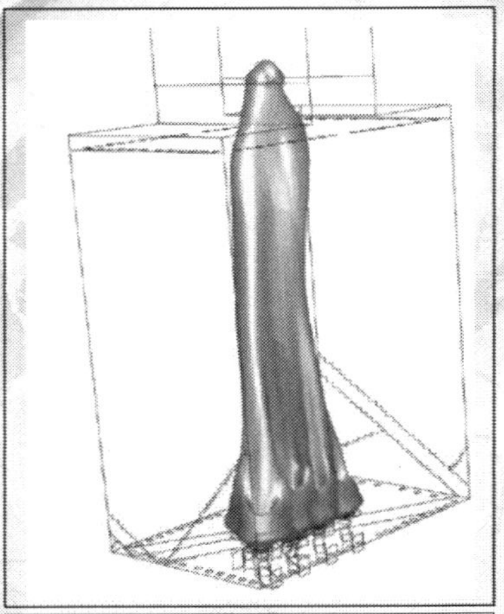

Figure 9: Flame envelope of burner leaning toward center of cell and merging together.

- Ultra-low NO_x burners were modeled in the heater
- The flames were interacting with flames from adjacent burners
- The end burner flames leaned towards the center of the heater and into the convection back

JOHN ZINK CLEAN COMBUSTION. CLEAR SOLUTIONS.™

The asymmetry of the cell reinforced the recirculation along the ends which push the flames together

Effects:

- Poor entrainment of flue gas by center burners
- The flames merged together into one large, elongated flame that exceed the maximum flame length requirement.

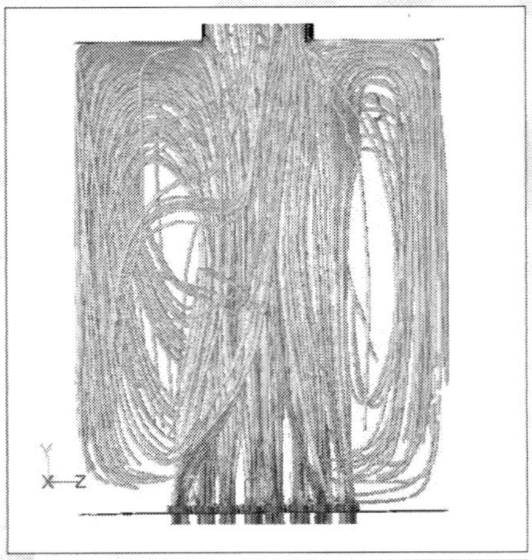

Figure 10: Flow pathlines show recirculation taking place at the ends of the burner rows instead of around each burner.

JOHN ZINK

CLEAN COMBUSTION. CLEAR SOLUTIONS.™

Burner Modifications

- Changing the burner fuel distribution to reduce flame length
  - Reducing the staged fuel to decrease the amount of flue gas recirculation
- Previous single burner testing had shown 20% reduction in flame length with similar modifications
- Burner spacing was not modified however overall flame length was acceptable

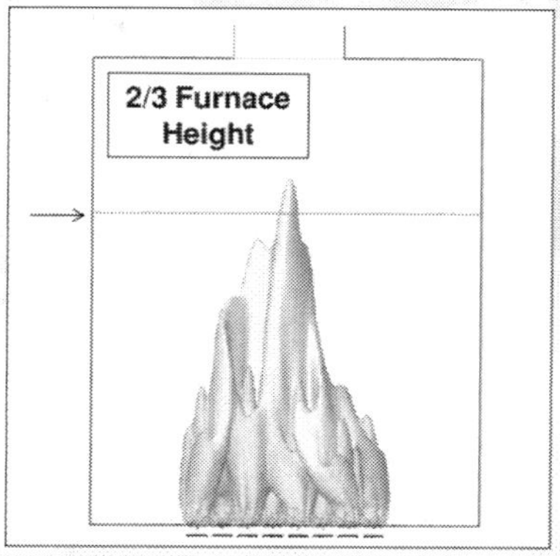

Figure 11: Flame envelope of shortened flames.

JOHN ZINK

CLEAN COMBUSTION. CLEAR SOLUTIONS.™

Results

Positive: Straightened and shortened the flame envelope.

Trade-Offs: There is still some flame to flame interaction present.

JOHN ZINK

CLEAN COMBUSTION. CLEAR SOLUTIONS.™

Case Study #3 – Single Circle Vertical Cylindrical Heater

JOHN ZINK

CLEAN COMBUSTION. CLEAR SOLUTIONS.™

Single Circle Vertical Cylindrical Heater

- Five ULNBs in a single circle in new heater
- Initial burner circle diameter (BCD) of the heater resulted in burner tip to burner tip spacing of 1.12 inches per MMBtu/hr
- CFD model revealed that the zone in the middle of the heater was oxygen deficient
  - Gas tips inside the burner circle could not properly entrain flue gas
- Flames merged and elongated

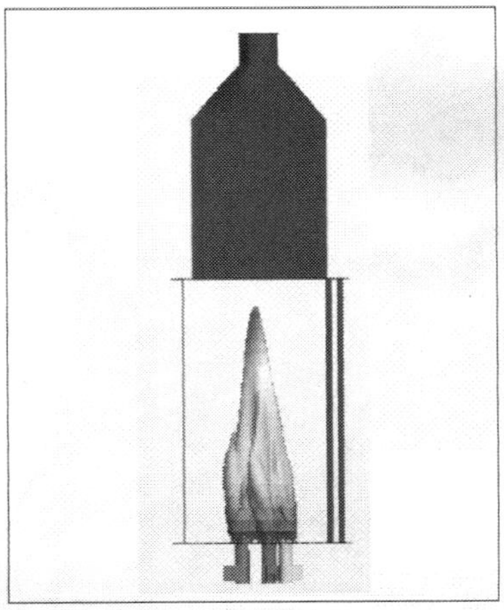

Figure 12: Flame envelope of ULNBs with initial BCD.

JOHN ZINK

CLEAN COMBUSTION. CLEAR SOLUTIONS.™

Results

The heater was re-modeled with a BCD that was increased to provide 2.37 inch per MMBtu/hr spacing - double the original case

- Individual flames were significantly shorter
- An increase in oxygen concentration in the center of the furnace was also shown
- No burner modification required

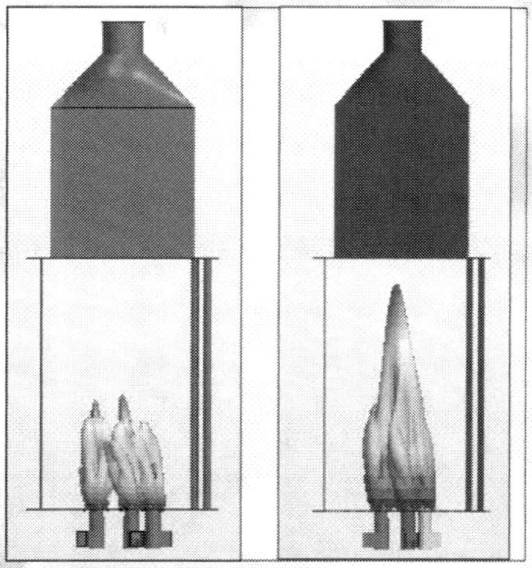

Figure 13: Effect of larger vs.. smaller BCD on flame envelope

JOHN ZINK

CLEAN COMBUSTION. CLEAR SOLUTIONS.™

Conclusion

- Ultra-low NO_x burners are more susceptible to furnace flow patterns than conventional burners.
- Several case studies were presented which revealed problems such as flame impingement on process tubes and undesired flame envelope.
- In each case study, CFD modeling was used to identify the problem. Once the problem was known, the correct modification was applied and the most cost effective solution was obtained.

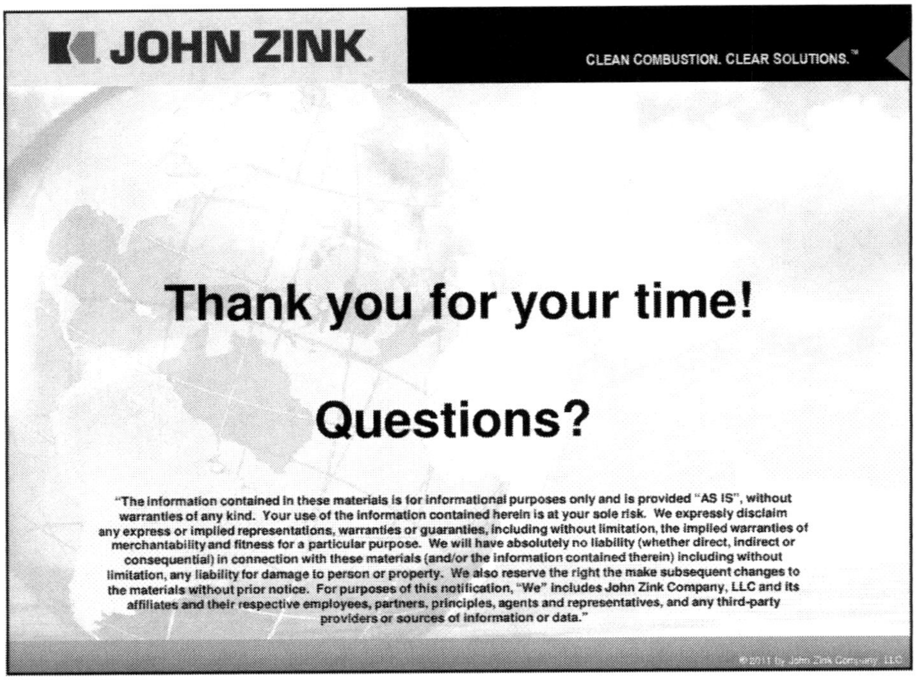

JOHN ZINK
CLEAN COMBUSTION. CLEAR SOLUTIONS.™
Thank you for your time!
Questions?
"The information contained in these materials is for informational purposes only and is provided "AS IS", without warranties of any kind. Your use of the information contained herein is at your sole risk. We expressly disclaim any express or implied representations, warranties or guaranties, including without limitation, the implied warranties of merchantability and fitness for a particular purpose. We will have absolutely no liability (whether direct, indirect or consequential) in connection with these materials (and/or the information contained therein) including without limitation, any liability for damage to person or property. We also reserve the right the make subsequent changes to the materials without prior notice. For purposes of this notification, "We" includes John Zink Company, LLC and its affiliates and their respective employees, partners, principles, agents and representatives, and any third-party providers or sources of information or data."

Exceptional Events Management Applied to Continuous Pharmaceutical Manufacturing: A Strategy to Detect, Diagnose and Mitigate Multiple Exceptional Events

Hamdan, I. M., Purdue University
Reklaitis, G. V., Purdue University
Venkatasubramanian, V., Purdue University
Originally presented on: 10/7/2011 0:50:00 - 1:15:00

Conventional regulatory control systems enforce process set points in individual pieces of equipment via implementation of PID and MPC controllers. Supervisory control systems, in contrast, encompass a network of these regulatory control systems, thus resulting in higher levels of process intelligence. The main purpose of the supervisory control layer is to reduce deviations of the entire process from the desired operating conditions. Though the supervisory layer is also able to detect abnormal process behavior, it is unable to diagnose and alleviate the problem. Thus, despite the design and implementation of sophisticated control systems, process anomalies persist. We designate these anomalies that control systems cannot handle as exceptional events, and propose a framework that can handle these occurrences.

The exceptional events management (EEM) framework is designed to detect, diagnose, and mitigate exceptional events as they manifest in process variable measurements. The detection and diagnosis is first performed using signed directed graph, followed by a higher resolution method to pinpoint the exceptional event using trend analysis; the framework is complemented with fast Fourier transform analyses to deal with events that manifest as periodic oscillations in measurements. Once the exceptional event has been determined, a mitigation strategy is either advised to the operator or automatically performed. This framework is also designed to run in conjunction with existing control layers; in essence, the EEM framework provides an added layer of security beyond the control systems to ensure that the plant/process remains in normal operating conditions.

Earlier versions of EEM were introduced and demonstrated on one piece of equipment – the roller compactor [1,2]. The most recent version of EEM incorporates the ability to handle multiple occurrences of abnormality, including simultaneously occurring events [3]. In this talk, we once again explore the EEM functionality by applying it to a continuous process, and compare it to a PCA-driven method to fault detection and diagnosis.

References:

- 1. Hamdan, I.M., Reklaitis, G.V., Venkatasubramanian, V. Exceptional Events Management for Continuous Pharmaceutical Manufacturing: Application and Integration into POPE and DeltaV, AIChE Annual Meeting. 2009
- 2. Hamdan, I.M., Reklaitis, G.V., Venkatasubramanian, V. Exceptional events management applied to roller compaction of pharmaceutical powders. Journal of Pharmaceutical Innovations. 2010; 5:147-160

- 3. Hamdan, I.M., Reklaitis, G.V., Venkatasubramanian, V. Exceptional Events Management for Continuous Pharmaceutical Manufacturing: Feeder, Blender, & Roller Compactor in Series, AIChE Annual Meeting. 2010

Improving Ethylene Charge Gas Compressor Performance

Young, W., GE Water & Process Technologies

Originally presented on: 10/7/2011 1:15:00 - 1:40:00

Feedstock to ethylene units is moving toward more ethane versus historical. Cracking ethane yields much fewer longer-chain and aromatic molecules that aid in solubilizing fouling components. As a result, an increase in downstream fouling issues can be expected. Compressor design & operational methods to reduce fouling will be discussed.

OSHALiANG
Intellectual Property Law
Houston · Paris · Silicon Valley · Tokyo · Austin

An Introduction to Intellectual Property & Patent Law

Aron T. Griffith
Osha Liang LLP
October 2011

OSHALiANG
Intellectual Property Law
Houston · Paris · Silicon Valley · Tokyo · Austin

Intellectual Property

- An intangible asset
- Intellectual Property - creations of the mind
  - Inventions
  - Literary and artistic works
  - Symbols, names, and images used in commerce.
- Tool for converting human capital into value by defining and capturing new knowledge

OSHA LiANG
Intellectual Property Law
Houston • Paris • Silicon Valley • Tokyo • Austin

Why Does IP Matter?

- More than defensive act against potential competitors: increasing market value and future profitability
- Generate income
  - Through sale, licensing, commercialization of IP-protected products and methods
- Provides expectation of future profits
- Obtain, maintain, improve market share
- Create brand and protect brand value
- Enhance value and image
  - In eyes of customers, investors, and financial institutions
  - In event of sale, merger, or acquisition

OSHA LiANG
Intellectual Property Law
Houston • Paris • Silicon Valley • Tokyo • Austin

Types of Intellectual Property

- Patents
- Trademarks
- Trade Secrets
- Trade Dress
- Copyrights

OSHA LiANG
Intellectual Property Law
Houston • Paris • Silicon Valley • Tokyo • Austin

TRADEMARKS

- Indication of source or origin of goods or services
- Rights flow from ***use***
  - Registration is confirmation of rights
  - Gives various advantages
- Rights are tied to specific type or types of products in a specific market
- No rights in the absence of actual use of mark in the relevant market

OSHA LiANG
Intellectual Property Law
Houston • Paris • Silicon Valley • Tokyo • Austin

TRADEMARKS

- Duty to Police Marks
  - Scope of right will lessen if competitors allowed to use similar marks
  - Rights may be lost if mark allowed to become generic term
    - Aspirin
    - Cellophane
    - Nylon
    - Thermos
    - Escalator

OSHALiANG
Intellectual Property Law
Houston • Paris • Silicon Valley • Tokyo • Austin

TRADEMARKS

- Less descriptive mark is stronger, more descriptive mark is weaker
  - Strongest marks are coined or arbitrary
    - "Apple" for computers
  - Suggestive marks are more difficult to protect
    - Netscape, Microsoft

OSHALiANG
Intellectual Property Law
Houston • Paris • Silicon Valley • Tokyo • Austin

Trademarks

- Almost every enterprise has a trade name and one or more trade or service marks
- Effective use allows your products and services to be differentiated and more easily recognizable
- Promote products / services and create loyal clientele

OSHA LIANG
Intellectual Property Law
Houston • Paris • Silicon Valley • Tokyo • Austin

TRADE DRESS

- "The visual image by which the product or service is presented to the relevant consuming public. It can include shape, appearance, and color of the product itself or packaging which is distinctive enough to identify the source of the goods or services."
  1 J. Thomas McCarthy, Trademarks and Unfair Competition § § 8:1-8:7, 7:23-7:33 (2d ed. 1984).
- Limited in scope
  - Closely related to trademark
  - Must indicate source of goods to the consumer

OSHA LIANG
Intellectual Property Law
Houston • Paris • Silicon Valley • Tokyo • Austin

TRADE SECRET

- General Definition:
  - (1) a secret which gives its owner an actual or potential advantage in business, and
  - (2) which the owner exercises reasonable measures to maintain as a secret

OSHALiANG
Intellectual Property Law
Houston • Paris • Silicon Valley • Tokyo • Austin

COPYRIGHT

- Right attaches to a creative work fixed in a tangible medium
  - Book, plans, drawings, sculpture, musical work, *etc.*
- Copyright attaches
  - Regardless of presence or lack of copyright notice
  - Regardless of whether or not copyright is registered
- Still good practice to include copyright notice
  - Use © or the word, "copyright", and the year

OSHALiANG
Intellectual Property Law
Houston • Paris • Silicon Valley • Tokyo • Austin

PATENTS

- Utility Patent
  - Protects functional aspects of products and methods
    - the way something works, is used, or is made
  - Usually what is meant by "a patent"
- Design Patent
  - Protects the ornamental design of an article
- Plant Patent
  - Protects asexual reproduction of new plant variety
- In some countries: Utility Model (China)

OSHA LiANG
Intellectual Property Law
Houston • Paris • Silicon Valley • Tokyo • Austin

What is a Patent?

- The grant of an exclusive right of a limited duration as a reward for development of new technology
- Technology must be new and non-obvious
- Inventor gets exclusive right for limited time, and public gets full disclosure of the technology, right to use when patent expires
  - Facilitates sharing of research

OSHA LiANG
Intellectual Property Law
Houston • Paris • Silicon Valley • Tokyo • Austin

Why Patent?

- Offensive Purposes
  - Exclusivity/Competitive Advantage
  - Licensing: negotiating power and income
  - Litigation
  - Positive Image
- Defensive Purposes
  - Protect freedom-to-operate
  - Cross-licensing

What Can Happen if You Don't

- Lose competitive advantage
  - Unable to prevent copycat competitors from "borrowing"
- Someone else may patent
  - Excluded from practicing your own invention
    - New AIA 2011 defense for prior commercial use**
  - Forced to license from competitors

OSHA LIANG
Intellectual Property Law
Houston • Paris • Silicon Valley • Tokyo • Austin

Deciding What to Patent

- Understand your technology and how it fits in the marketplace and your business plan
- What is the potential commercial value of the invention?
- How easily can a competitor design around?

OSHALIANG
Intellectual Property Law
Houston • Paris • Silicon Valley • Tokyo • Austin

Look Beyond "Typical" Subject Matter

- New products – of course!
- Don't forget methods
  - Methods of making
  - Methods of using (new uses of old products)
- Don't forget intermediates and components of products

OSHALIANG
Intellectual Property Law
Houston • Paris • Silicon Valley • Tokyo • Austin

Considerations in What to Patent

- Who is the target infringer?
- Is infringement easy/difficult to detect?
- Consider potential design-arounds
- Do not limit to commercial product or what is monetarily feasible today
- Claiming an appropriate breadth
  - Specific compound vs. family of similar compounds

OSHA LiANG
Intellectual Property Law
Houston • Paris • Silicon Valley • Tokyo • Austin

Where Should We Patent?

- A patent is granted and enforced in a single country
- To possess exclusivity in a country, you must file a patent application there
- File first application in home country, then file foreign within 12 months

OSHA LiANG
Intellectual Property Law
Houston • Paris • Silicon Valley • Tokyo • Austin

Foreign Protection of IP

- Global protection is costly (over $500K for single invention in 50 major countries)
- Any place you do not file, no claim to exclusivity
- Should balance need for exclusivity with:
  - Cost
  - Ability to obtain and enforce
  - Return on investment

OSHA LIANG
Intellectual Property Law
Houston • Paris • Silicon Valley • Tokyo • Austin

When to File – To have a Valid Patent

- Promptly after the inventor has conceived or determined how to achieve the results desired
- Not necessary to wait for construction of a working embodiment to disclose or patent

- Patents Filed in United States
  - A patent application must be filed within 1 year of any disclosure event of the invention described.
  - [Currently] First to Invent – Can swear behind disclosure event within 1 year
    - March 16, 2013 – US moving to First to File with grace period**
- Patents Filed in Foreign Countries
  - Generally, a patent application must be filed prior to any disclosure event of the invention described.
  - First to File – Race to the patent office

OSHA LIANG
Intellectual Property Law
Houston • Paris • Silicon Valley • Tokyo • Austin

What are Considered Disclosure Events?

- Publication or presentation of a technical paper or other description
- Display at trade shows
- Postings on the Internet
- Outsourcing prototype manufacture without a non-disclosure agreement
- Sale or offer to sell
- Use of the invention in public (no trade secret protection enforced)

OSHALiANG
Intellectual Property Law

Houston • Paris • Silicon Valley • Tokyo • Austin

How - What to Include in Invention Disclosure

- A complete written disclosure of the invention
- Attach drawings or sketches, reaction pathways
- Disclose various embodiments, when possible
- ALWAYS include the "best mode" (do not hold back)
- Inventor and two witnesses sign and date

OSHALiANG
Intellectual Property Law

Houston • Paris • Silicon Valley • Tokyo • Austin

The Patent Document

- Patent rights flow from patent application
  - Written disclosure or "specification"
  - Drawings when appropriate
  - "Claims"
    - Claims define the scope of patent monopoly
    - Each claim stands independently
    - Validity of each claim determined separately

OSHALiANG
Intellectual Property Law
Houston • Paris • Silicon Valley • Tokyo • Austin

Requirements of a Patent Specification

- Enablement
  - Teach how to make and use invention without undue experimentation
- Written description of invention
  - Explains what the applicant believes the scope of the invention to be
- "Best mode"
  - Duty to describe the best mode of carrying out the invention known to the inventor at the date of filing

OSHALiANG
Intellectual Property Law
Houston • Paris • Silicon Valley • Tokyo • Austin

Note on America Invents Act 2011

- New prior art rules
- Third party submissions
- Post-grant proceedings
  - Opposition period
  - Inter partes reexamination
  - Supplemental Examination
- Microentity
- Commercial Use Exception
- Many other changes

OSHALIANG
Intellectual Property Law
Houston • Paris • Silicon Valley • Tokyo • Austin

Practical Considerations

- For the U.S., we MUST be able to prove date of invention
- Lab notebooks are ESSENTIAL!!
- Mover to first-to-file with grace period may lessen impact on proving date of invention, but still important to document (derivation, best mode, etc.)

OSHALIANG
Intellectual Property Law
Houston • Paris • Silicon Valley • Tokyo • Austin

Lab Notebooks

- The goal is to maintain records in a provable form that may be admitted in a court of law
- Detailed and accurate
- Primary record of the achievements of an individual
- A source for preparation of invention disclosures
- 2 Colleagues qualified to understand the data should witness each page with his/her signature and date

Assignment of IP Rights

- Only individual inventors can file for a US Patent – corporations cannot directly file
- Invention is owned by the inventors – they are the presumed owners
- Inventors must sign an Assignment transferring rights to an employer/ corporation/university

OSHA LiANG
Intellectual Property Law
Houston • Paris • Silicon Valley • Tokyo • Austin

Appropriate Agreements in Place

- Employee
- Consultant
- Vendor/Vendee

OSHA LiANG
Intellectual Property Law
Houston • Paris • Silicon Valley • Tokyo • Austin

Employee / Consultant Agreements

- Be aware of employee mobility
- Be proactive – require all employees and consultants to sign agreements at start of employment

OSHA LiANG
Intellectual Property Law
Houston • Paris • Silicon Valley • Tokyo • Austin

Employee / Consultant Agreements

- Confidentiality Provisions
  - Covers BOTH use and disclosure
  - Extends after end of employment
- Invention assignment provisions
  - Assist in the preparation and enforcement of patent applications
- Copyright assignment provisions
- Non-compete provisions
- Non-solicitation clauses

OSHALiANG
Intellectual Property Law
Houston • Paris • Silicon Valley • Tokyo • Austin

Vendor / Vendee or Collaborative Agreements

- Define who owns what
- Limitations on field of use
- Confidentiality Provisions
  - Covers BOTH use and disclosure
  - Extends after end of cessation of agreement
- Employee education on necessity of NDA (specific) prior to collaborative work

OSHALiANG
Intellectual Property Law
Houston • Paris • Silicon Valley • Tokyo • Austin

Tips to Help Spur Creation of Patent Rights

- Focus on individual and specific lines of business or technologies
- Monitor competitors' filings and activity
- Conduct regular brainstorming sessions
- Employee education (all levels)
  - Who, what, where, when, why, how of patenting
- Invention reward system
- Appoint a technical person as IP Manager

OSHA LIANG
Intellectual Property Law
Houston • Paris • Silicon Valley • Tokyo • Austin

What happens after the application is filed?

- Application is:
  - Assigned to examining group based on class/subclass
  - Assigned to appropriate examiner within group
  - Examined in order of date assigned, unless special case
- Timing for U.S. PTO action (Roughly):
  - 14 - 18 months for initial examination
  - 4 months for responsive examination
  - 4 months for issuance
  - 2.5 - 3 years filing to issue

OSHA LIANG
Intellectual Property Law
Houston • Paris • Silicon Valley • Tokyo • Austin

When Not to Patent?

- Trade Secret
  - a secret that gives its owner an actual or potential advantage in business or trade,
  - which the owner exercises reasonable measures to maintain as a secret
  - rights can last indefinitely, but there is no protection if separately discovered
- Defensive Publication
  - Publish to ensure freedom-to-operate and prevent patent attempts by others

OSHALIANG
Intellectual Property Law
Houston • Paris • Silicon Valley • Tokyo • Austin

Potential Liabilities & Minimizing Risks

OSHALIANG
Intellectual Property Law
Houston • Paris • Silicon Valley • Tokyo • Austin

Freedom to Operate

- Always keep eventual commercialization in mind
- Must conduct patent & TM searches before commercializing
- Patent Grant ≠ Freedom to Operate

OSHALIANG
Intellectual Property Law
Houston • Paris • Silicon Valley • Tokyo • Austin

When Presented with Notice of Patent by Third Party

- Request specifics (*e.g.*, claim charts and accused products) from the patentee if letter is received
- Request opinion if
  - Useful to make informed business decision
  - Close case – answer not clear
  - Invalidity/unenforceability position only

OSHALIANG
Intellectual Property Law
Houston • Paris • Silicon Valley • Tokyo • Austin

Withholding Information from Attorney

- Provide all known prior art that is relevant to invention
  - The inventor and any individual within the company who are involved in the patent application
  - Must submit any known information that an Examiner would be likely to consider important in examining an application
- Provide “best mode” for carrying out invention

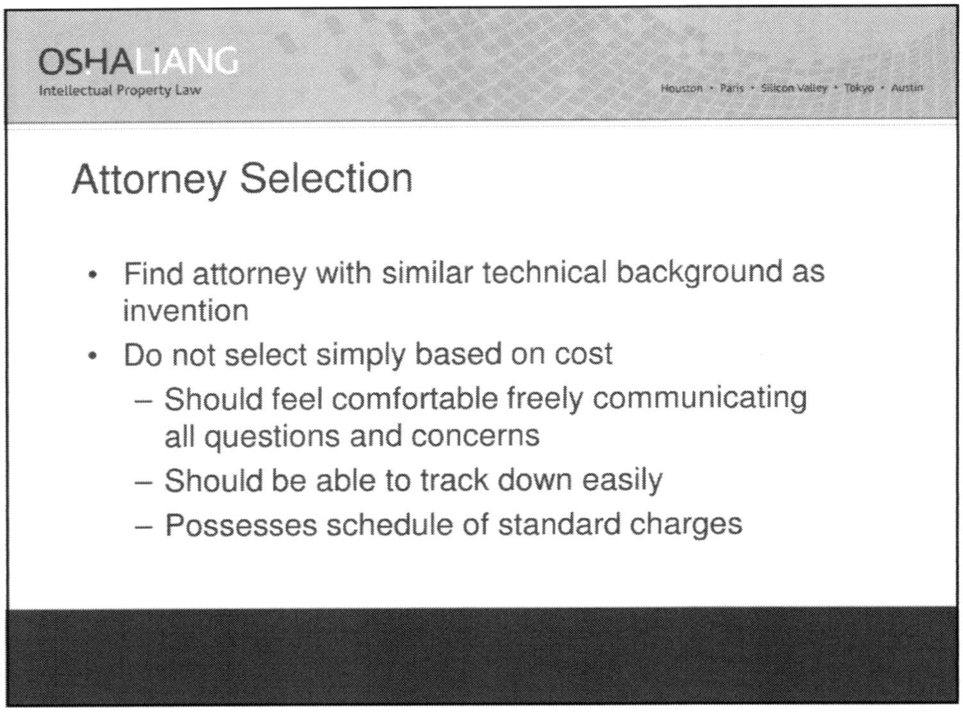
OSHALiANG
Intellectual Property Law
Houston • Paris • Silicon Valley • Tokyo • Austin
Attorney Selection
• Find attorney with similar technical background as invention
• Do not select simply based on cost
– Should feel comfortable freely communicating all questions and concerns
– Should be able to track down easily
– Possesses schedule of standard charges

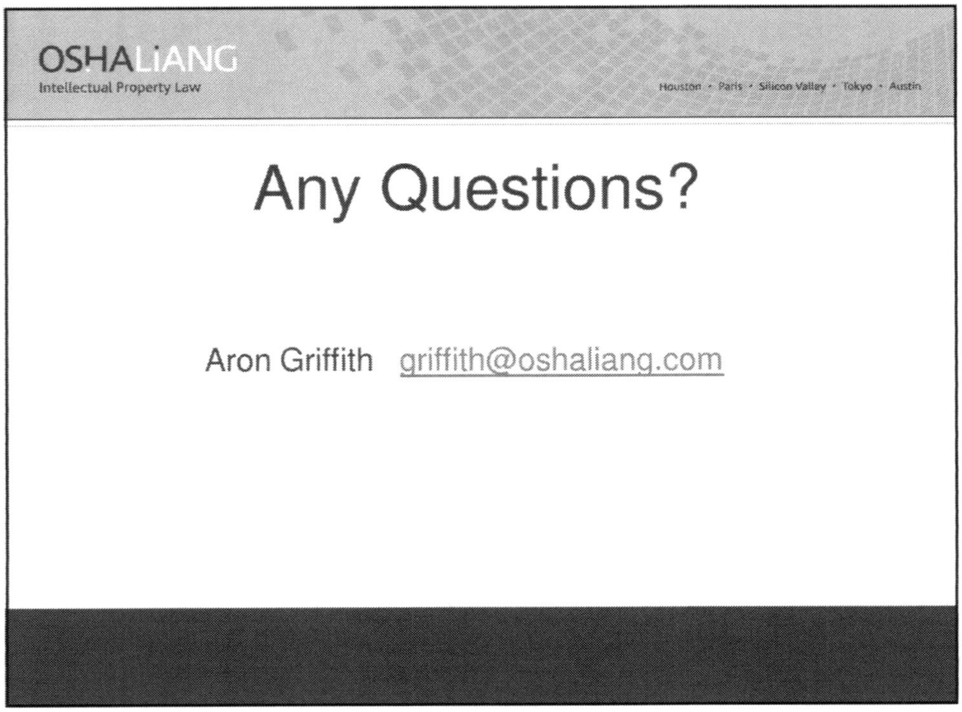
OSHALiANG
Intellectual Property Law
Houston • Paris • Silicon Valley • Tokyo • Austin
Any Questions?
Aron Griffith griffith@oshaliang.com

The Current Situation

- Volatility of diesel to gasoline prices
- More difficult crudes and tightening product specifications

→Customization of catalytic solution required
→Criterion/Zeolyst Proprietary Zeolite Manufacturing Technology

ZEOLYST CRITERION Page 2

A Hydrocacker Diagram

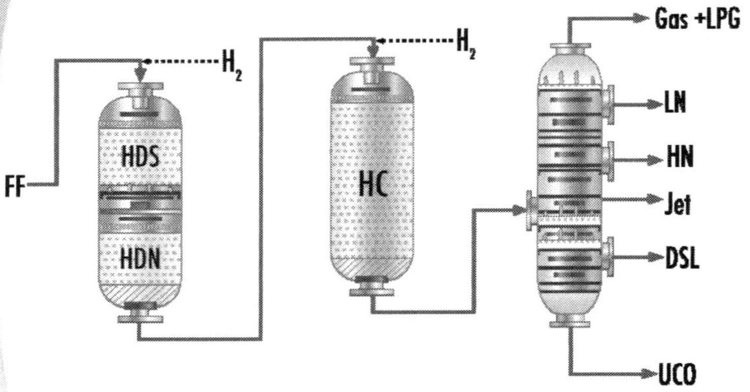

ZEOLYST INTERNATIONAL CRITERION

Page 3

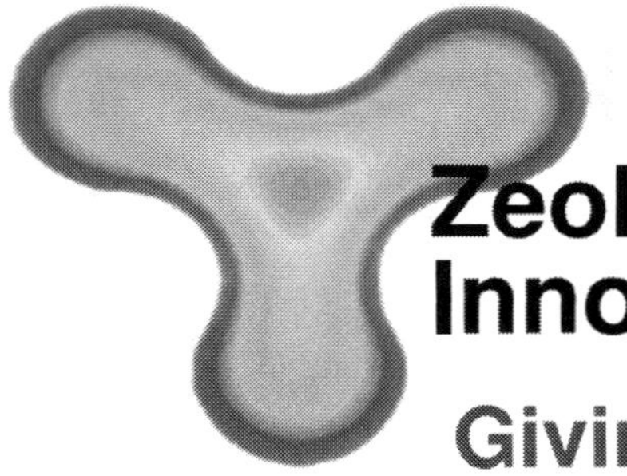

Zeolite Innovations.

Giving you the Flexibility to Thrive!

ZEOLYST INTERNATIONAL CRITERION

Page 4

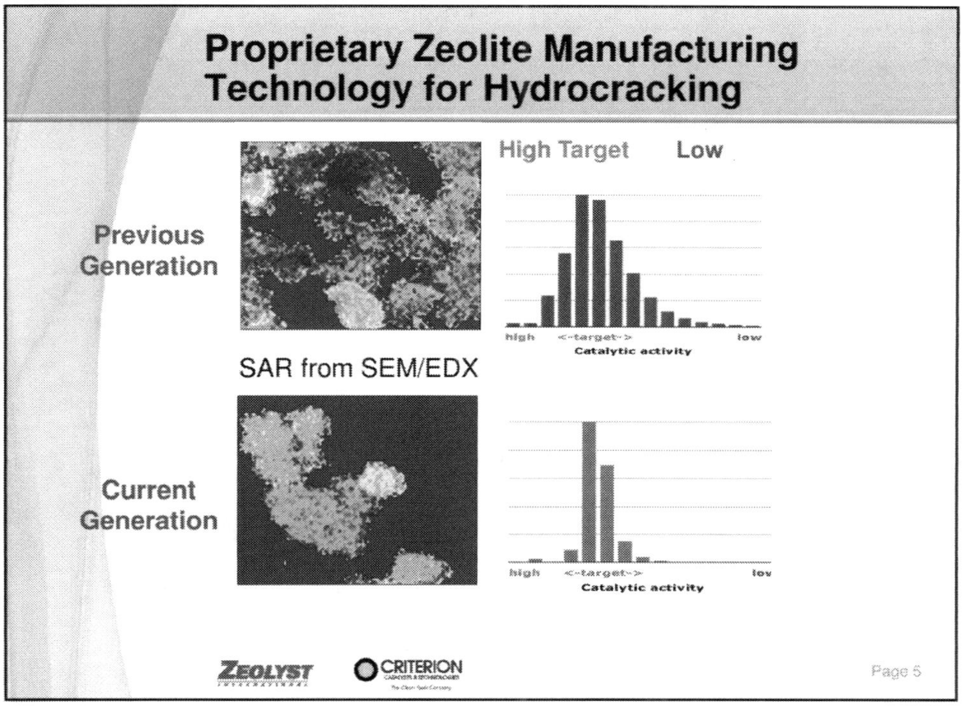
Proprietary Zeolite Manufacturing Technology for Hydrocracking
Previous Generation
SAR from SEM/EDX
Current Generation
High Target Low
high
<--target-->
low
Catalytic activity
high
<--target-->
low
Catalytic activity
ZEOLYST
CRITERION
Page 5

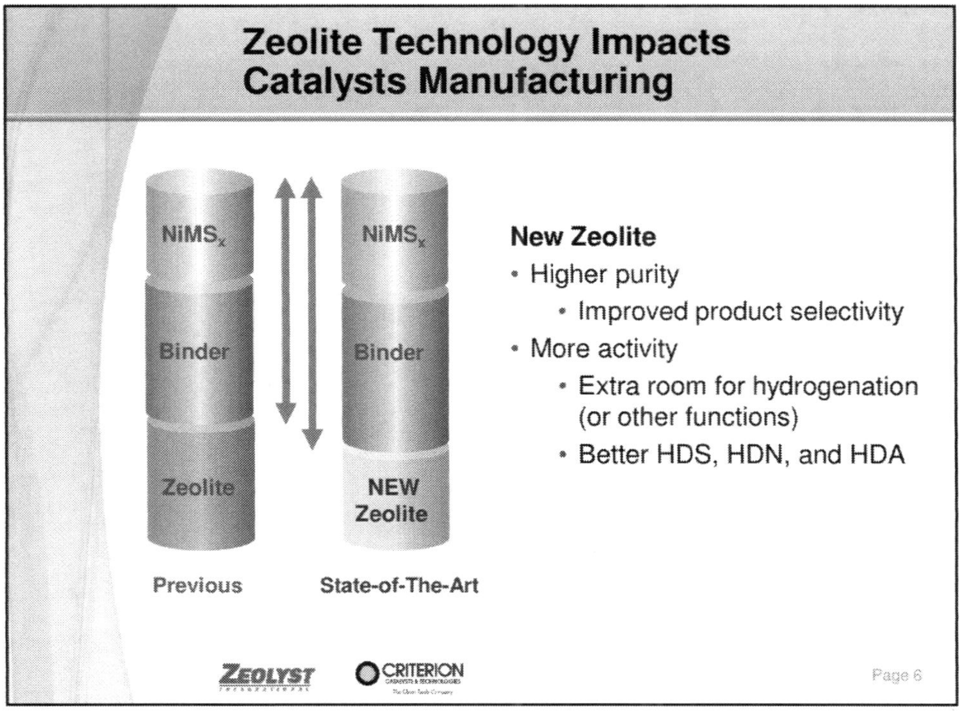
Zeolite Technology Impacts Catalysts Manufacturing
NiMS$_x$
Binder
Zeolite
NiMS$_x$
Binder
NEW Zeolite
Previous
State-of-The-Art
New Zeolite
• Higher purity
• Improved product selectivity
• More activity
• Extra room for hydrogenation (or other functions)
• Better HDS, HDN, and HDA
ZEOLYST
CRITERION
Page 6

Advanced Trilobe eXtra (ATX) Shape

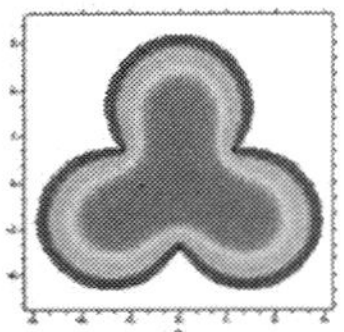

TL Trilobe Shape

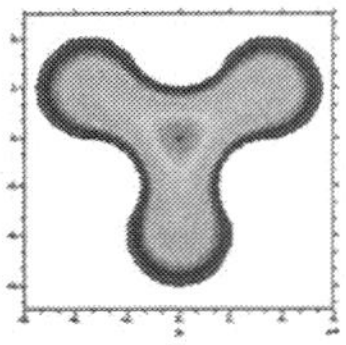

ATX Trilobe Shape

- Advanced Trilobe eXtra
- Now available for our HC catalyst
  - Lowers SOR ΔP (ΔP ATX = 0.75 x ΔP TL)
  - Improves diffusion
  - Delays onset to rapid ΔP increase

ZEOLYST CRITERION

Criterion/Zeolyst Hydrocracking Catalysts

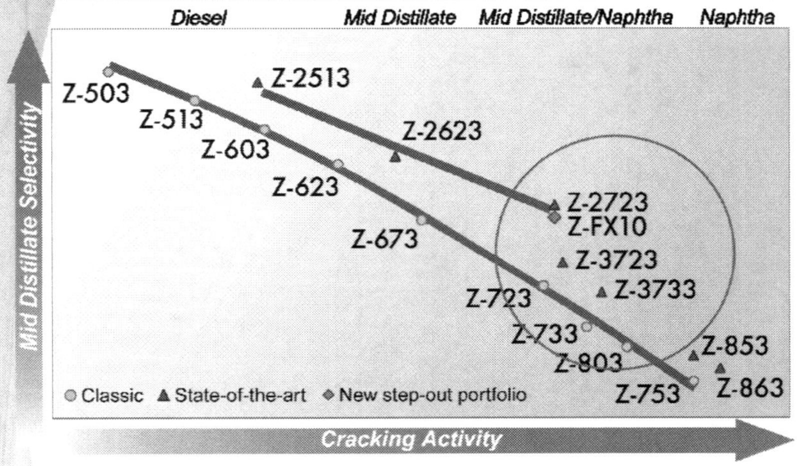

ZEOLYST CRITERION

New Z-FX10 Catalyst (Flexible)

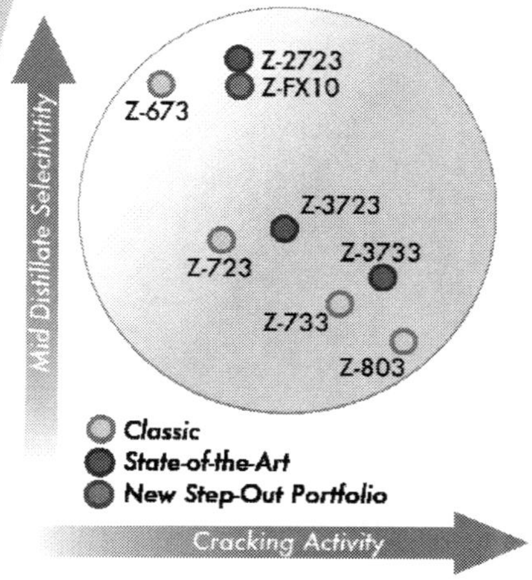

Z-FX10 Performance

	Z-3723	Z-FX10	
Feed		Canadian Synthetic VGO (Ring Rich)	HVGO
Middle Distillate selectivity (%)	Base	**Base +4**	**Base +3**
Diesel Cetane	Base	**Base +1**	Base

Z-FX10 Hydrogenation Activity

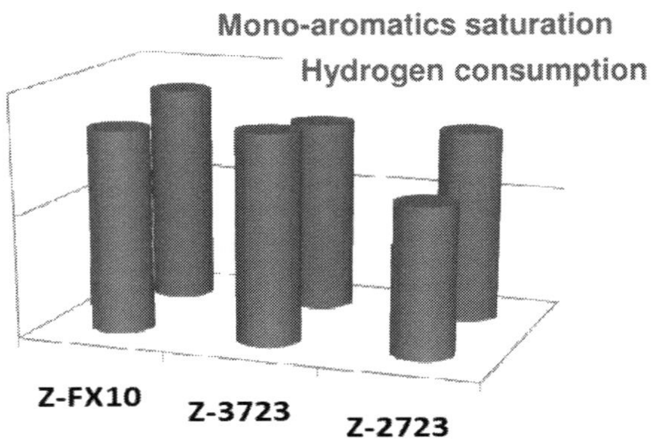

Optimizing your Hydrocracker Operation Thro' Customized Catalyst Solutions

Example 1: Z-853 (Naphtha) vs Z-2723 (Flexible) Catalyst

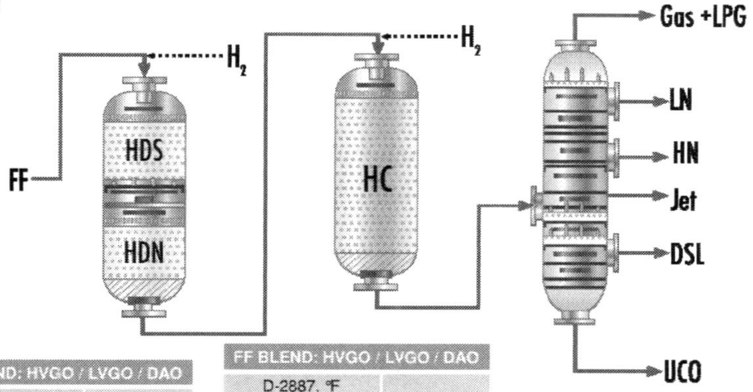

FF BLEND: HVGO / LVGO / DAO	
API	26
Sulfur, wt%	0.60
Nitrogen, ppm	1000

FF BLEND: HVGO / LVGO / DAO	
D-2887, °F	
10%	580
50%	820
FBP	1370

ZEOLYST INTERNATIONAL   CRITERION CATALYSTS & TECHNOLOGIES The Clean Tech Company

Example 1: Yields and Product Properties

	Z-853	Z-2723	Z-853	Z-2723
Conv to 343-°C, vol% ff	55		70	
Product Yields, vol% ff	Multiplier			
LPG	Base	0.65	1.69	1.08
Total Naphtha	"	0.91	1.54	1.37
Jet	"	**1.03**	**1.15**	**1.26**
Diesel	"	**1.15**	**0.91**	**1.13**
C5+ Yield	"	1.00	1.01	1.02
Chemical H2 Cons, scfbff	"	0.98	1.25	1.19
	Delta to Base			
HC Catalyst WABT, °C	"	**+12**	**+7**	**+19**
ULSD Cloud Pt., °C	"	-5	-0.5	-6
ULSD Sulfur, ppm	"	-2	0	-3

ZEOLYST INTERNATIONAL   CRITERION CATALYSTS & TECHNOLOGIES The Clean Tech Company

Example 1 Summary

- Catalyst System Considerations
  - LPG constraints
  - Product Sulfur or Cetane constraints
  - Cold Flow constraints
  - Value and Definition of "Volume Swell"
    - C3+ or C5+ yield
  - Value of Unconverted Oil
  - Reactor Temperature impact on Cycle Life

ZEOLYST CRITERION

Example 2: Processing more difficult feed Z-FX10 (Flexible) vs Z-673/Z-623 (Diesel)

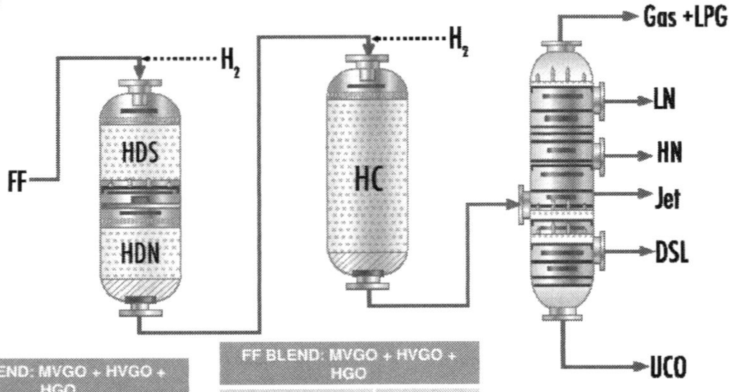

FF BLEND: MVGO + HVGO + HGO	
API	26
Sulfur, wt%	3.05
Nitrogen, ppm	815

FF BLEND: MVGO + HVGO + HGO	
D-2887, °F	
10%	680
50%	818
95%	1018

ZEOLYST CRITERION

Example 2 – Increasing pretreat and post treat catalyst with Z-FX10

	Z-673/Z-623	Z-FX10
Pretreat SV, hr^{-1}	1.8	1.6
Cracking SV, hr^{-1}	3.7	5.1
Post treat SV, hr^{-1}	20	15
Middle Distillates Yield, LV%	Base	0.96 x Base

Unit runs to constant UCO VI target

- Flexibility of running higher N (10% above base case) or heavier feedstock
- Option of increasing throughput (15% above base case)
- Longer cycle length

ZEOLYST CRITERION

Page 17

Example 2 – Summary

- Improved activity and selectivity of Z-FX10 creates more options for profitable operation
- A unique solution depends on the careful consideration of the constraints.

ZEOLYST CRITERION

Page 18

Conclusion

→Customization of catalytic solutions using Proprietary Zeolite Technology and new catalyst shape

Allows refiners to maximize benefits while handling the challenges of:

- Volatility of diesel to gasoline prices
- More difficult crudes and tightening product specifications

Criterion/Zeolyst

Zeolite Innovations.

Giving you the Flexibility to Thrive!

CRITERION/ZEOLYST DISCLAIMERS

ZEOLYST CRITERION

Processing Heavy Crudes in Existing Refineries

Kevin Turini, James Turner, Arthur Chu, and Shankar Vaidyanathan

FLUOR

Presentation Goals

When considering running heavy crude…

- What are the issues related to the nature of the heavy crude and the overall refinery configuration?
- What units are the most affected?
- What are the major challenges in these units? Are there any available low cost solutions for these challenges?

Caveats…

- Some techniques discussed work with incremental heavy crude.
- There is no one size fits all solution.

FLUOR

Agenda

Design Basis

- Type of Heavy Crude, Dilbit vs. Synbit
- Naphthenic Acid
- Sulfur Content
- Vacuum Tower Cut Point

Unit Specific Issues

- Crude/Vaccuum
- Coker/Resid Processing
- Hydrotreaters
- Amine
- Sour Water
- SRU/TGTU
- Naphtha Processing

Q&A

FLUOR®

2

Canadian Crude Oil Forecast

	2010	2015	2020	2025
Total Canadian Crude Oil Production (million BPD)(including oil sands)	2.8	3.3	3.9	4.3
Canadian Oil Sands Production (million BPD)	1.5	2.2	2.9	3.5

Data from Canadian Association of Petroleum Producers (CAPP)

FLUOR®

3

Yield of Crude Oil

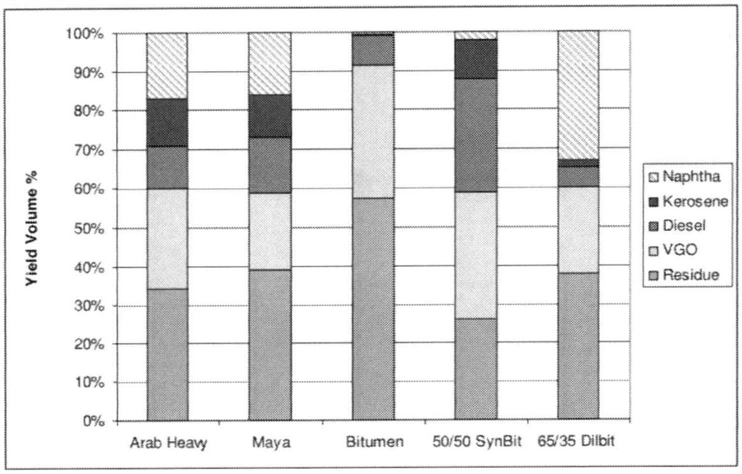

FLUOR

Nature of the Crude

- Dilbit versus Synbit
- Naphthenic Acid of Crude
- Sulfur Content
- Aromatics Content
- Asphaltenes
- Nitrogen

FLUOR

Vacuum Unit Cut Point

Refineries always have the mindset:

Deeper Vacuum Tower Cut Point = Better

- This is true for existing assets.
- May not be true if…
  - Building a new Vacuum Unit
  - Building a new Coker or Other Resid Processing
  - Not constrained on Resid sales (asphalt)

FLUOR.

Less Deep Vacuum Tower Cut Point Comparison

ADVANTAGE	DISADVANTAGE
✓ Operability of the Crude/Vacuum Unit	✗ More Coke Make
✓ Less Metals entrained in HVGO to Gas Oil	✗ Higher Cost Coker
✓ Lower Cost Crude/Vac. Unit	
✓ Lower Cost Gas Oil Hydroprocessing	

FLUOR.

Unit Impacts

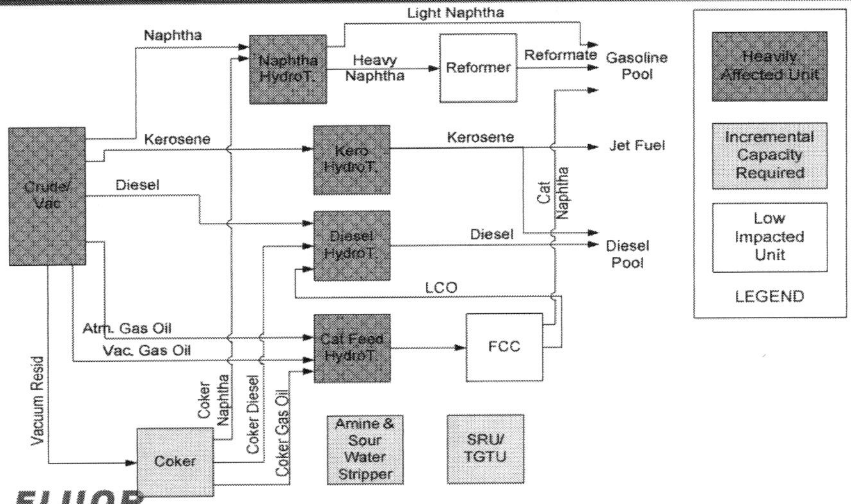

FLUOR.

8

Crude/Vacuum Unit

- Metallurgy
- Cracked Gas Make
- Coking and Vacuum Unit Run Length
- Crude Preheat Exchanger Train
- Desalting
- Kerosene Product Sulfur Specification

FLUOR.

9

Coker/Vac. Resid Processing

- Incremental Vac. Resid In Feed
  - Increase Coker Capacity Required (or more VTB sales)
- Naphthenic Acid corrosion can be a concern at elevated temperatures on Coker Feed
  - Coker Furnace Temperatures decompose Naphthenic Acid

FLUOR.

16

Hydrotreaters - Challenges

- Incremental Difficult Feed Sulfur
  - Additional Catalyst Volume or Shorter Cycles
- More Metals and Silicon with Native Crude
  - Additional Trap Catalyst is Required
- Additional Coker Products
  - These are the most challenging liquids in the refinery
  - Highest in sulfur
  - High in Nitrogen
  - Contain Silicon from the antifoam
  - High Aromatics

FLUOR.

17

Hydrotreaters – Major Design Issues

- Metallurgy
  - Naphthenic Acid Corrosion
- Wash Water Rates
  - Higher Nitrogen
- Amine System Loading
  - Higher Sulfur

FLUOR.

12

Amine System & Sour Water System

- Amine
  - Additional Sulfur Load can Overwhelm Amine System
  - Consider switching Amines – MDEA or other proprietary blends
- Sour Water System
  - Make sure that units are not overwashing. Recycle within the unit to the limits of NH_4HS.
  - Look at recycling stripped sour water as wash water makeup.

FLUOR.

13

SRU / TGTU

- Bebottleneck the existing units – O_2 Enrichment
- Consider Sparing Philosophy
  - Many Refineries Operate with a full or partial "Spare" Sulfur Train
  - Consider using load shedding instead of "Sparing" for unplanned shutdowns

FLUOR.

Desalter Wastewater Treating

- Water from Desalters have a larger "rag layer" emulsion
- Consider proprietary technologies for separation
- Fluor has developed a technology for the separation of the oil and water in this emulsion

FLUOR.

Naphtha Processing

What do you do with all the naphtha that came in the with the Dilbit crude?

Incremental Naphtha appears in Crude unit naphtha.

Options:

- Process into Gasoline
  - Some can likely fit within the existing pool.
  - Requires additional Naphtha Hydrotreater / Reformer Capacity
- Return to Production Site
  - Can it be returned?
  - What are the logistics involved?
  - VERY site specific.

FLUOR

16

Conclusion

- Design Basis Issues
  - Dilbit vs Synbit vs Native Heavy Crude
  - Vacuum Tower Cut Point
- The Most Affected Units
  - Crude/Vac
  - Hydrotreaters
  - Naphtha Processing
  - Sulfur Complex (Sour Water Stripping, Amine Regeneration, SRU/TGTU)
  - Desalter Water Processing

FLUOR

17

Questions

?

FLUOR

lyondellbasell

Gasoline Blend Property Control and Optimization

Clint Dolby
Houston Refinery LP

Robert Hawkins
Lyondell Chemical Company

2011 AIChE Regional Process Technology Conference
October 7, 2011

LyondellBasell

59 plants in 18 countries	More than 14,000 employees worldwide	Sales in more than 100 countries

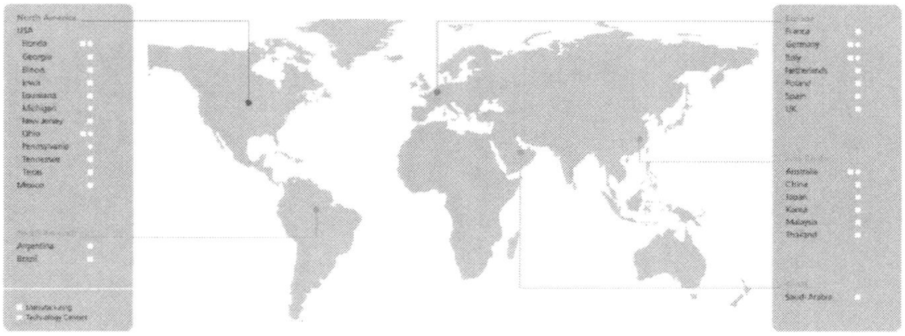

lyondellbasell.com

lyondellbasell

LyondellBasell Houston Refinery

- Two identical blend headers
  - Capable of 168,000 BPD each
  - 10 blending component loops each
  - Blend ratio control was developed in-house to manage start-up, shutdown, and recipe changes
  - The blenders share RVP, sulfur and NIR analyzers
- Gasoline is blended to on-site tanks and batched to pipeline tanks

lyondellbasell.com lyondellbasell

The Gasoline Blending Control Requirements

- Get the tank on-specification at the end
- Minimize the cost of the overall blend
  - Minimum giveaway on constraining specifications
- Make the actual blend recipe consistent with:
  - The overall refinery operating plan
  - The near-term volumetric constraints

- Do all of this in spite of uncertainties in the qualities and flows of the blend components, and gaps in the measurements of the blend properties

lyondellbasell.com lyondellbasell

VOC % Reduction

- Very complex calculation defined in 40 CFR Part 80
- Based on distillation, vapor pressure, and aromatic content of gasoline with ethanol
- Property measurements available without ethanol
- Too difficult for direct manual control
- Economically significant

Implementation Alternatives

Product Choices

- Buy an off-the-shelf commercial product or develop a custom in-house solution?

Technical Choices

- Simple volumetric averages or complex nonlinear models for blend property estimation?
- Manipulate individual component flow rates or total blend rate and component ratios?
- Dynamic control of line analyzer, estimated tank properties, or both?

Observations on Optimization

- # Active Constraints = # Components (at most)

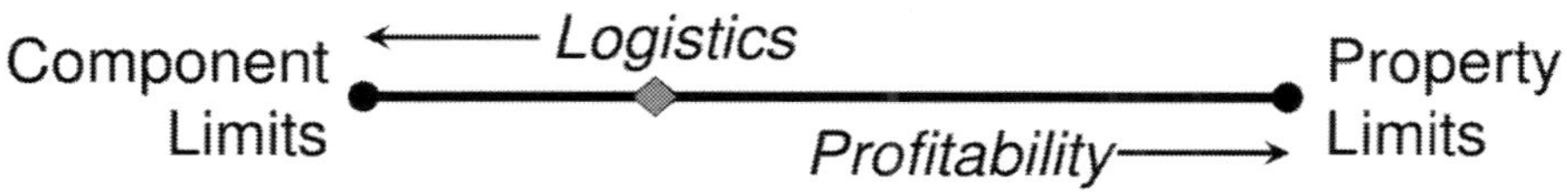

- Economic benefit comes from:
  - Reducing average cost of components
  - Increasing volume blended
  - Increasing average price of blended gasoline
- Making money requires changing the overall mix of components and/or products

Controller Model Matrix

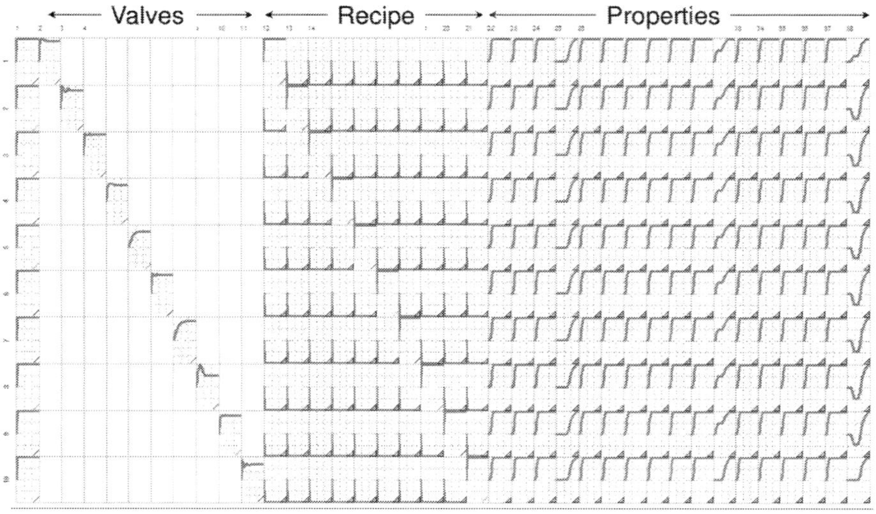

Typical Blend Results

- Controller takes ~ 1 hour to "fix" initial recipe
- Very little change after initial adjustment
- Small changes in property gains through blend
- Cycling is possible if gains change too much

lyondellbasell.com   lyondellbasell

Increase Blend Rate

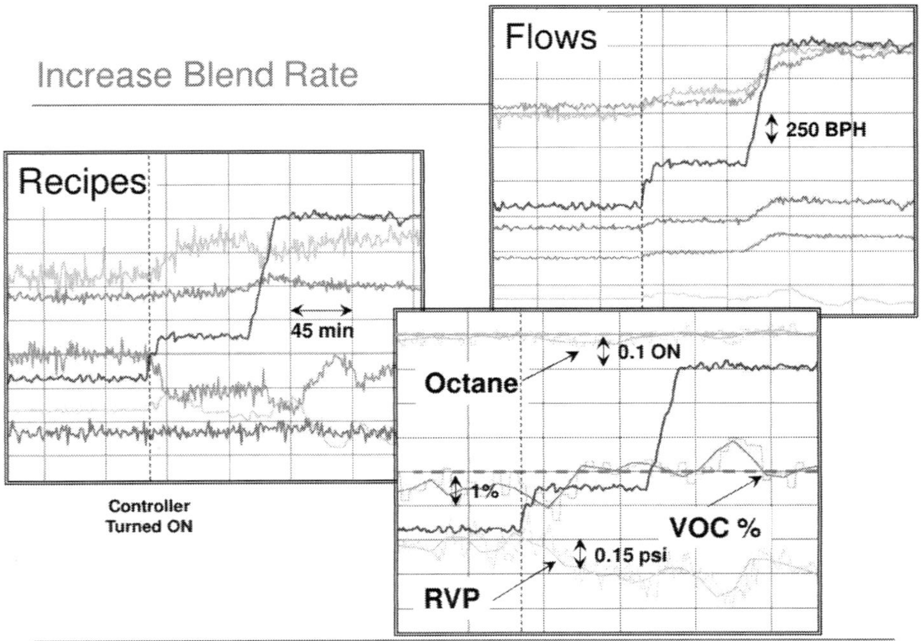

lyondellbasell.com   lyondellbasell

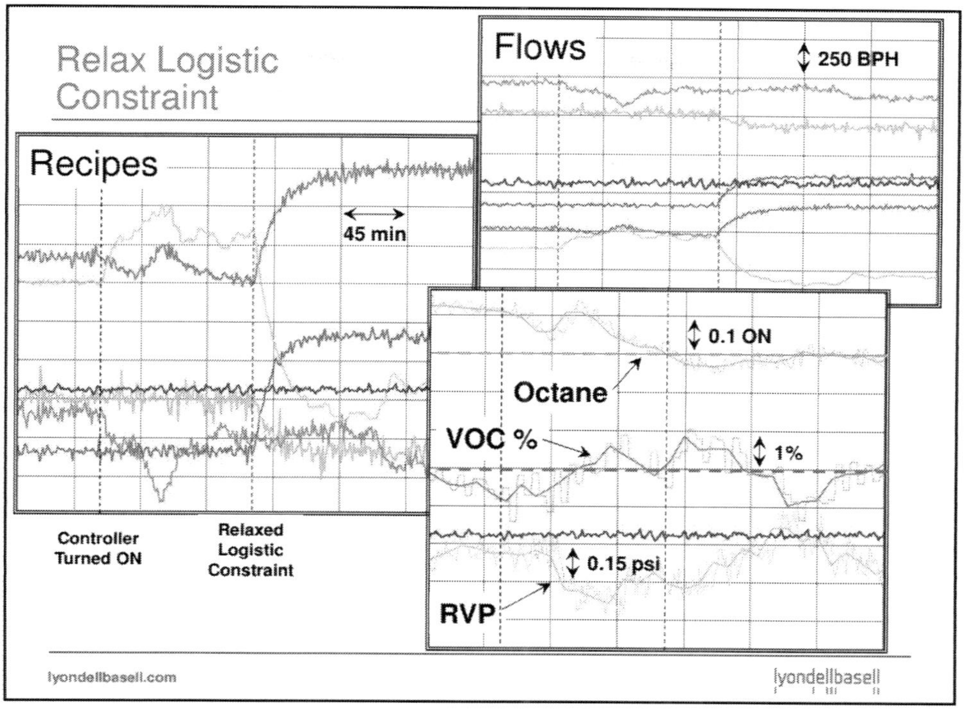

Relax Logistic Constraint
Flows
250 BPH
Recipes
45 min
0.1 ON
Octane
VOC %
1%
0.15 psi
RVP
Controller Turned ON
Relaxed Logistic Constraint
lyondellbasell.com
lyondellbasell

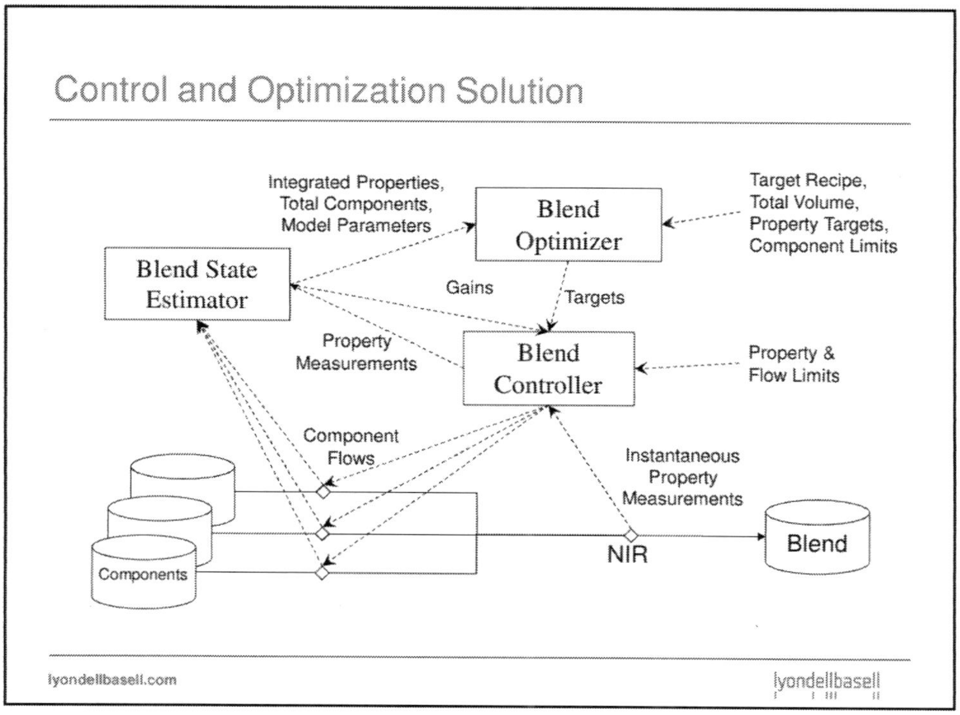

Control and Optimization Solution
Integrated Properties, Total Components, Model Parameters
Blend Optimizer
Target Recipe, Total Volume, Property Targets, Component Limits
Blend State Estimator
Gains
Targets
Property Measurements
Blend Controller
Property & Flow Limits
Component Flows
Instantaneous Property Measurements
Components
NIR
Blend
lyondellbasell.com
lyondellbasell

Blend Controller Gains

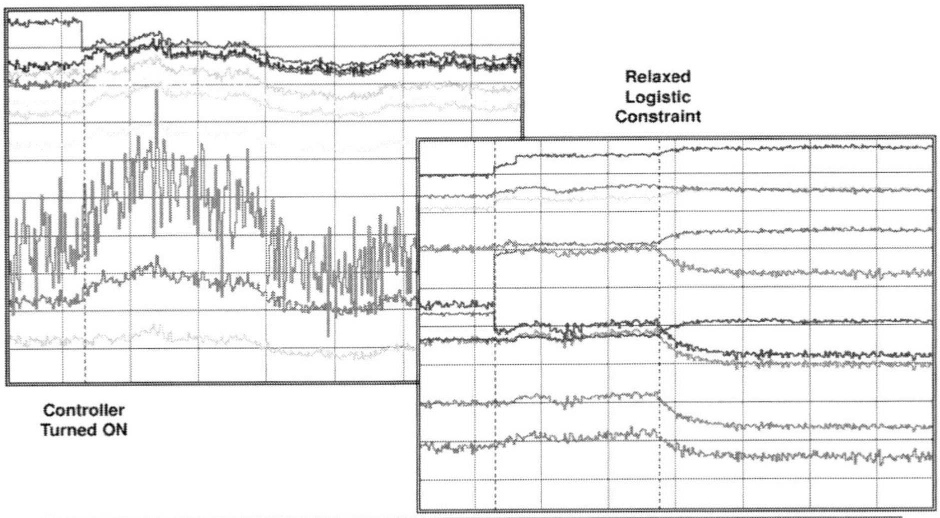

lyondellbasell.com

Gasoline Blending Information Flow

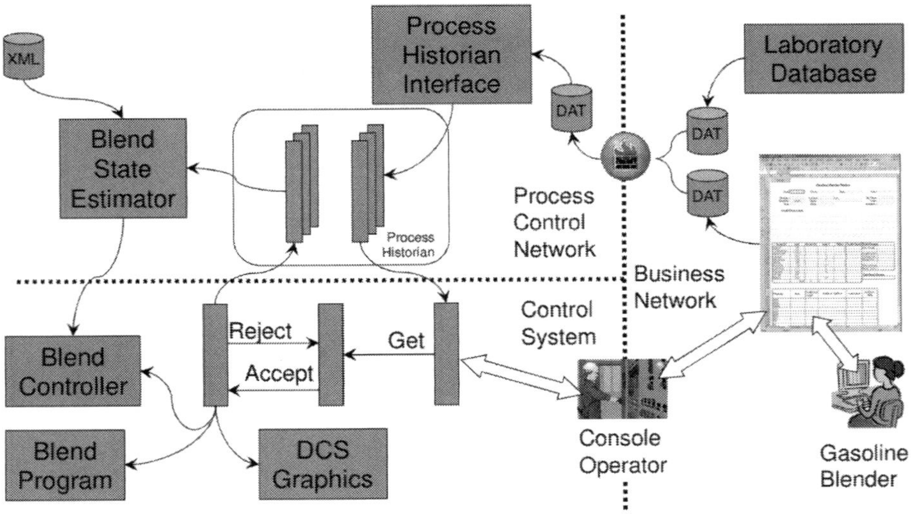

lyondellbasell.com

"Open" Design Decisions

- Nonlinear optimizer / Blend controller:
  1. Pass recipe targets to blend controller
  2. Pass property targets to blend controller
  3. Control tank properties directly in blend controller
- Property estimator model reconciliation:
  1. Use analyzer to adjust component properties
  2. Use analyzer to adjust blending model parameters
  3. Do both at same time
  4. Do neither

lyondellbasell.com lyondellbasell

Future Work

- Controller/Optimizer Interface
  - Property targets or ratio targets?
- Interaction with Planning and Scheduling
  - Update LP model from blend results
  - Evaluate blend component purchases and sales
- Investigate blending direct to pipeline
  - Reduced testing and intermediate storage

lyondellbasell.com lyondellbasell

lyondellbasell

Disclaimer

All information ("Information") contained herein is provided without compensation and is intended to be general in nature. You should not rely on it in making any decision. LyondellBasell accepts no responsibility for results obtained by the application of this Information, and disclaims liability for all damages, including without limitation, direct, indirect, incidental, consequential, special, exemplary or punitive damages, alleged to have been caused by or in connection with the use of this Information. LyondellBasell disclaims all warranties, including, but not limited to, the implied warranties of merchantability and fitness for a particular purpose, that might arise in connection with this information.

Step-Out Improvements in Heavy Oil Conversion

Irv Wiehe

Soluble Solutions

AIChE Process Technology Conference, Galveston, TX

Polynuclear Aromatics, The Barrier to Conversion

"Good Guys" ← → "Bad Guys"

Phenanthrene

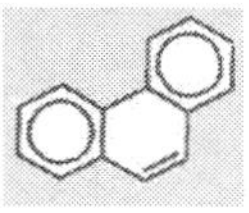

BP = 644ºF

Pyrene

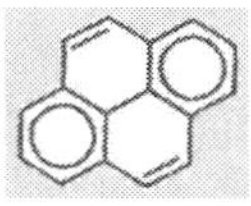

BP = 739ºF

Benzo-pyrene

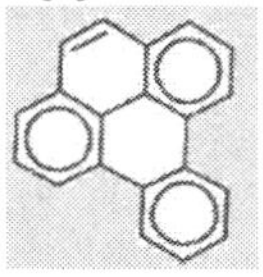

Benzo-perylene

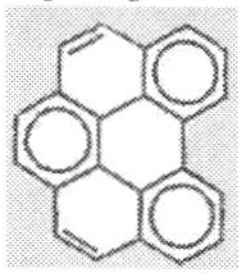

Coronene

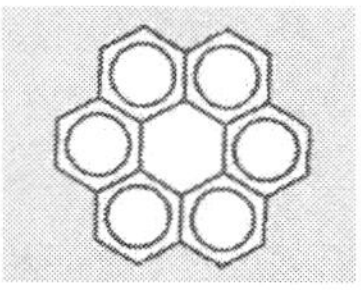

BP = 977ºF

- PNA's Only Part of Molecule (Asphaltenes: 50% Aromatic C)
- PNA's of 5 Rings or More: Not Economical to Convert PNA Part (Jury Out for 4 Rings)

AIChE Process Technology Conference, Galveston, TX

Polynuclear Aromatics, The Barrier to Conversion

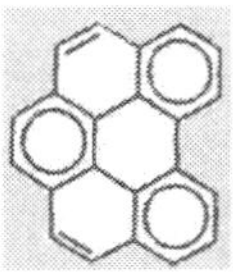

- PNA's Are Thermally Stable
- PNA's Cause Insolubility
- Nonvolatile PNA's Cause Coke Formation
- Many PNA's Are Formed During Conversion
  - Dehydrogenation
  - Combination of Smaller Aromatics

AIChE Process Technology Conference Galveston, TX

Upgrading Options for Heavy Oils & Resids

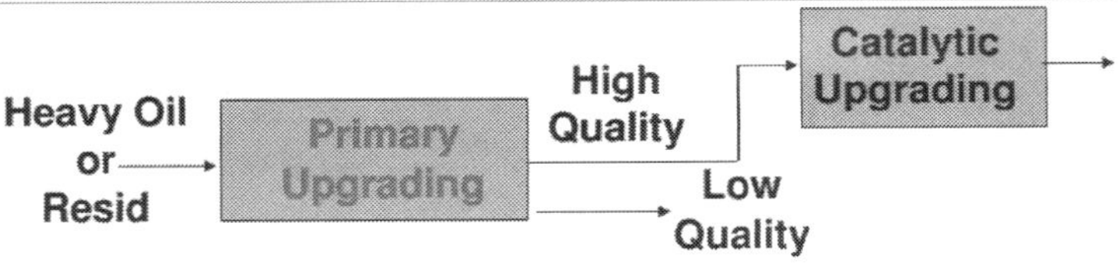

Primary Options	Low Quality Options	Catalytic Options
1. Coking	1. Residual Fuel Oil	1. Hydrotreating
2. Hydroconversion	2. Fuel Coke	2. Fluid Cat. Cracking
3. Deasphalting	3. Gasification Feed	3. Gas Oil Hydrocracking
4. Visbreaking	4. Bury or Pile	

AIChE Process Technology Conference Galveston, TX

Pendant-Core Building Block Model: Conradson Carbon Residue Predicted To Be Linear Function of Hydrogen Content

Resid Reactant		Distillable Liquids		Conradson Carbon Residue
P-P				
P-P-P				
P-●-P				
P-P-●-P-●	⟶	nP	+	-●-●-●-●-
P-●-●-P-●-P-P				
P-●-P-●-P-●-P-●				
H-Content: H		H_P		H_C

H Balance: $100\,H = (100-CCR)\,H_P + CCR\,H_C$

$CCR = 100\,(H_P - H) / (H_P - H_C)$ Linear Function

AIChE Process Technology Conference Galveston, TX

Solu ble tions

Data Are in a Linear Band

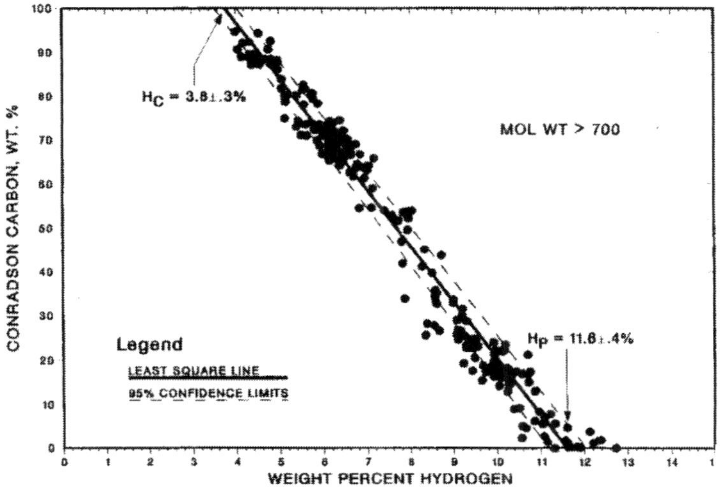

AIChE Process Technology Conference Galveston, TX

Solu ble tions

P-C Model Compounds Show Large PNA's Not Required to Make Coke

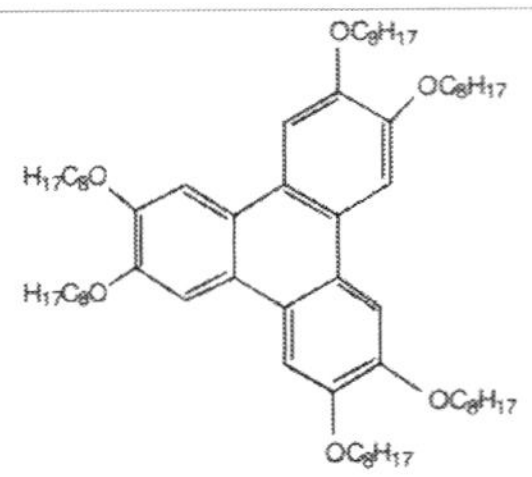

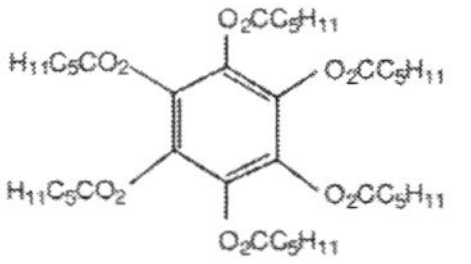

Benzenehexa-n-hexanoate

2,3,6,7,10,11-Hexaoctoxytriphenylene

Compound	CCR (wt%)	C (wt%)	H (wt%)	O (wt%)	Mol Wt (amu)
Hexaoctoxytriphenylene	24.1	79.46	10.91	9.62	998
Benzenehexa-n-hexanoate	14.0	65.66	9.35	24.99	768

- Part of CCR (Less than 5 Rings) Can be

AIChE
Process Technology Conference
Galveston, TX

Solu ble tions

Increase Liquid Yield from Coking

- Inhibit Formation of Extrinsic Coke

INTRINSIC EXTRINSIC

REDUCE COMBINATION FREQUENCY OF AROMATIC FRAGMENTS

- Reduce Secondary Reactions of Desired Liquids

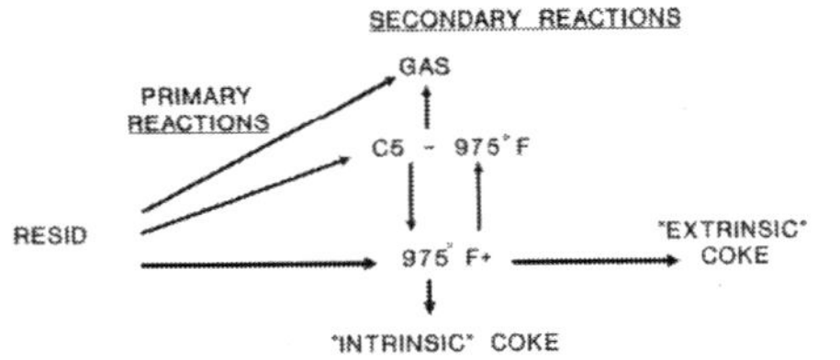

AIChE
Process Technology Conference
Galveston, TX

Solu ble tions

Short Vapor Residence Time Increases Yield

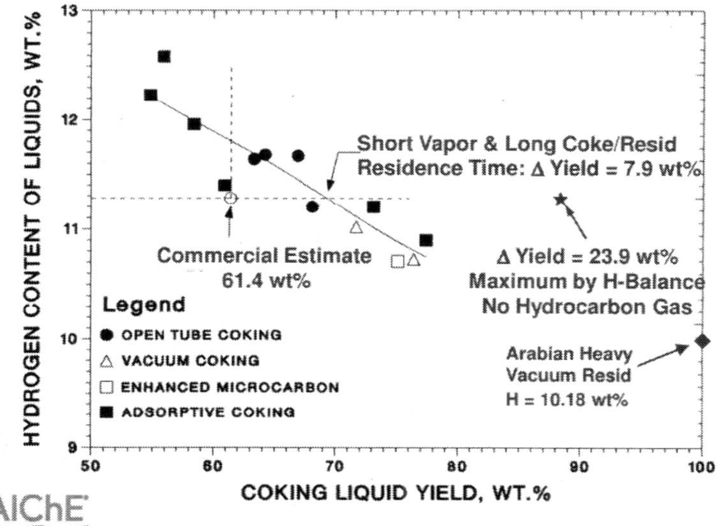

AIChE
Process Technology Conference
Galveston, TX

Soluble Solutions

Add Concept to Reduce Extrinsic Coke

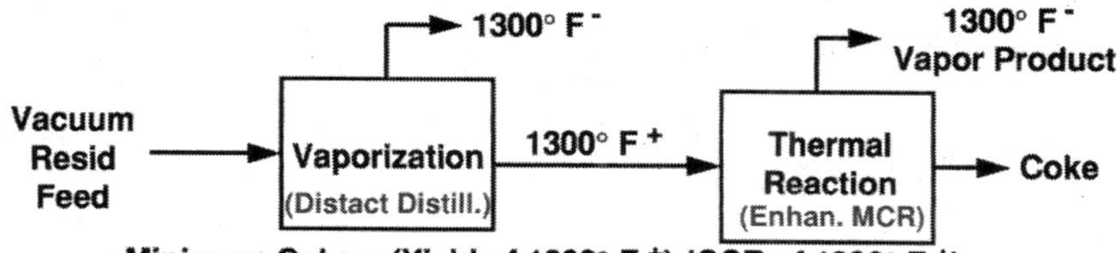

Minimum Coke = (Yield of 1300° F ⁺) (CCR of 1300° F ⁺)

Berri Vacuum Resid (CCR = 18%):	Coke = (.382)(35.1) = 13.4 wt%
Arabian Heavy Vac. Resid (CCR = 22%):	Coke = (.546)(35.9) = 19.6 wt%
Syrian Light Vac. Resid (CCR = 23%):	Coke = (.541)(31.9) = 17.3 wt%

Coke = 0.8 CCR
Commercial Cokers: Coke = 1.2- 1.6 Times CCR

AIChE
Process Technology Conference
Galveston, TX

Soluble Solutions

Separation Further Increases Liquid Yield

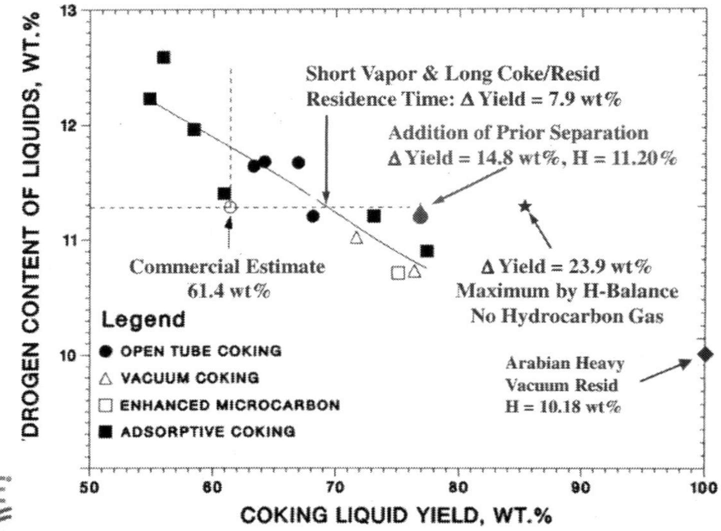

ETX Upgrader (Cross Flow) Under Development
Short Vapor – Long Resid Residence Time

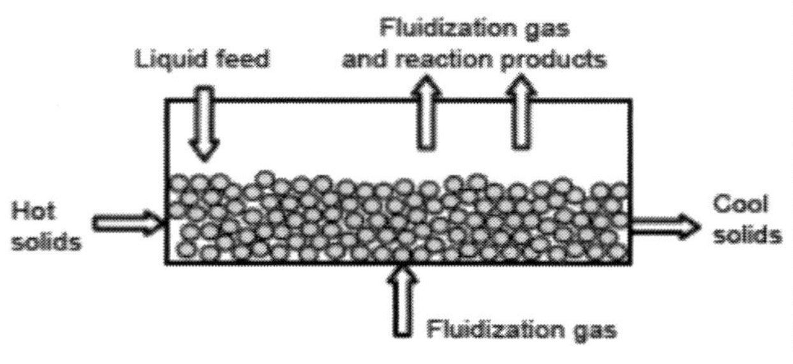

ETX Systems, Inc., Calgary, Canada
Pinchuk et al. (2005)

Holding Volatiles in Reactor Hastens Coke Formation

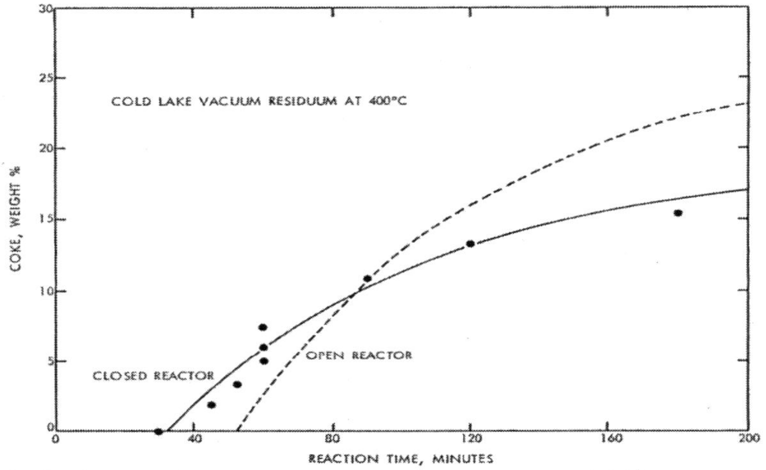

AIChE
Process Technology Conference
Galveston, TX

Soluble Solutions

Hydroconversion

- **Advantages**
  - Minimizes PNA Growth: Greater Yield (50-98%)
  - Greatly Increases Coke Induction Period
  - Removes Sulfur, Nitrogen, and Metals: Ni, V, Fe
- **Limitations**
  - PNA Aromatic Cores
    - Incompatibility (need to Strip with Hydrogen)
    - Last Conversion: Slow, Waste Hydrogen, Aromatic Product
  - Cost of Hydrogen, Catalyst, and High Pressure Reactors

AIChE
Process Technology Conference
Galveston, TX

Soluble Solutions

Dispersed (Slurry) Catalyst Processes

- Microcat-RC Process (ExxonMobil)
- HCAT Technology (Headwaters)
- Veba Combi-Cracking Process
- VRSH (Chevron)
- UOP Slurry-H.T. Process
- Eni Slurry Technology
  - 1200 B/D Demonstration Plant Since 2005
  - 23,000 B/D Industrial Plant by 2012

▶ Catalysts: Mo Sulfide (100 ppm) & Fe Sulfide (Dispersed in Feed)
▶ Conversion: Over 95% by wt., Over 100% by Volume

AIChE
Process Technology Conference
Galveston, TX

Soluble Solutions

Microcat-RC Process (ExxonMobil)

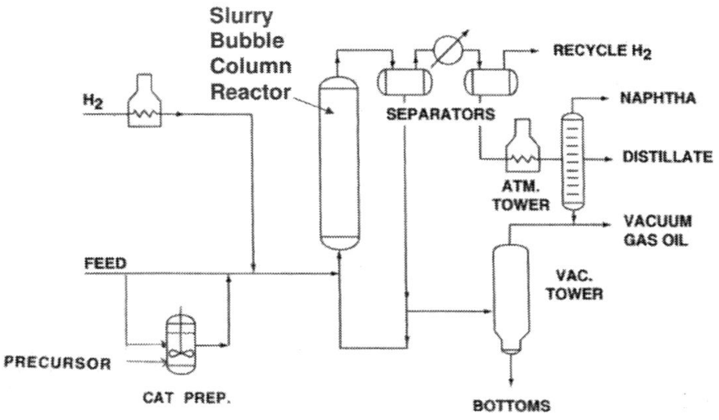

Soluble Solutions

- Operates at 440°C under 1500 – 2200 psi of H_2 with 10 ppm Mo
- Attains 90-95% Conversion of 1050°F$^+$ Bottoms (≈ 85% Liquids)
- Requires Stripping out Volatile Liquids to Get >80% Conversion

AIChE
Process Technology Conference
Galveston, TX

90+ % Conversion May Not be Optimum

- **Requires Hydrogenating PNA's, Cracking Saturated Rings to Form Hydrocarbon Gas & Smaller PNA's**
  - Cracking Saturated Rings is a Slow Reaction (Can Double Reactor Volume)
  - Hydrocarbon Gas Byproduct Consumes Hydrogen
  - Liquids Produced are Aromatic (4 Ring Aromatics)
- **Does Reduce But Not Eliminate Bottoms Disposal**

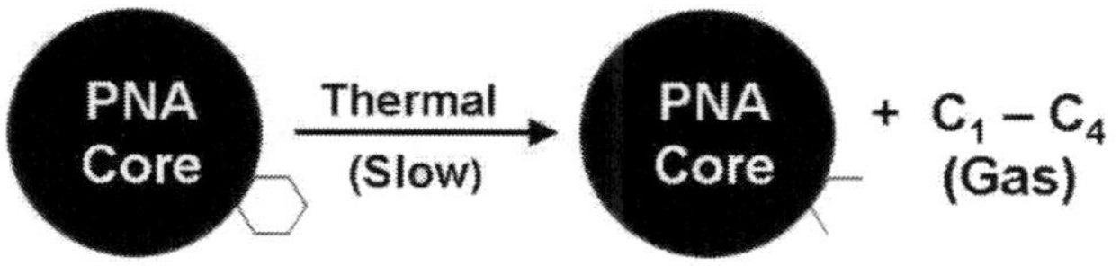

Resid Conversion Processes Can Be Improved (Arabian Heavy Vacuum Resid)

Process	Liquid Yield (wt%)
Conventional Visbreaking (HSC)	28 (44)
Recycle Visbreaking + Selective Separation*	62
Conventional Delayed Coking	59
Once-Thru Delayed Coking	61
Conventional Fluid Coking	61
Once-Thru Fluid Coking	66
Short Vapor Res. Time Coking*	69
Sep.+ Short Vapor Res. Time Coking*	76
Ebullating Bed Hydroconversion	65-70
ENI Slurry Hydroconversion	89

* Laboratory Scale

Details in IA Wiehe, "Process Chemistry of Petroleum Macromolecules", CRC Press, June, 2008; Code = 802DC

When Things Don't Go Right

James Turner

FLUOR

H2S Stripper that didn't Strip H2S

- In 1990, I was Lead Process Engineer on two hydrotreater revamps for a midwest refinery, as part of an overall refinery re-configuration
- One hydrotreater revamp was converting a 3 MBPD Naphtha Hydrotreater to a 7 MBPD Kerosene Hydrotreater
- After startup, the product H2S Stripper was not successfully removing all of the H2S

FLUOR

H2S Stripper that didn't Strip H2S

- At start-up, the Stripper was not removing all the H2S.
- With Fluor assistance the stripper could meet the specifications at reduced rate "most of the time" - but was very sensitive to operational upsets.

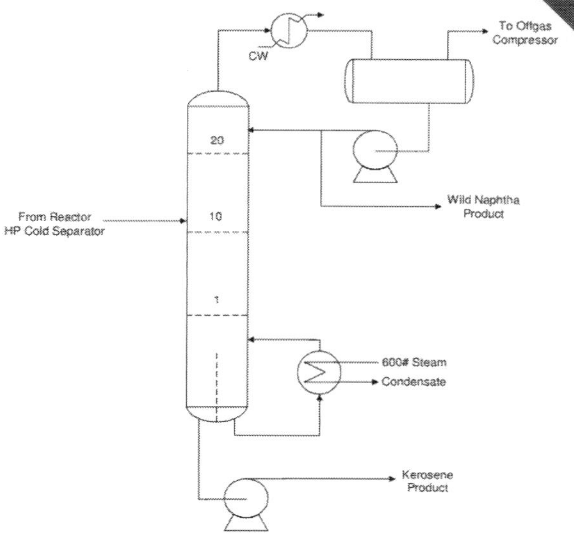

FLUOR.

H2S Stripper that didn't Strip H2S

- Design Basis had a minimum steam pressure of 600 psig – but actual pressure was 585 psig.
- Tower overhead pressure drop was slightly higher than predicted, also raising the reboiler pressure.

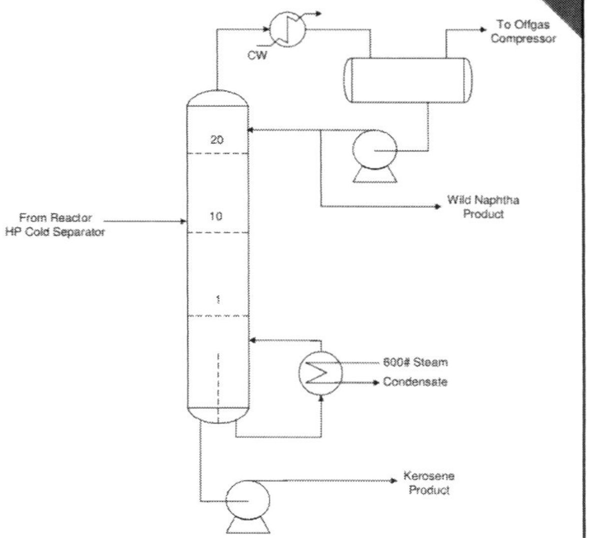

FLUOR.

H2S Stripper that didn't Strip H2S

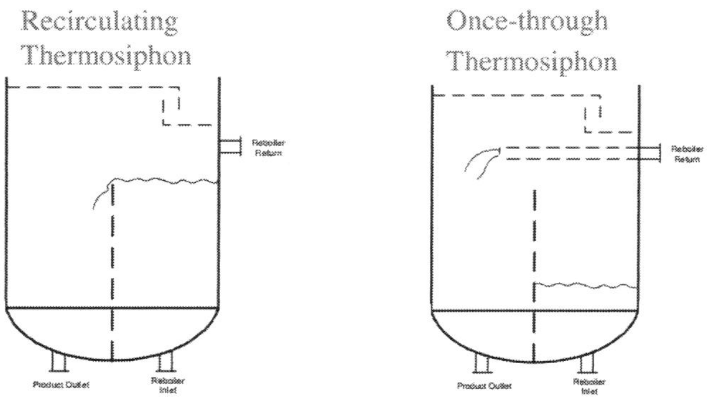

Existing Tower/Reboiler design was a recirculating thermosiphon – this means that the reboiler is at a higher pressure than the tower due to the liquid static head.

FLUOR

H2S Stripper that didn't Strip H2S

The three factors caused a "pinch" on exchanger temperatures, making the available exchanger surface area inadequate

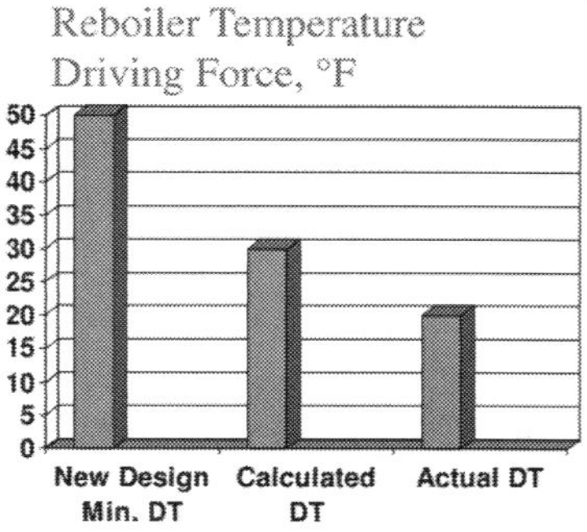

The existing trays were checked during design by the vendor, and considered OK, but in reality, they were very sensitive to vapor rate, and would weep at rates just below the design rate.

FLUOR

H2S Stripper that didn't Strip H2S

Possible Solutions:

- Scrap the exchanger and install a new fired heater
  - Costly – and would take a long time to implement.
- Stick with exchanger, but consider the following:
  - Replace trays
  - Modify baffle to convert reboiler to once-through (lower reboiler pressure)
  - Replace reboiler bundle with finned bundle to improve heat transfer

FLUOR®

H2S Stripper that didn't Strip H2S

- Client decided to stick with exchanger and implement all three proposed modifications.
- Stripper worked fine after modifications installed.

FLUOR®

"Lessons Learned"

- Sometimes, everything that can go wrong, will go wrong.
- Sometimes little things can make a big difference. Details matter.
- Need to properly document possible design issues during Design Phase. The tightness of the reboiler design was known in advance by Fluor, but not documented in writing to client – we could have pulled the client into the design decisions in advance.
- Hydraulics on thermosiphon reboiler designs on low pressure columns should be checked for impact to equilibrium flash, not just for hydraulic driving force.

FLUOR.

Hydrotreater Hot Separator Problem Solved

James Turner

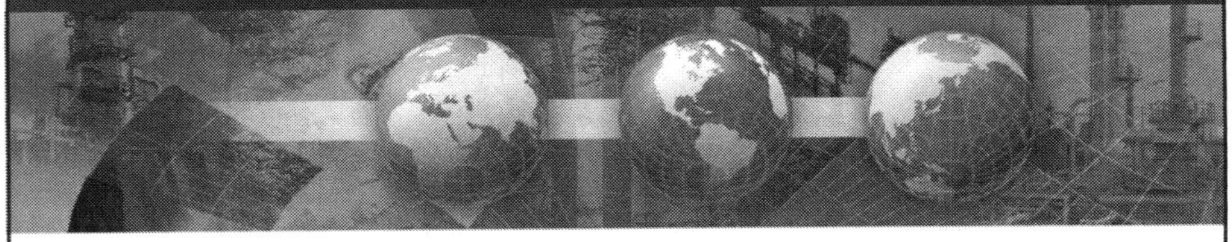

FLUOR.

Hot Separator Problem - 1998

- Designed a gas oil hydrotreater Hot Separator for a major unit Gulf Coast refinery
- Unfortunately, it didn't separate!
  - Major hydrogen carry-under out the bottom with the liquid
  - Vessel was full of dense foam – therefore level guage was incorrect
  - Operations didn't know it, raised the level above the vessel inlet, causing massive carry-over
  - Operations failed to respond to the rising level in the downstream vessel, leading to a "series of unfortunate events" culminating in shutting the whole refinery down
- On top of unplanned refinery shutdown, problem was costing refiner ~$0.5 MM/month in lost profits

FLUOR.

Hot Separator Problem

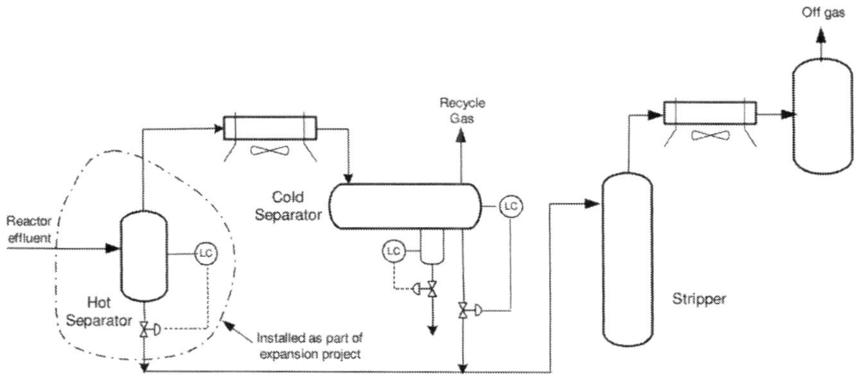

FLUOR.

Hot Separator Problem

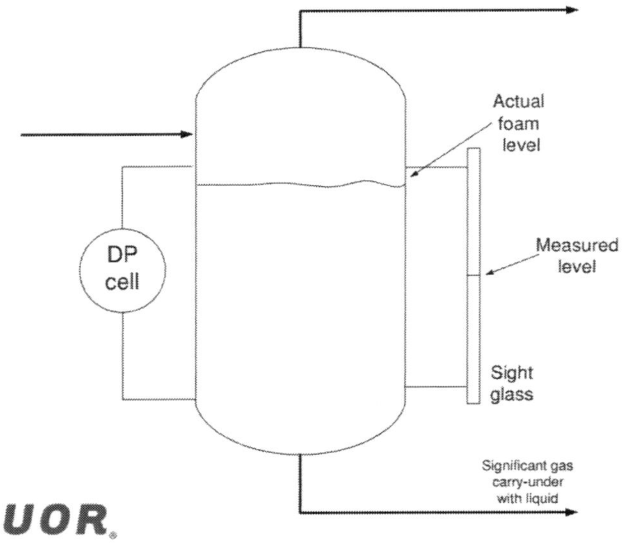

FLUOR

Hot Separator Problem

- This led to a high level collaboration between Fluor and Client Management, resulting in my relocation to the refinery “until an agreeable path forward could be determined”
- This was a “high profile” opportunity, so I was motivated to perform
- I led a three pronged study to determine:
  - Impact of Process Variables on Separator Performance
  - Survey of Other Refineries with Similar Units
  - Potential Alternative Vessel Designs/Modifications

FLUOR

Data Correlation "solutions"

Remove resid material in feed	Not economic to remove
Lower liquid velocity	Would require significant throughput reduction
Increase operating temperature	Small improvement – higher increase not practical for heat integration and metallurgy impacts
Remove trace impurities	Trace impurities not present in unusual levels
Lower piping velocity	No improvement observed – not economical

FLUOR.

Hot Separator Problem

- Result of data gathering – no practical solution by manipulating process conditions
- Industry survey found no problems with refiners that have horizontal separators
- one refiner that had solved a similar problem by install vendor inlet internals
- My recommendation was to install vendor internals to improve the separation
  - Industry experts that were consulted said it wouldn't work

FLUOR.

"Vortex Cluster"

FLUOR

Reference Article

"Stop foaming on hydrotreater hot separator", J. Turner, J. Asquith, R. Atkinson, Hydrocarbon Processing, June 1999

FLUOR

Hot Separator Problem

- This was the first time that Fluor had tried this type of internals in this service -
  - Fortunately, it worked very well!
- Fluor has done many other projects with this client since (including more than one multi billion project)
- These internals have been subsequently used in several similar situations, all with good success
- Recently, these internals were used for the first time in a Crude Unit Preflash Column
  - They also worked very well in this application.

FLUOR®

Lessons Learned

- Hydrogen and Heavy Oil can cause difficult foaming issues
- "Experts" may not know all there is to know
- By surveying other operations, good solutions may be found
- Vendor systems typically used in one industry may be very useful in other industries
- "crisis" = dangerous opportunity

FLUOR®

Questions / Comments

???

FLUOR®

Safety Moment

NEWSFLASH – On November 8, 1999, a refinery storage tank near Catlettsburg, KY ruptured, spilling ~285,230 gallons of petroleum distillate (gas oil) into a tributary of the Big Sandy river. Some of the gas oil was entrained with steam escaping the tank and "rained" down on the surrounding community.

Why did this happen?
What made the situation worse?
How could it have been prevented?

FLUOR®

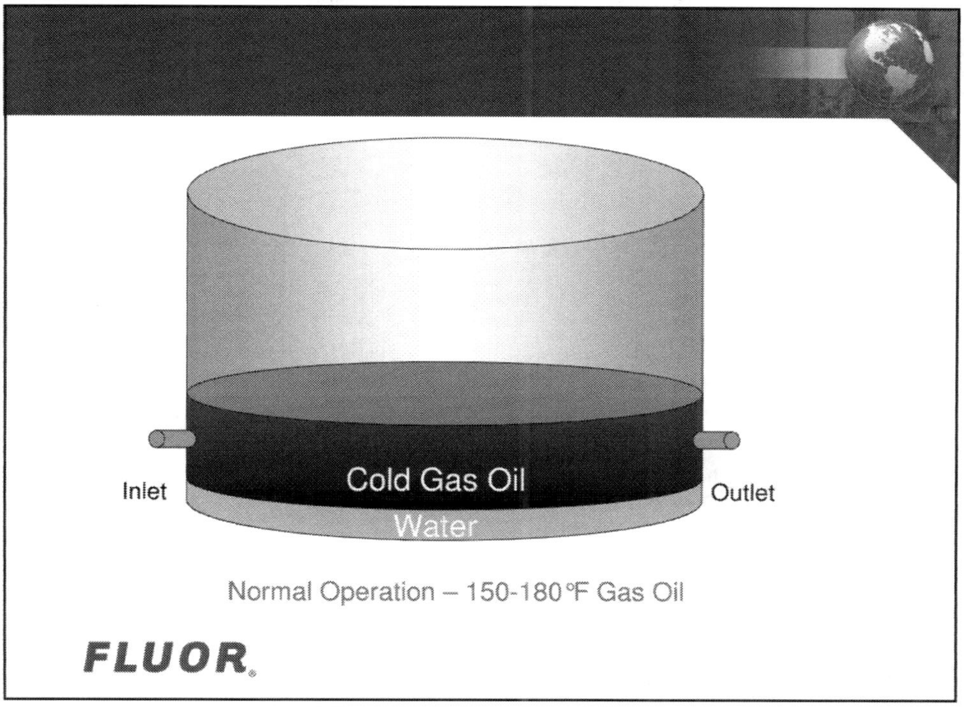
Cold Gas Oil
Inlet
Outlet
Water
Normal Operation – 150-180°F Gas Oil
FLUOR

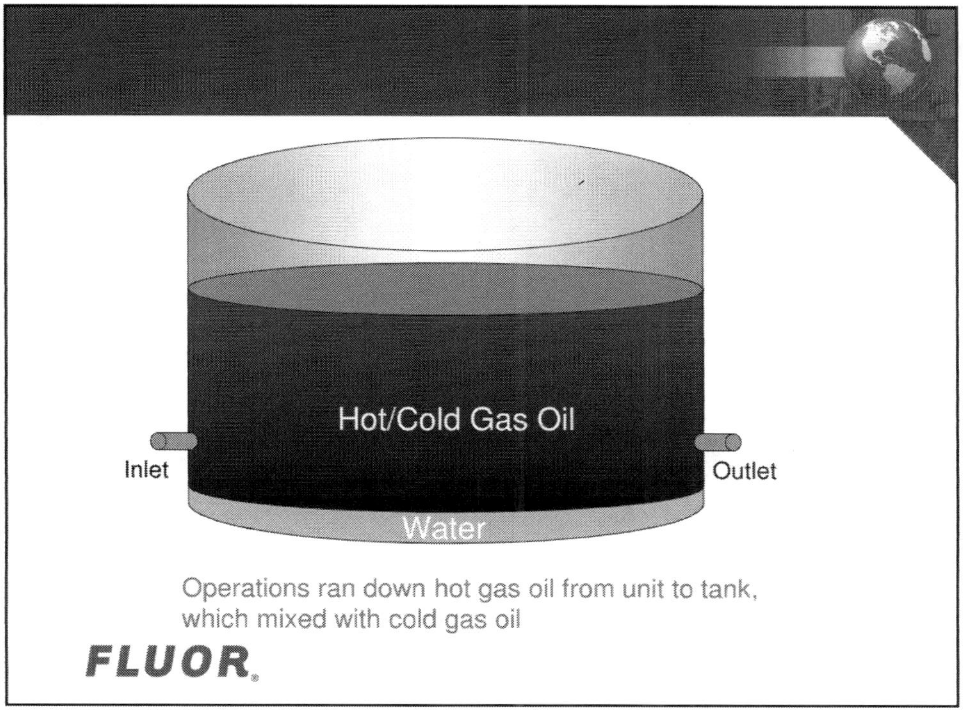
Hot/Cold Gas Oil
Inlet
Outlet
Water
Operations ran down hot gas oil from unit to tank, which mixed with cold gas oil
FLUOR

As gas oil was pumped out, hot gas oil caused water to vaporize, raising pressure in tank, causing the side the tank to rupture.

FLUOR

Summary – Hot oil and water do not mix well!

Takeaway – Good understanding and proper execution of operating procedures is required for safe operation.

FLUOR

Advances in Digital Automation: Lowering Refinery Operating Costs

Tim Olsen

Emerson Process Management

AIChE

EMERSON.
Process Management

Agenda

- Background: Advances in Computing
- Examples: Advances in Automation within Refining
  - Energy Efficiency
  - Reliability
  - Operator Effectiveness
  - Safety

Typical Process Computer's Performance

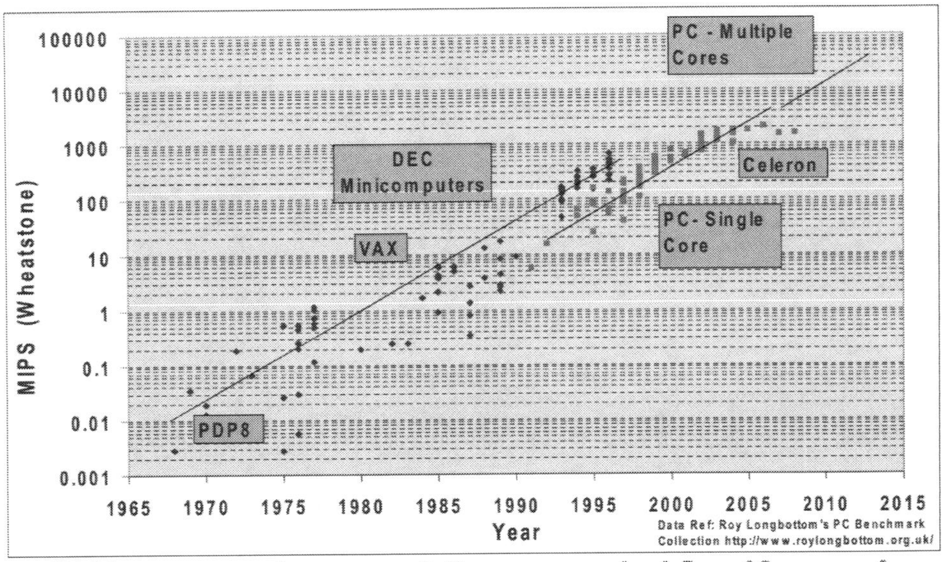

50% per year increase! Compounded for 40 years!

Computer Advances – Access to Performance

Cray 2 Supercomputer
1 Gigaflop
1985 - $23 million
Ref *Wired*, February 22, 2008, by Rob Beschizza

Cray CX1: 4 Teraflops

2008 Launch Price: $25,000

Now Starting Price: $8,239

Ref: www.CX1.Cray.com, October 2011

Cell Phone Advances – Access to Information

The "Brick"
Phone only
Late 1980s

I-Phone

Applications, Real-time information

Available Today

Memory Cost

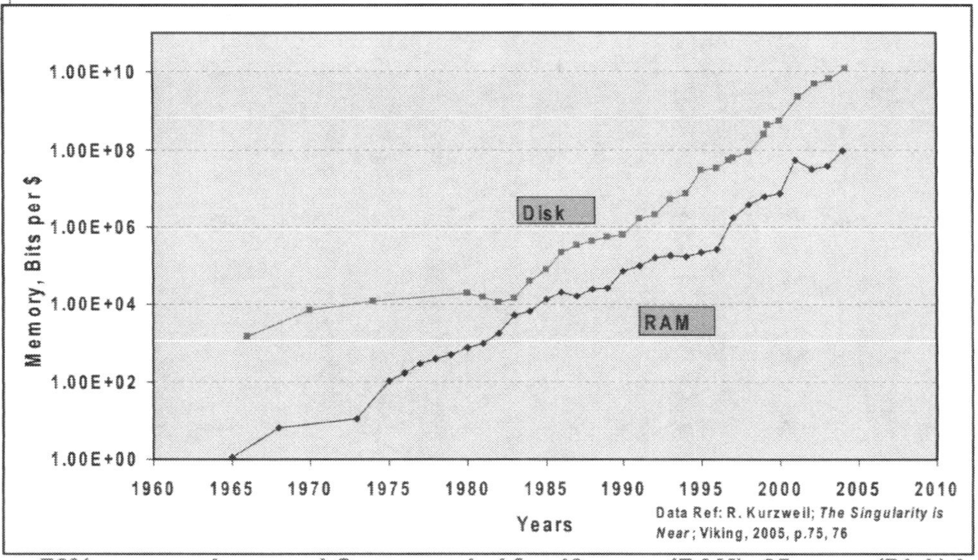

70% per year increase! Compounded for 40 years (RAM); 25 years (Disk) !

100 gigabytes – Basic Demographic information on every human on earth!

IP Network Communication – Wired
(Max - Bandwidth x Distance (wo amp))

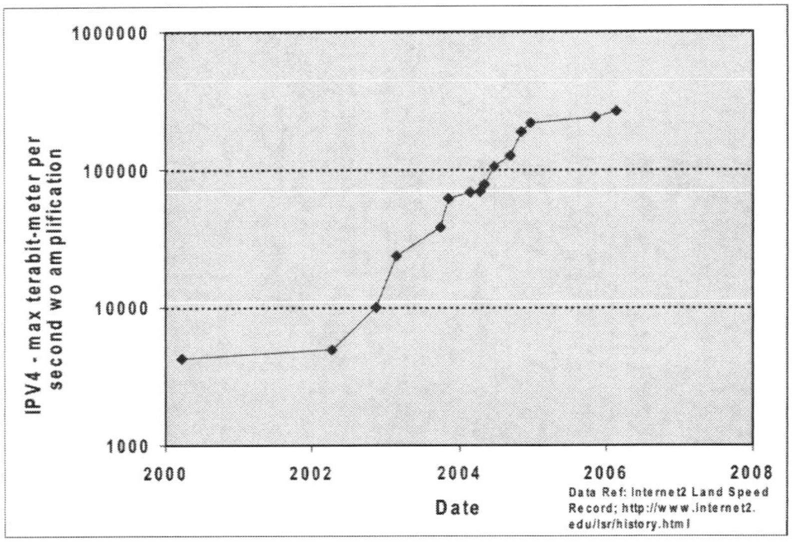

Any information will be available anywhere!
Bandwidth will substitute for local memory!

Network Communication - Wireless

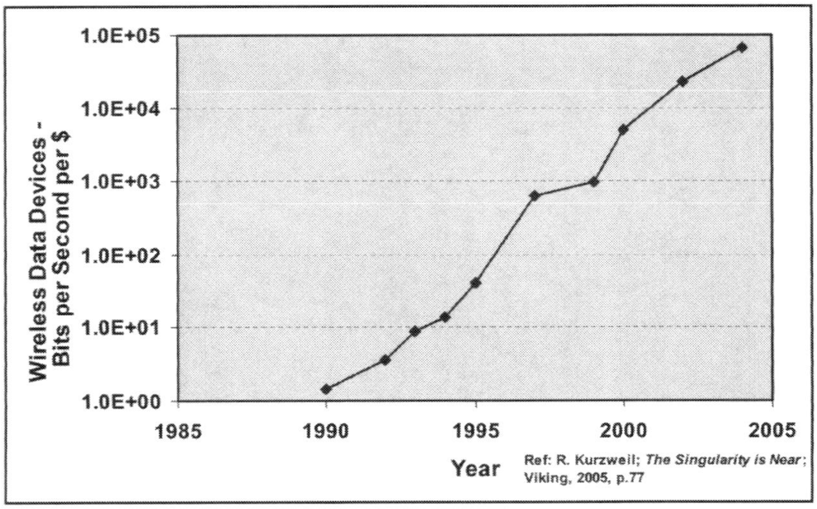

200% per year compound increase!

The Exception - Software Productivity Growth

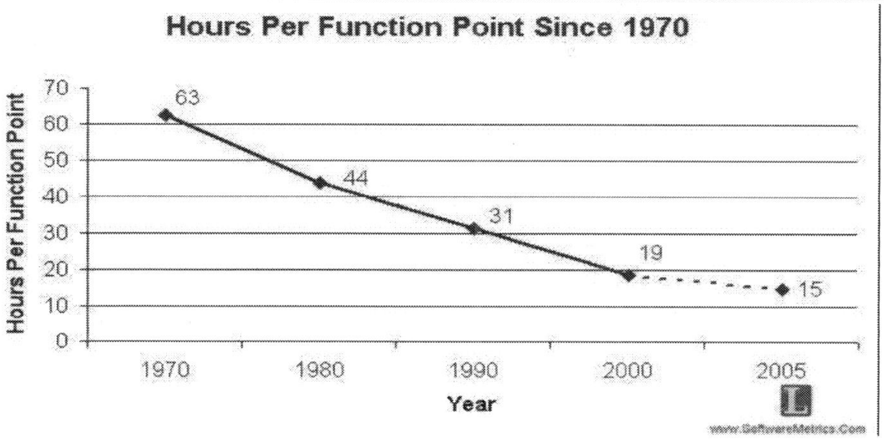

Only 4% per year increase from 1970 to 2000; less thereafter! Salary increases similar! No net cost reduction.

Software is dominant cost in system development/ implementation!

However, software has much longer life than hardware.

Implications of Computing Trends

- Low cost computing / communications replaces higher cost "other"
- "Distance Doesn't Matter" - Physical location of individuals irrelevant with regard to communication and information availability – virtual organizations
- Computational limits are receding
- Everything that can be digitized will be digitized
- No "practical limits" on data / information storage

What does this mean for the Refining Industry?

Advances in Digital Automation: Lowering Refinery Operating Costs

EMERSON.
Process Management

Older Automation Field Networks – Analog 4-20 ma

Only 1 value is transmitted
Only one direction

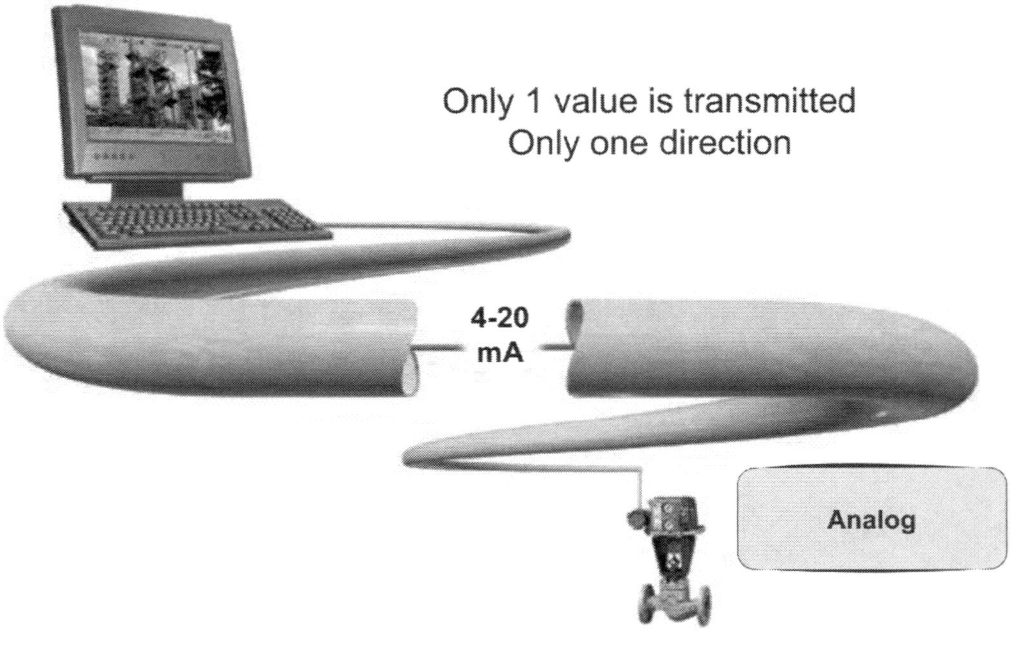

But Computers and Communication and Calculations have migrated to the Field Devices…

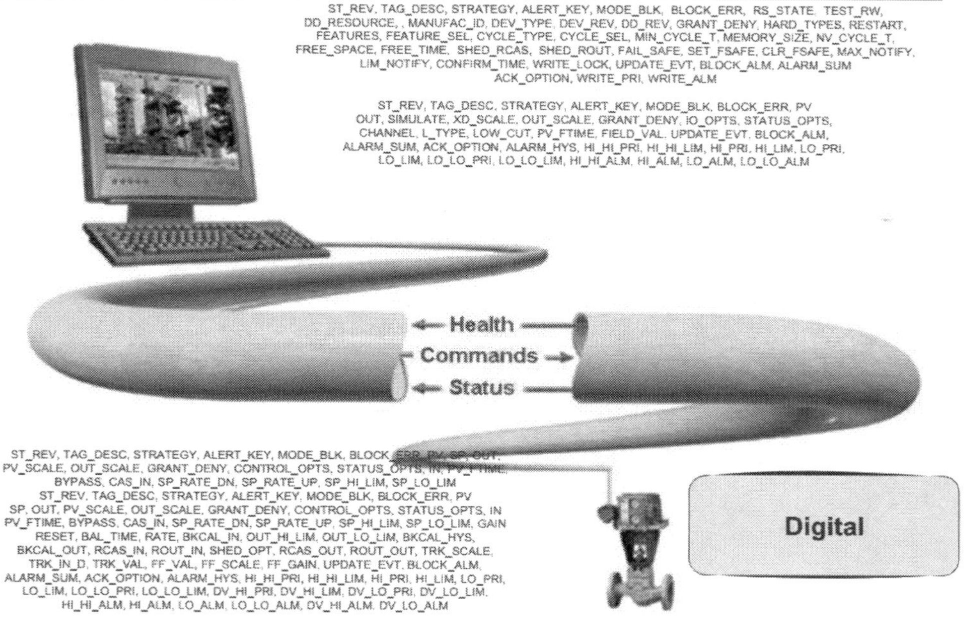

Control Rooms have also Modernized

Site Wide Control Room

Coordinated Multiple Loop Digital Control

Computer Screen Interface

Photo Ref: Emerson Process Management

Photo Ref: Emerson Process Management

Each Unit had a separate Control Room

Single Loop Analog Control

Panel Board Interface

The Manufacturing Balance Has Changed

It is time to revisit technology justifications

YESTERDAY:

- Cheap Energy
- Cheap Water
- Cheap Waste
- 'Free' Emissions
- Predictable Raw Materials
- Available Talent Pool

- Expensive Controls & Automation Technology
- Expensive Software and IT

TODAY:

- Energy cost swings
- Water cost rising
- Expensive Waste
- Expensive Emissions
- Unpredictable Raw Materials
- Limited Talent Pool

- Cheap Controls & Automation Technology
- Cheap Software and IT

ARC Advisory Group

Thought Leaders for Manufacturing & Supply Chain

Ref: A. Chatha; "Current Business Drivers and Trends;" ARC 2009 Orlando Conference; February, 2009

Advances in Digital Automation in Refining

Key areas that have benefited from advances in digital automation

- Energy Efficiency
- Reliability
- Operator Effectiveness
- Safety

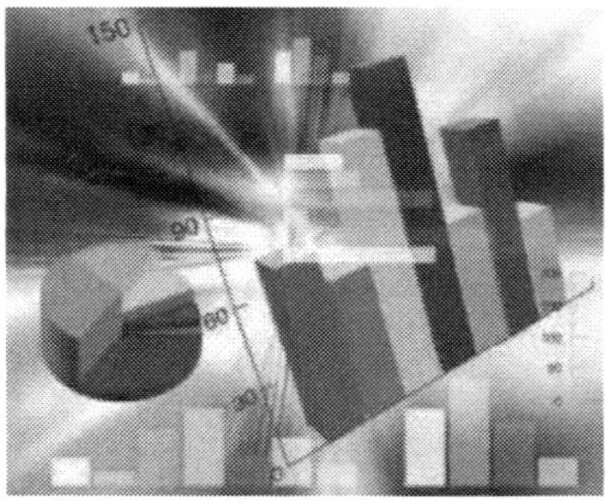

Examples…

Advances in Digital Automation in Refining

Key areas that have benefited from advances in digital automation

- Energy Efficiency
  - Key Performance Indicators (KPI)
  - Advanced Process Control (APC)
- Reliability
- Operator Effectiveness
- Safety

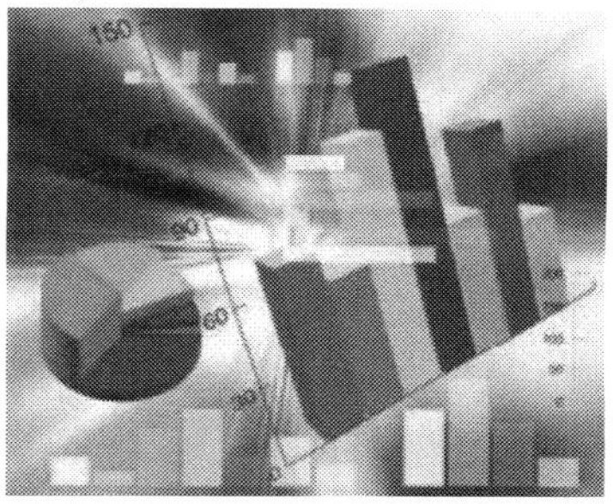

Key Performance Indicators, KPI

- Quickly at a glance, understand issues with the refinery.
- Poor performance does not go undetected.
- Energy use per barrel processed is tracked and trended.
- Predictor for poor heater operation or effects of heat exchanger fouling.
- Detect losses with heat and weight balancing

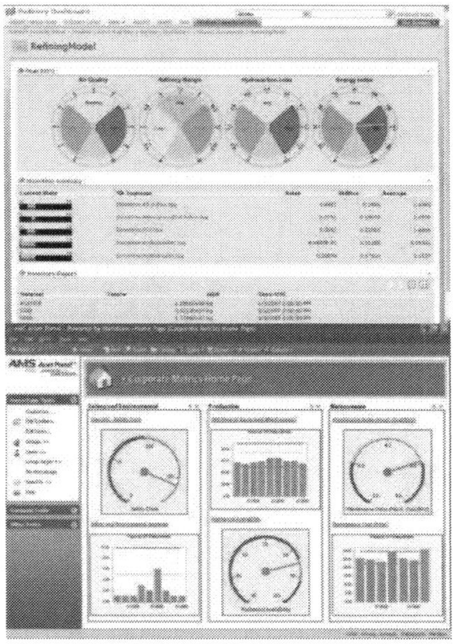

KPI Dashboard Case

*"After the installation of data reconciliation software and a KPI dashboard display, the model showed an incorrect process unit lineup. A recycle line was open that should have been closed. After closing the recycle we were able to increase hydrotreater throughput by 8 percent. This resulted in an energy savings of $400,000 per year and a product profit increase of $500,000 per year, for a **total benefit of $900,000 USD per year**."*

Planning Manager
Refinery, North America

Abnormal Operation does not go Undetected

Advanced Process Control, APC

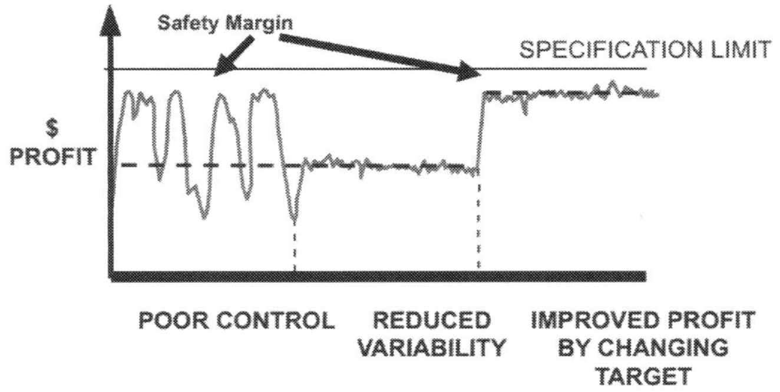

Embedded APC within Modern Automation Systems

- Better measurements and tighter control valve resolution
- Easy and Faster Startup
- Easy for Console Operators to understand, thus stays "ON"
- Easy to Maintain and Change as Needed

APC Case, Complex Distillation Column

- 2 weeks configuring and commissioning
- 1 week proof testing
- Average overehead product loss reduced by ~22%
- Impurities in bottoms were maintained within specifications.
- Steam usage reduced ~7%
- Estimated value over $700,000 USD per year

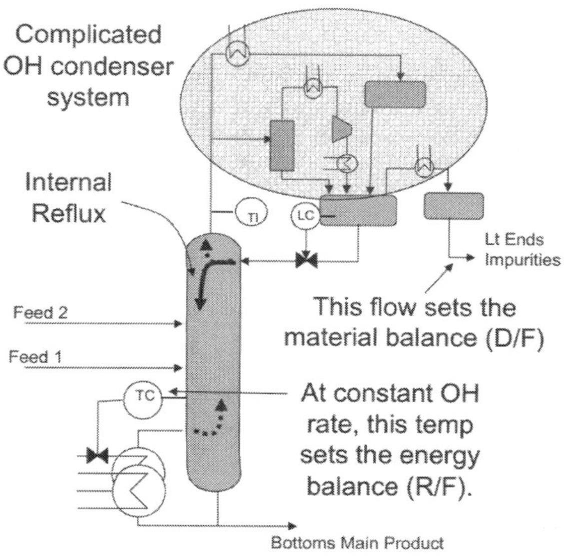

Advances in Digital Automation in Refining

Key areas that have benefited from advances in digital automation

- Energy Efficiency
- Reliability
  - Asset Health Monitoring
  - Modernization
- Operator Effectiveness
- Safety

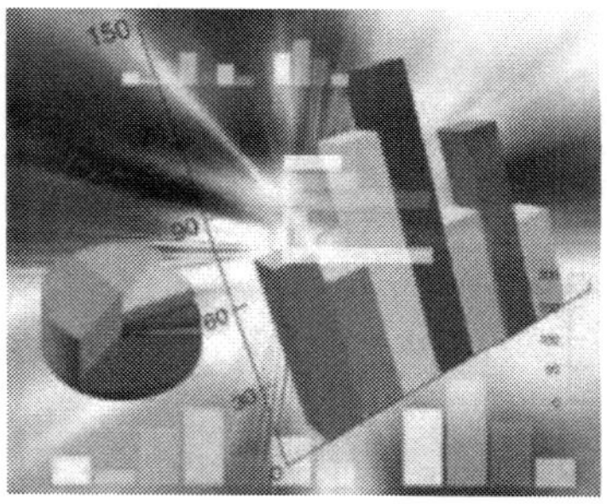

Predictive Diagnostics and Analytics

Future Trends – Predictive Analytics - How do we avoid this?

Photo Ref: Emerson Process Management

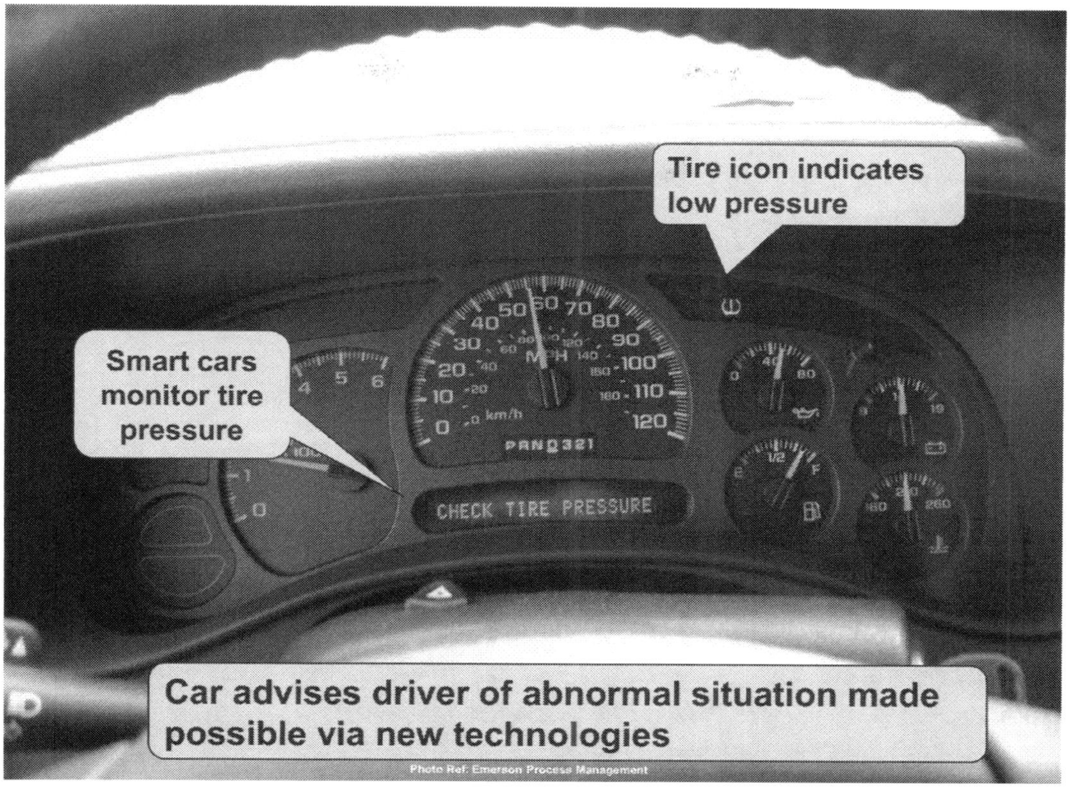
Tire icon indicates low pressure
Smart cars monitor tire pressure
CHECK TIRE PRESSURE
Car advises driver of abnormal situation made possible via new technologies
Photo Ref: Emerson Process Management

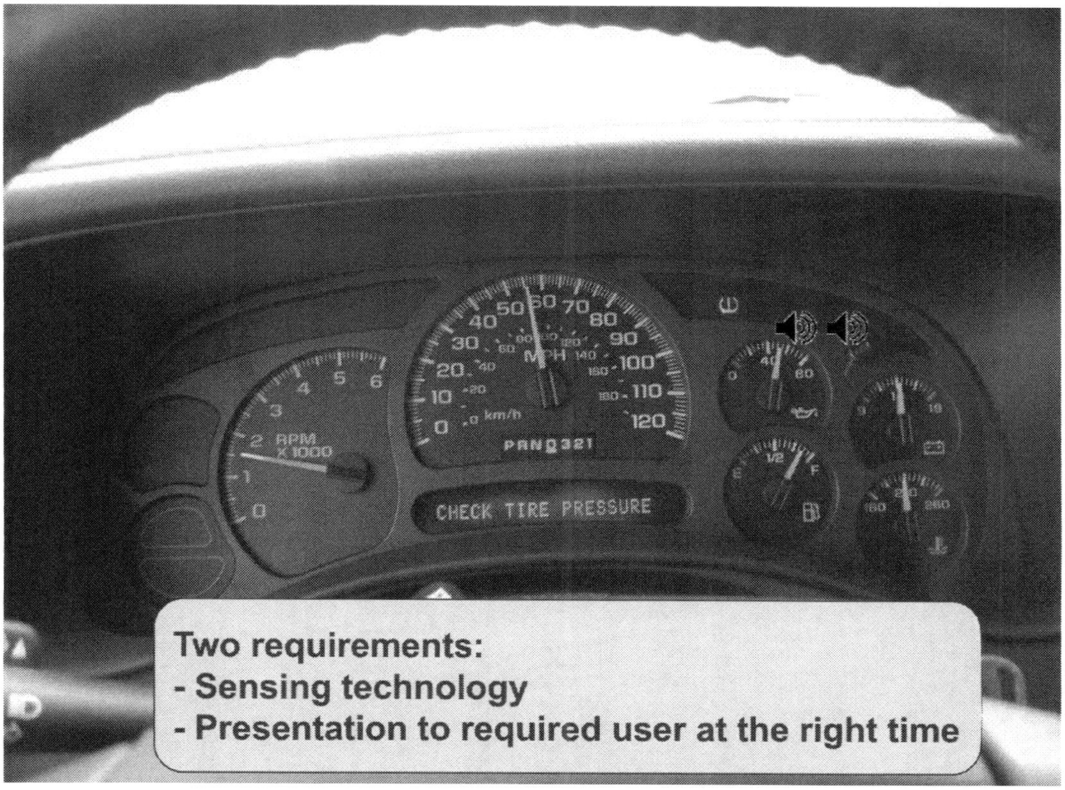
CHECK TIRE PRESSURE
Two requirements:
- Sensing technology
- Presentation to required user at the right time

Right Person, Right Time, Take Action

Asset Health

- Maintenance / Turnaround Planning
  - Normal wear and tear
  - More information and details
  - Predict and plan maintenance
- Operations
  - Abnormal operation or imminent failure
  - Alert
  - Immediate action
- Safety System
  - Instruments have HART for diagnostics
  - Improved
    - Availability – safety system works when it needs to
    - Reliability – no trips due to instrumentation

Hydrocracker Quench Control Valve

Low instrument air pressure to quench valve actuator – poor response to control

- Maintenance and Operator alerted
- Potential exothermic reaction runaway avoided
- No emergency depressuring / shutdown required
- No safety incident

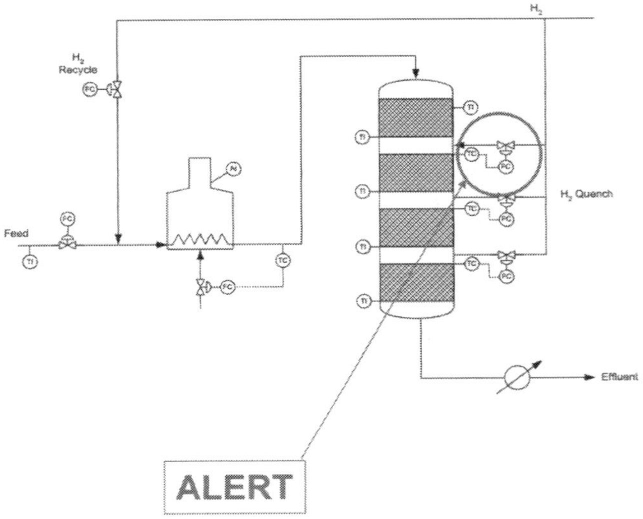

Smart Turnaround Case

- Three-year study with Gulf Coast Refiner to reduce maintenance spending utilizing predictive instrument and valve diagnostics
- Average maintenance savings $100,000 USD per year
- Additional benefits a magnitude greater than maintenance savings per year operations with better process control and plant availability

Fix only what needs maintenance, nothing more

Pump Health Monitoring Case

NA Refiner

- Significant number of process pumps monitored manually, once a month.
- Implemented asset health monitoring (wireless). Both Maintenance and Operations now have access to information previously unable to justify with wired solution.

"I only have two seconds worth of manually collected data each month to base decisions on maintenance. We don't catch all pump failures."

System Modernization Case

"Poor reliability and frequent shutdowns were undermining the refinery's efforts to serve the market effectively."

- Automation System Upgrade on Gulf Coast Hydrocracker
- Results showed sustained 50% reduction in reliability risk
- Estimated Savings, $1-2 Million per year

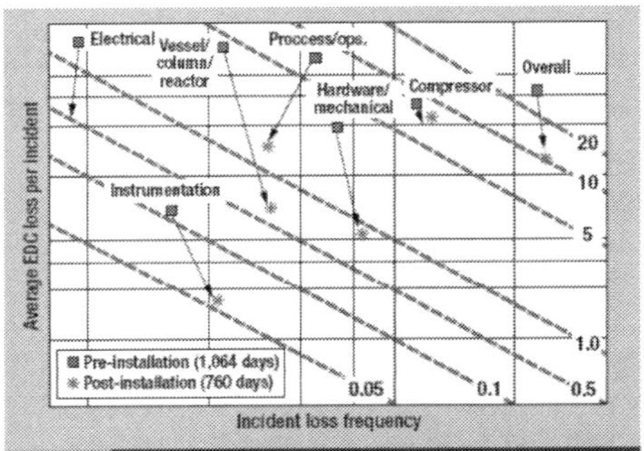

FIG. 3 Hydrocracker incident frequency and severity EDC loss.

Advances in Digital Automation in Refining

Key areas that have benefited from advances in digital automation

- Energy Efficiency
- Reliability
- Operator Effectiveness
  - Consolidated Control Rooms
  - Operator Training Simulations
- Safety

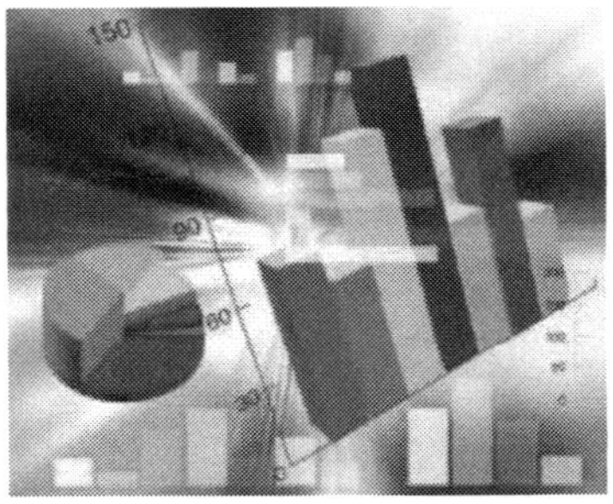

Multi-Site Control Center / Remote Support

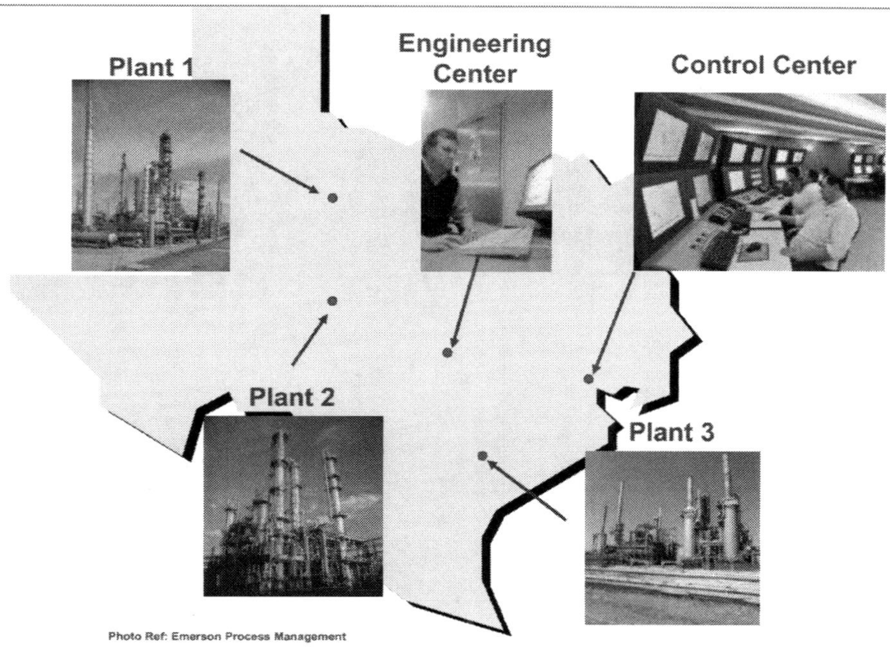

Photo Ref: Emerson Process Management

Operator Training Simulation

Training for both new and existing operators on infrequent events such as:

- Startup
- Shutdown
- Emergency Shutdown
- Abnormal Events

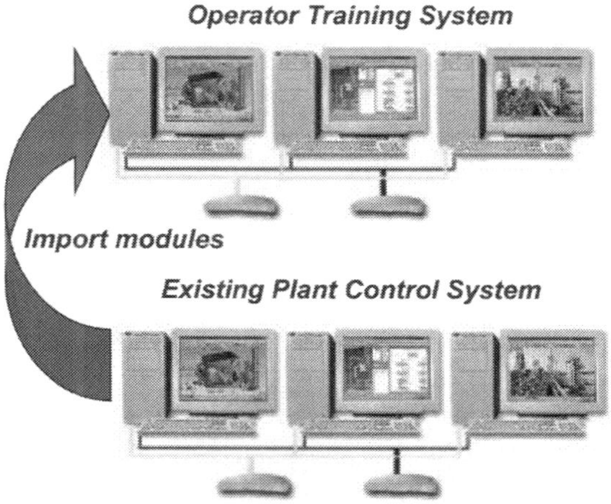

Modern Systems Utilize Exact Plant Control System Configuration

Advances in Digital Automation in Refining

Key areas that have benefited from advances in digital automation

- Energy Efficiency
- Reliability
- Operator Effectiveness
- Safety
  - Accumulation effect of above

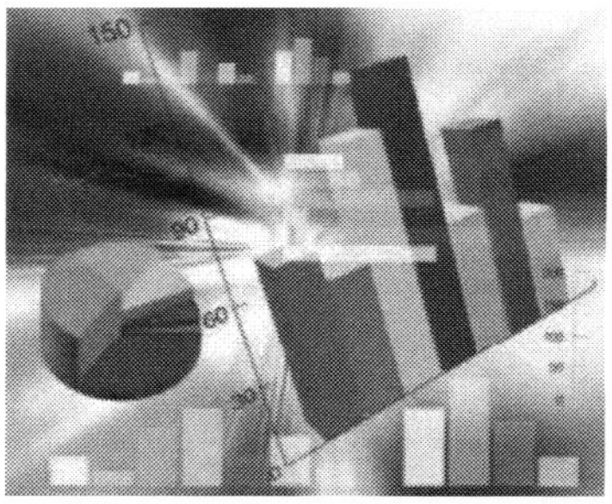

US Oil Refining Industry Employee Productivity

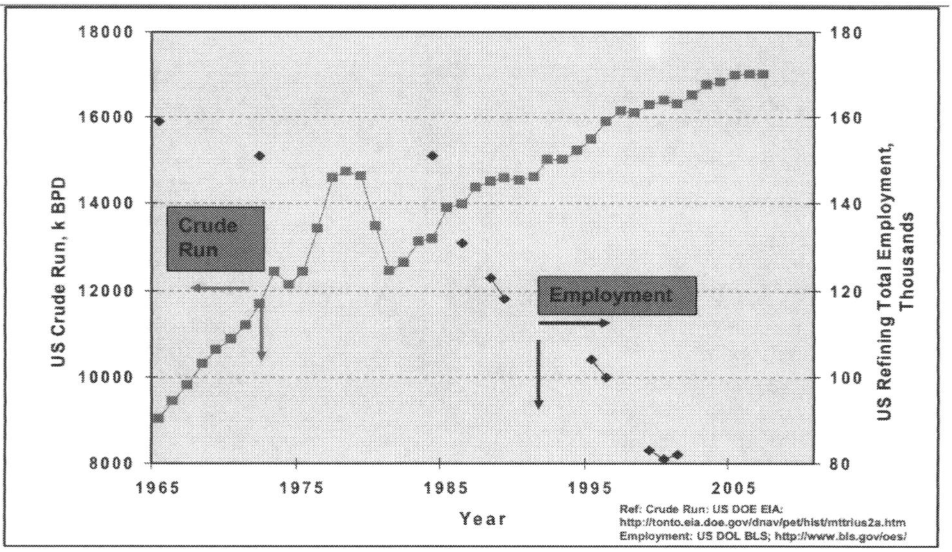

Employee Productivity Up 300%, Even With:
- Increased Complexity - More Crudes/ Product Grades
- Increased Regulatory Requirements

Refining Safety Impact

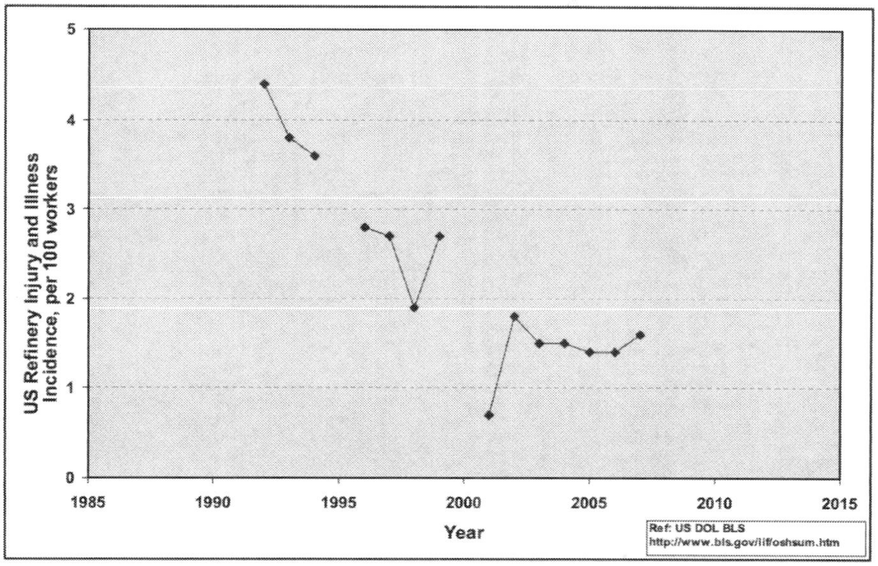

Conclusion

The continuing evolution in digital computing and communication capabilities and the application of these technologies have led to fundamental differences in the way plants operate and will continue to change the way they operate in the future.

Key areas that have benefited from advances in digital automation:

- Energy Efficiency
- Reliability
- Operator Effectiveness
- Safety

Questions?

Tim.Olsen@Emerson.com

AIChE

EMERSON.
Process Management

Examining the Value Chain in Production of Renewable Diesel

Riley, B., Grace

Originally presented on: 10/7/2011 0:25:00 - 0:50:00

Production of diesel and higher hydrocarbon fuels from natural oil feedstocks has been an active area of interest for the petroleum refining community, with much of the published work focusing on the conversion processes for these oils. A holistic view of the entire value chain, from oil preparation to finished fuel production, reveals areas of significant value that have been traditionally overlooked. This presentation will examine both the technical and economic aspects associated with the entire process of production of renewable diesel, with a particular aim to examine the processes prior to conversion necessary to remove impurities from feedstocks. In addition, we will discuss the issue of scale, as well as consideration of non-vegetable oil feedstocks and the complications that arise.

UOP FCC Innovations Developed Using Sophisticated Engineering Tools

Wolschlag, L., UOP

Originally presented on: 10/7/2011 0:50:00 - 1:15:00

Fluid Catalytic Cracking (FCC) technology has been a part of the petroleum industry since the 1940s. Yet, despite being a very mature technology, as many refiners move their FCC operations from fuels production to higher value products, continued development is vital to help reinvent the role of the FCC unit. Through the commercialization of world record scale FCC units, technical discoveries are emerging, which are providing opportunities for improvements across units of all size. As FCC technology is pushed into new areas of production, more is learned about controlling hydrocarbon species versus just a boiling range of fuels products.

Advanced diagnostics, design tools and pilot plant operations are accelerating process development and have resulted in several technical innovations. Through the combination of sophisticated engineering tools, Computational Fluid Dynamic (CFD) modeling, and radioactive tracer and tomography combined with physical inspection reports and commercial yields, technology advancements are being brought to market more quickly and with greater success.

This paper will highlight advancements in regenerator capacity, emissions reduction and riser feed distribution. It will demonstrate how innovative tools are being used to both improve existing FCC designs and move an aged technology towards true growth opportunities.

Maintain Peak HEN Performance under Fouling Conditions

Jimmy D Kumana*, Kumana & Associates, Houston, Texas
Simon J Pugh, IHS-ESDU, London, UK
Dr. Edward M Ishiyama, IHS-ESDU, London, UK

AIChE Regional Process Technology Conference, session 8
Galveston, Tx (6-7 October, 2011)

* SPEAKER

OUTLINE

- Cost of HX Fouling
- Causes of Fouling
- Fouling Control & Mitigation
- Optimum HEN Management
- Case study results

 2

REFINERY ENERGY COST ISSUES

- Energy inefficiency due mainly to
  - Poor operating practices
  - Sub-optimal heat recovery (HEN design)
  - HX fouling — Focus of this presentation
- Typ. energy costs = 59% of Opex
  ≈ 0.36 MMBtu/bbl
  = $3.65 /bbl @ $60/bbl

3

Impact of Fouling on Energy Efficiency

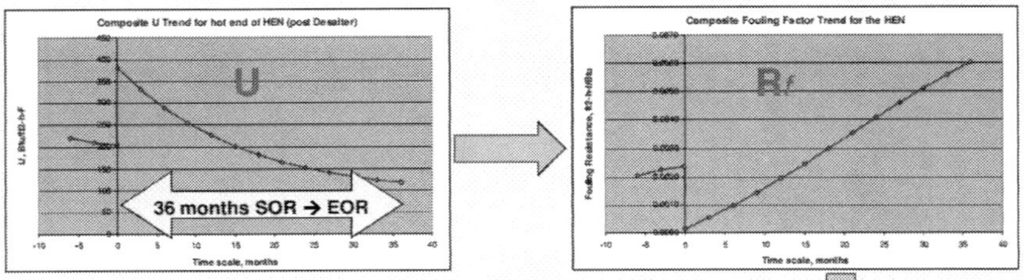

Simulation for 200 MBD Refinery, at $ 6/MMBtu (gas)

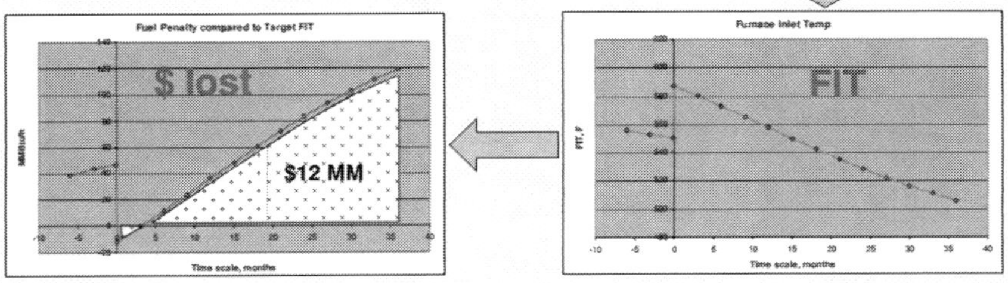

4

Fouling can also cost Production Loss

MM$

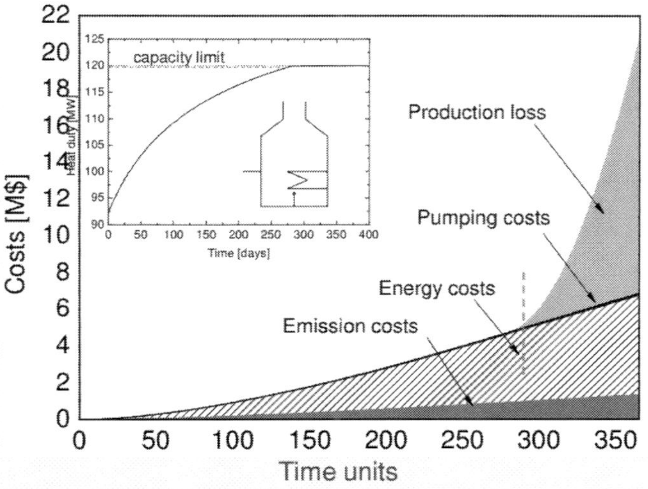

REF: F Coletti and S Macchietto, "Refinery preheat train network simulation undergoing fouling assessment of energy efficiency and carbon emissions", *Proc of Intl Conf on HX Fouling & Cleaning VIII*, Schladming, Austria (June 2009)

5

Total Cost of Fouling

BASIS:

- 200 MBD
- 340 days/yr
- $6 /MMBtu
- 20°F potential rise in FIT

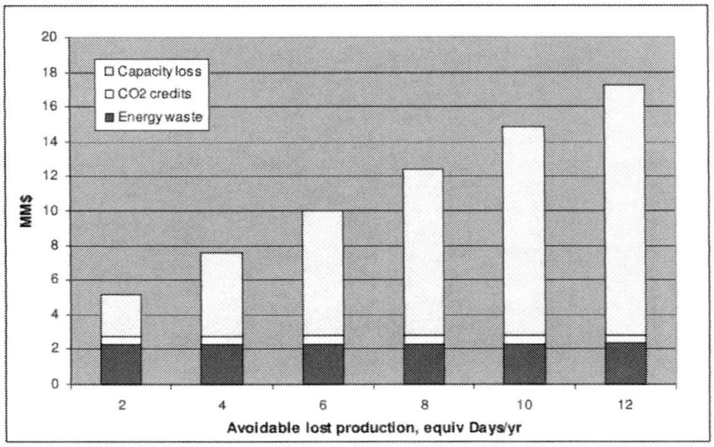

6

CDU FOULING: post-Desalter is worst

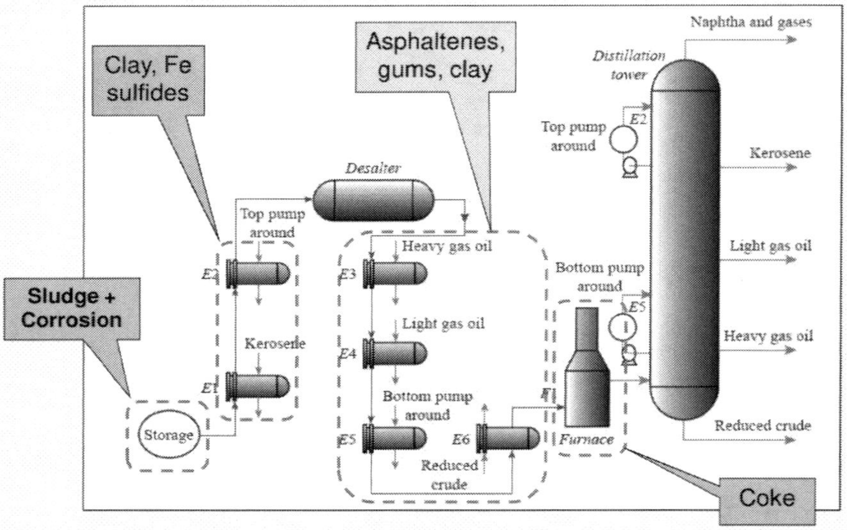

Fouling Rate = Fn. of Blend Composition

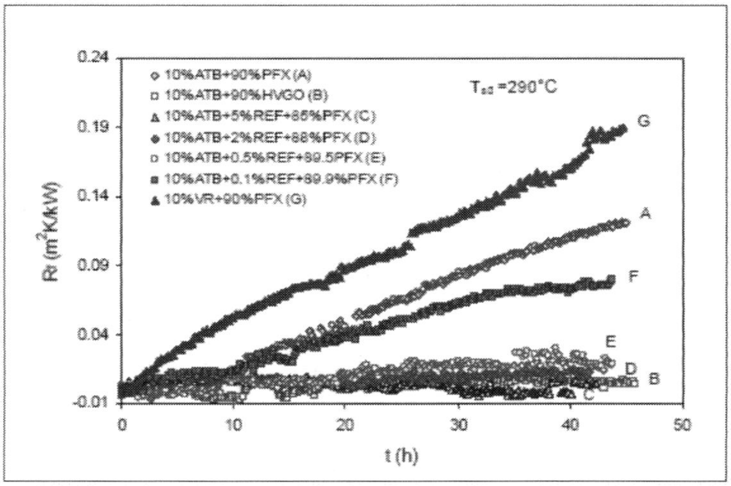

REF. E and Watkinson (2007)

Fouling Rates are Temp Sensitive

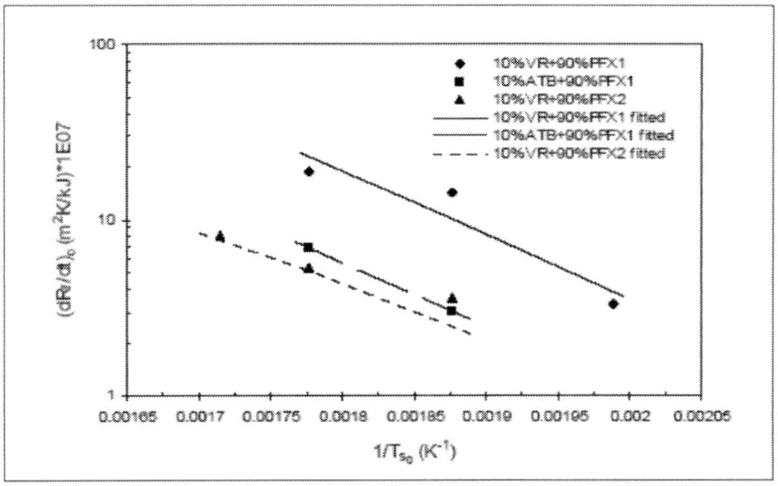

REF. E and Watkinson (2007)

Asphaltene Fouling

Well described by the deposition-removal model of Ebert & Panchal. Generalized 3-parameter model (Polley et al, 2009):

$$\frac{dR}{d\theta} = \frac{A}{\alpha}\exp\left(\frac{-E}{RT_f}\right) - \gamma\tau_w$$

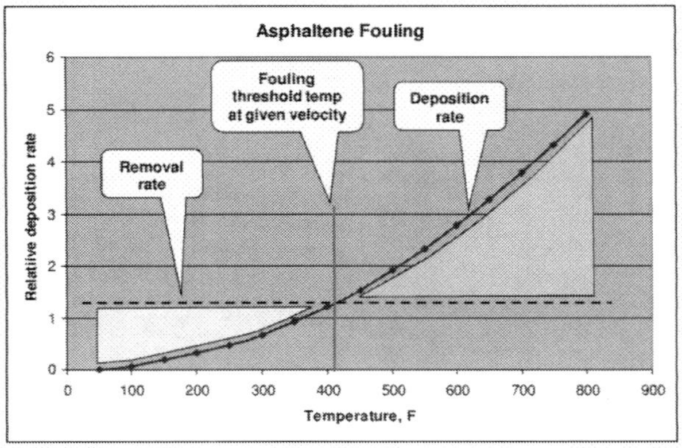

Fouling Threshold Concept

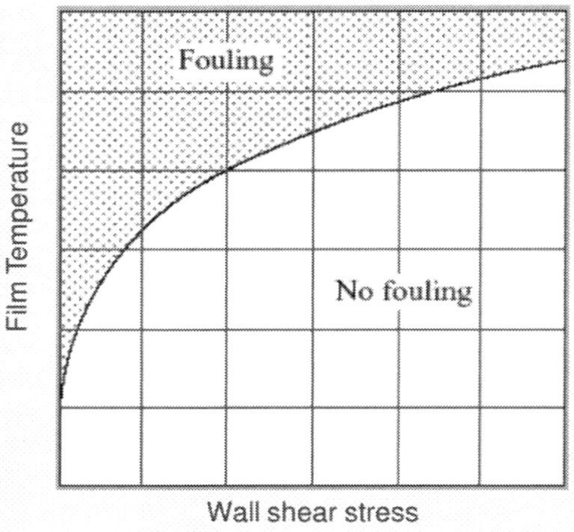

 11

FOULING MITIGATION & CONTROL

SEVERAL TOOLS & TECHNIQUES:

- Crude oil blending
- Parameter control for S&T hx (shear rate, temp)
- Tube inserts
- Low-fouling HX – twisted tube, compabloc, helical
- Anti-fouling additives
- Optimize HX cleaning schedule
- Restructure HEN configuration

Combined into single software

 12

New Methodology

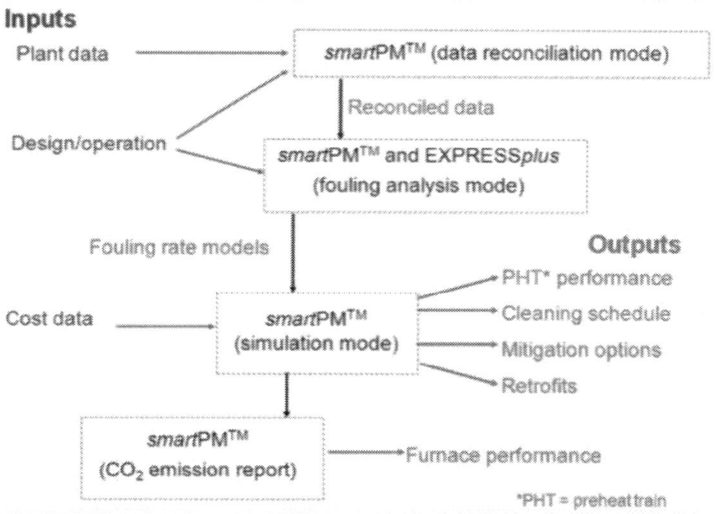

Results of Fouling Analysis

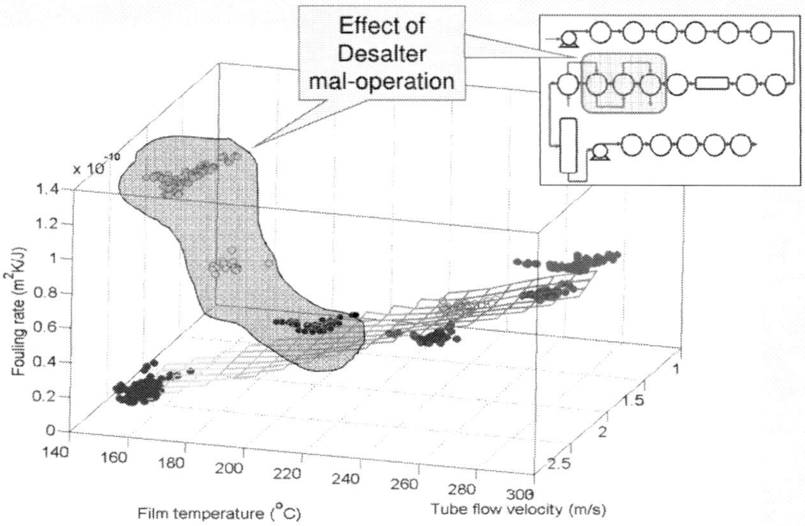

REF. Ishiyama, "Scheduling & desalter control in CDUs", Schaldming conf (2009)

HX Parameter Control, using Fouling Map

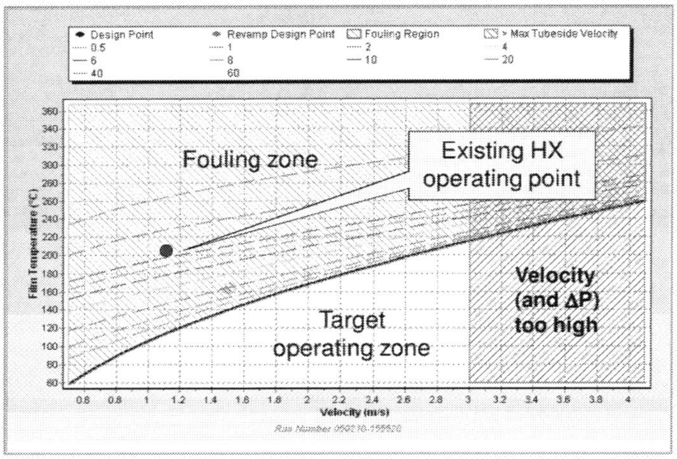

Black line = acceptable (target) fouling rate
Fouling rates increase rapidly at higher temps

REF. Screenshot, ExpressPlus® s/w from IHS-ESDU (2006)

15

HX modification strategy

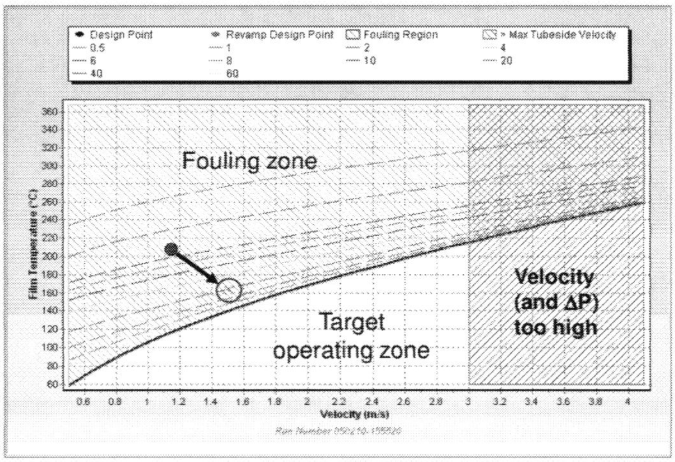

REF. Screenshot, ExpressPlus® s/w from IHS-ESDU (2006)

Goal = Move operating point towards Target Zone

16

HEN REVAMPS

- Re-sequencing of HX matches to keep tube-wall temps below fouling limit (goes against temp profile philosophy of Pinch Analysis)
- Generally not advisable unless done in conjunction with major process modification or capacity expansion
- Installed spares are expensive (and take up valuable real estate)

HEN Restructuring, based on DFP

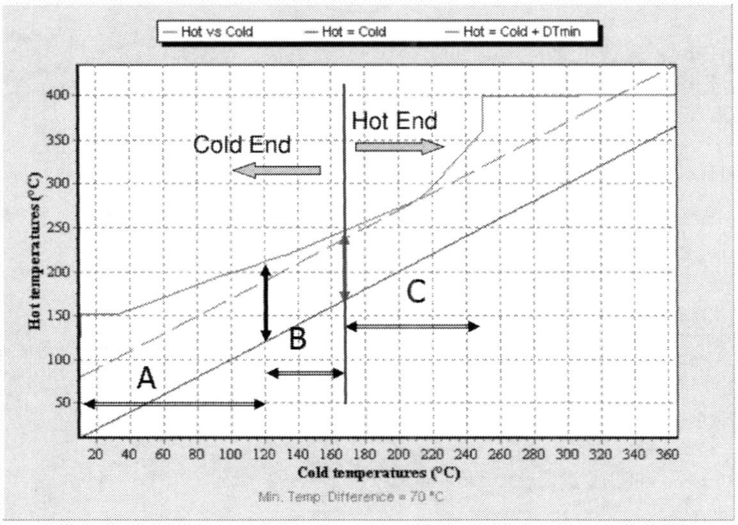

(Screen shot from *Integrity®* software, IHS-ESDU)

Design HEN and HX within Zones

- Section A – non-fouling regime, follow Pinch Design method (using DFP)
- Section B – velocity control (may require HX mods) to suppress fouling
- Section C – needs special consideration

Implications for HEN structure + economics

- Smaller HX area for high-ΔT cross-match
- More HX area for low-ΔT cross-match
- Reduced overall fouling rate over a larger heat transfer range, which helps to extend run time (extremely valuable)
- Tube inserts can reduce fouling rates even further
- No energy penalty
- Furnace capacity maintained

Optimizing HX Cleaning Schedule - 1

Traditional method:

- Calculate U for each HX from plant data
- Plot U vs time
- Clean HX (usually hydrodrilling) when U falls below some target value

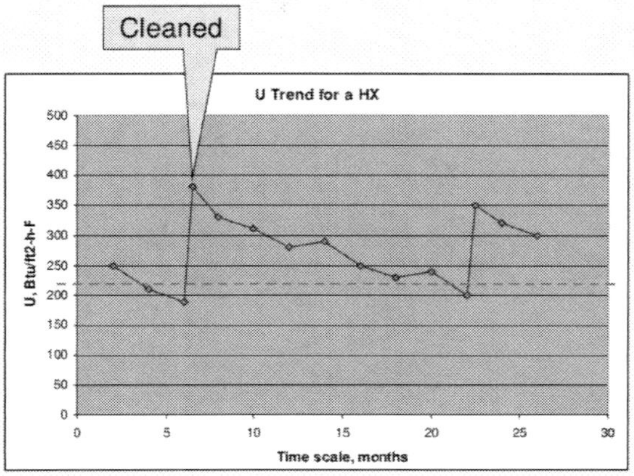

- U may be low for reasons other than fouling, eg. low flow rates, change in fluid properties, etc
- This HX may not be the main cause of low FIT (i.e. the "pinch")

Optimizing HX Cleaning Schedule - 2

Better method:

- Calculate U for each HX from reconciled plant data
- Calculate clean film h_t & h_s using correlations
- Estimate $R_f = 1/U - 1/h_t - 1/h_s$
- Plot R_f vs time
- Predict impact on FIT w/ hi-fidelity HEN sim model
- Monitor ΔP (both sides) across each HX
- Apply fouling suppression techniques
- Clean when cost of extra energy or lost production capacity exceeds Cleaning cost

CDU Case Study 1 (2009)

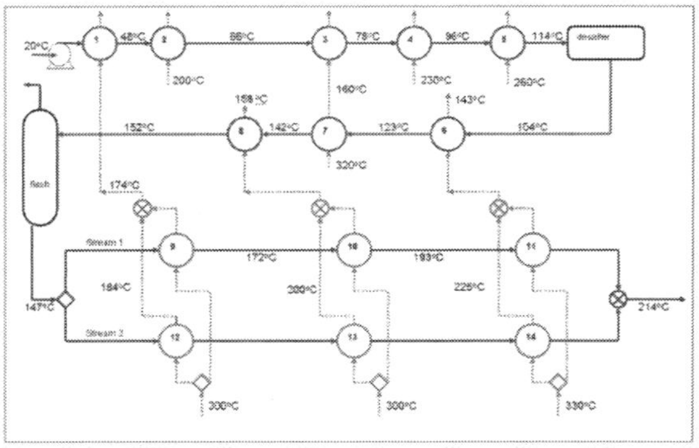

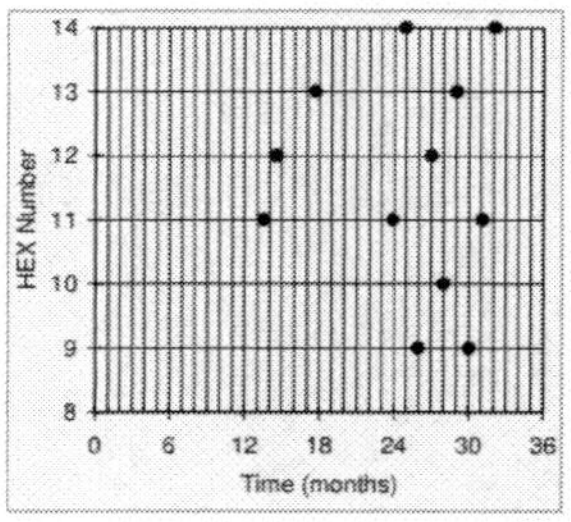

Furnace coil inlet temp (CIT) was maintained within Target range

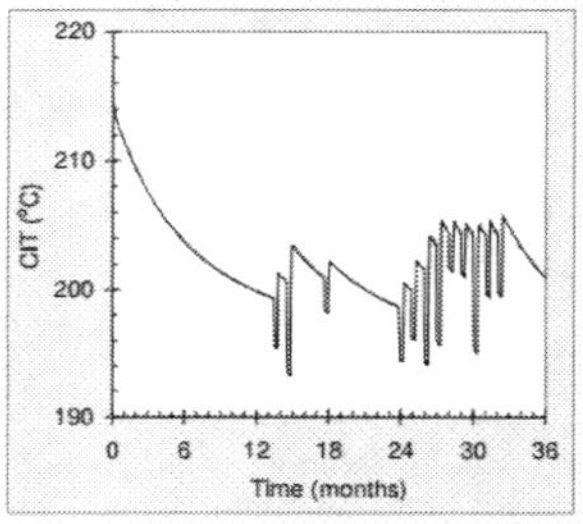

REF. Ishiyama, "Fouling mitigation options in ref preheat", Energy & Fuels (2009)

CDU Case Study 2 (2011)

- Capacity/Energy Revamp, India
- 3 parallel CDU trains of different capacities
- Multiple "opportunity" crudes (~50)
- Feedstock blend "optimized" by Business Planning group
- Typical run length of 3-10 days per blend
- Extremely high fouling rates

Hot End of Crude Preheat Train

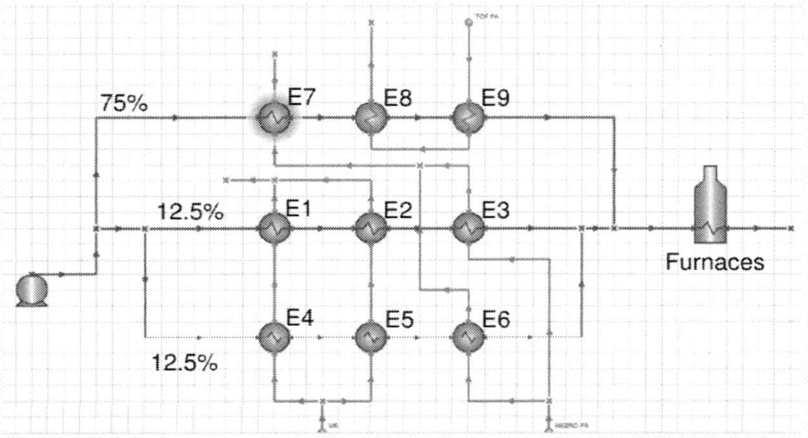

- Post-desalter HEN
- 3 branches on crude side
- Unequal % flow splits
- Reconciled data

Data Reconciliation is Critically Important

- How do we know if the measured data values are correct?
- Need some "anchor points" of data that can be trusted, e.g. using redundant measurements, or consistency checks.
- Rest can be filled in by HMB simulation

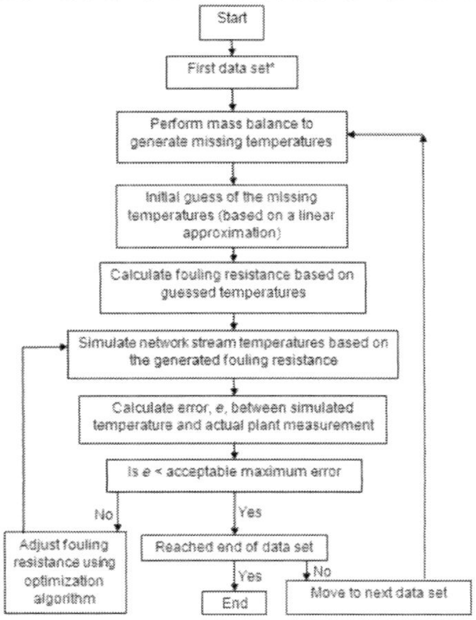

Test of Data Reconciliation Methods

- Well instrumented refinery, trusted data (all known)
- X axis = Rf based on trusted data
- Y axis = Rf based on simulated temps (assuming we did not know selected temps)
- Red squares → Conventional data reconciliation
- Blue diamonds → Smart data reconciliation

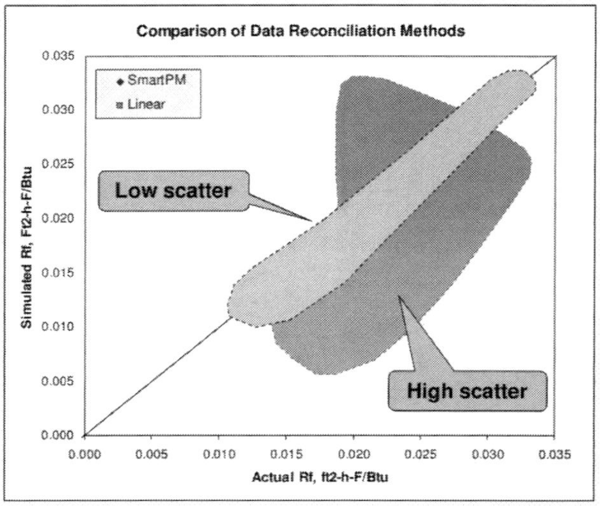

 27

Cleaning Strategy from Std Data Recon

- High R_f → frequent cleaning
- Low R_f → infrequent cleaning

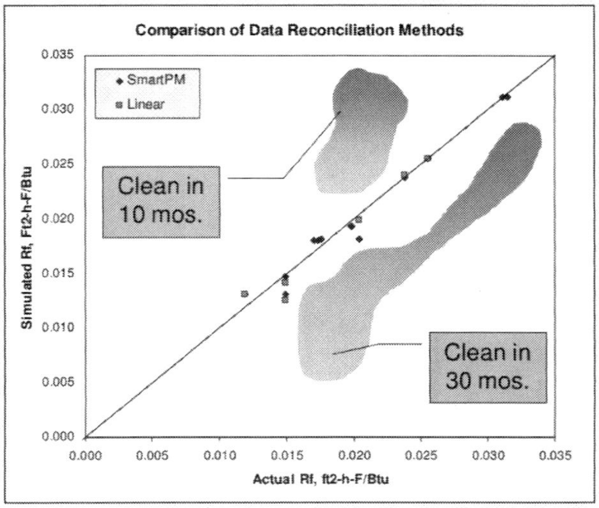

 28

Cleaning Strategy from Smart Data Recon

- The Cleaning strategy is radically different
- Std Data Recon cleans one HX needlessly and fails to clean the other HX when required

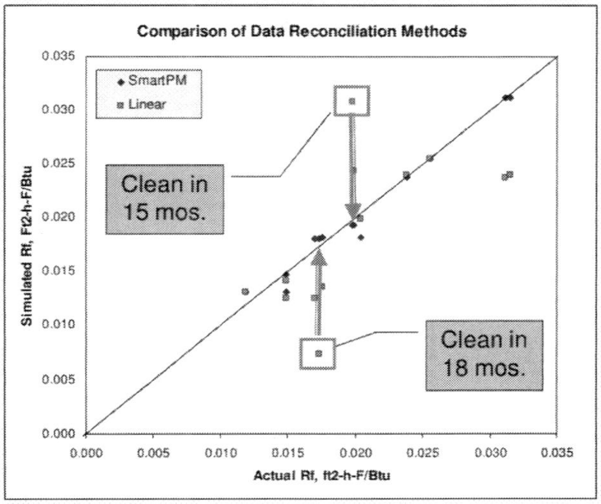

29

Case Study 2 – Hot End of PHT (2011)

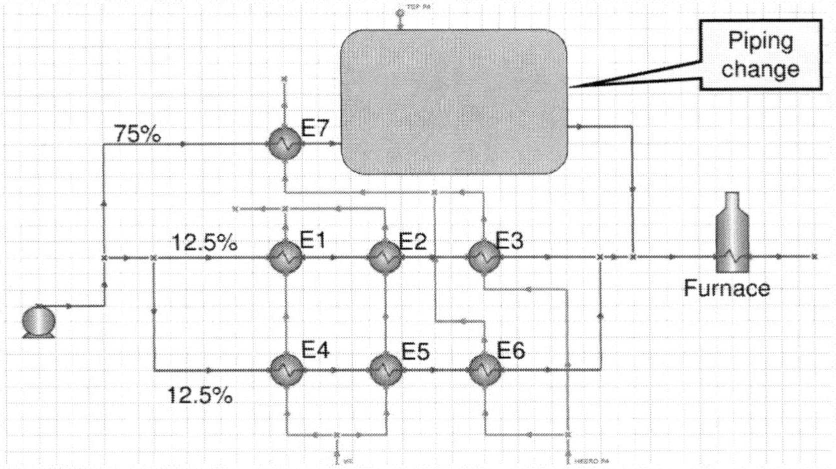

- Parallel vs series config. for E8 and E9, w/ flow splits
- Crude swapped from shell-side to tube-side

30

Performance Comparison – 3 year Run

Scenario	Description
1	Original network without any modification
2	Targeted cleaning, *i.e.* the network is simulated based on a predefined cleaning schedule. In this example unit 9 is cleaned every 6 months.
3	Optimized cleaning
4	Retrofitted network – arrangements considered to reduce rate of fouling

Scenario	Heater Q MMBtu/h	Fuel cost MM$	CO2 em. 1000 tons
1	63.8	10.80	90
2	54.4	9.21	80
3	41.1	6.95	60
4	4.1	0.70	10

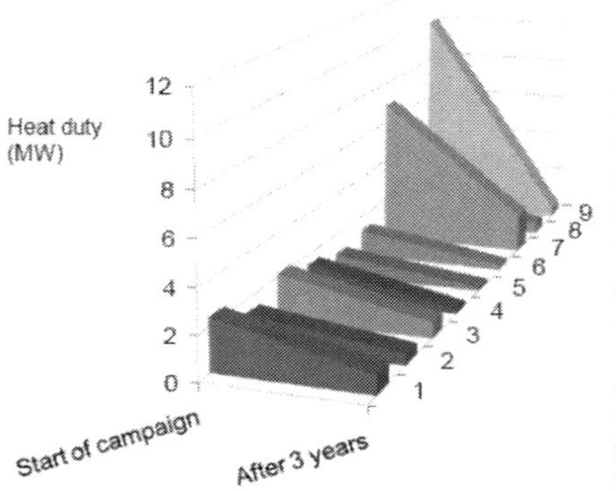

~ $10 MM fuel cost savings between SOR and EOR

Cautionary Notes

- Most plants today have a Plant Information (PI) system that stores hourly data on key measured process variables, but experience shows that they are seldom used to their full potential for optimization and process trouble-shooting.
  - Lack of expertise in Data Mining
  - Inconsistent and unreliable data (not reconciled)
- Fouling model depends upon feedstock properties. So if crude slate is frequently changed, model may not be reliable.
- Data Reconciliation is a resource-intensive activity (high-skill level, hi-fi models, advanced software algorithms), best outsourced to specialists.

SUMMARY & CONCLUSIONS

- Historically Refinery Fouling has been accepted as unavoidable & uncontrollable
- Research progress in past 10-15 years shows that it is mostly due to asphaltene deposition
- Portfolio of techniques for mitigation and control:
    - Crude oil blending
    - Parameter control for S&T hx (shear rate, temp)
    - Tube inserts
    - Low-fouling HX – twisted tube, compabloc, helical
    - Anti-fouling additives
    - Optimize HX cleaning schedules
    - Restructure HEN configuration
- Potential Energy savings ~**5 cents/bbl**

 33

Q&A

 34

ExxonMobil
Taking on the world's toughest energy challenges.™

The Outlook for Energy
a view to 2030

Ram Seetharam
3rd AIChE Regional Process Technology Conference
October 7, 2011

Economic and Energy Evolution

As societies and technologies develop over time…

… energy needs evolve as well

ExxonMobil
Taking on the world's toughest energy challenges.™

ExxonMobil 2010 Energy Outlook

3

ExxonMobil

Taking on the world's toughest energy challenges.

Ongoing Development Challenges

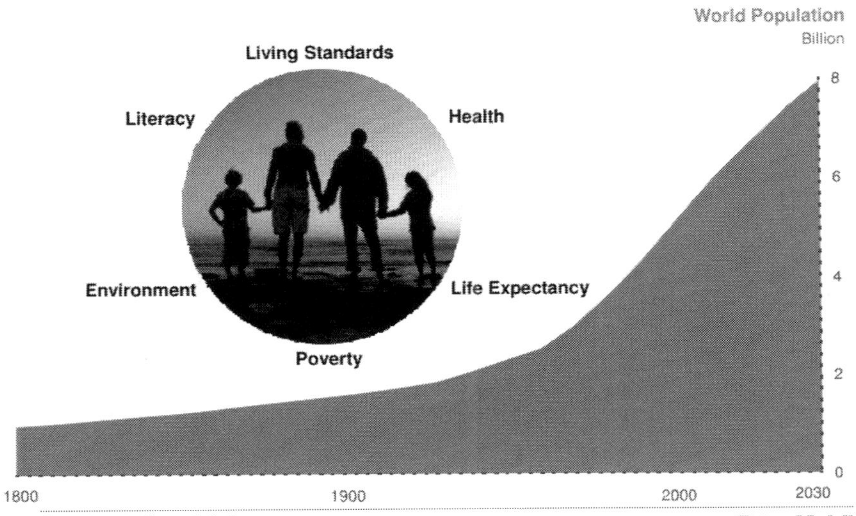

ExxonMobil 2010 Energy Outlook

4

ExxonMobil

Taking on the world's toughest energy challenges.

Source: United Nations; ExxonMobil

ExxonMobil 2010 Energy Outlook

5

ExxonMobil
Taking on the world's toughest energy challenges.

Economic and Energy Evolution

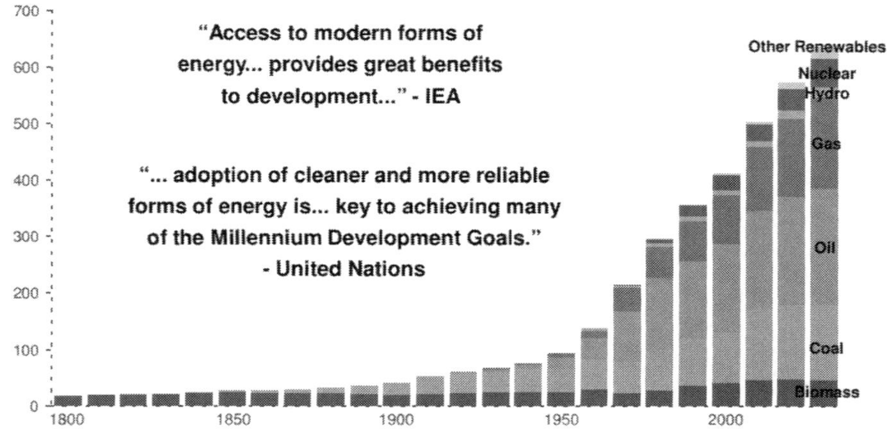

ExxonMobil 2010 Energy Outlook

6

ExxonMobil
Taking on the world's toughest energy challenges.

Source: Smil, *Energy Transitions*; ExxonMobil

ExxonMobil 2010 Energy Outlook

7

ExxonMobil

Taking on the world's toughest energy challenges.

Energy and Prosperity Linked

Energy per Capita

MBTU

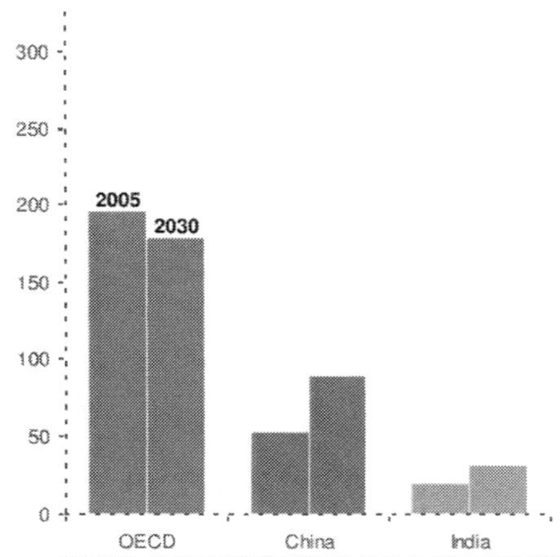

GDP per Capita

2005 $k

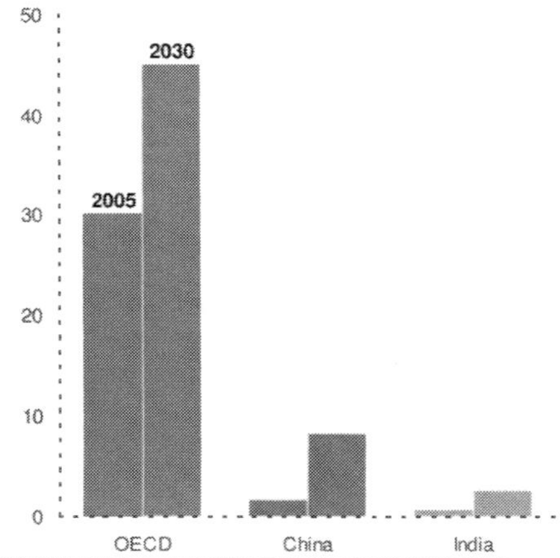

ExxonMobil 2010 Energy Outlook

8

ExxonMobil

Taking on the world's toughest energy challenges.

ExxonMobil 2010 Energy Outlook

9

ExxonMobil

Taking on the world's toughest energy challenges.

Parsepartationsportaiod Geowansltefactilyns

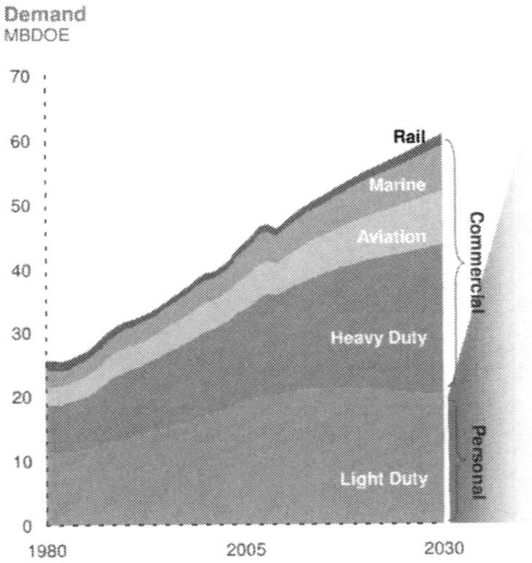

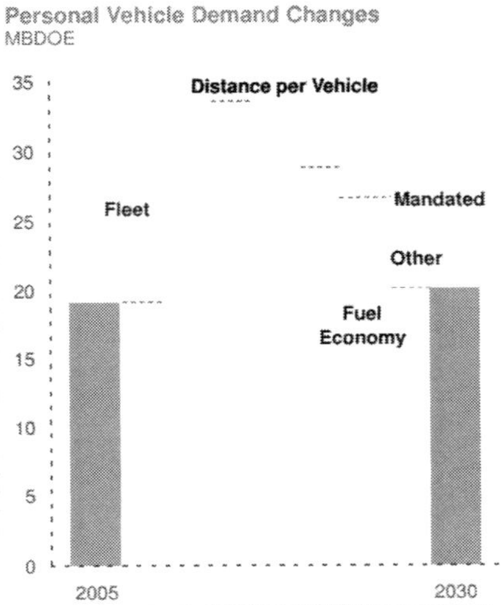

ExxonMobil 2010 Energy Outlook

10

ExxonMobil

Taking on the world's toughest energy challenges.

ExxonMobil 2010 Energy Outlook

11

ExxonMobil
Taking on the world's toughest energy challenges.

Electricity Demand Continues to Surge

By Sector
k TWh

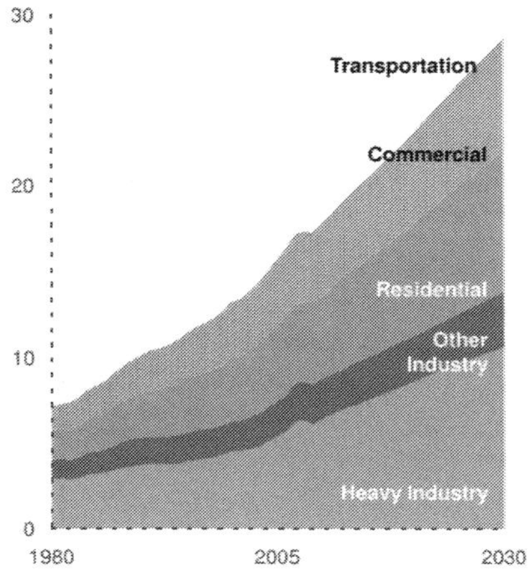

By Region
k TWh

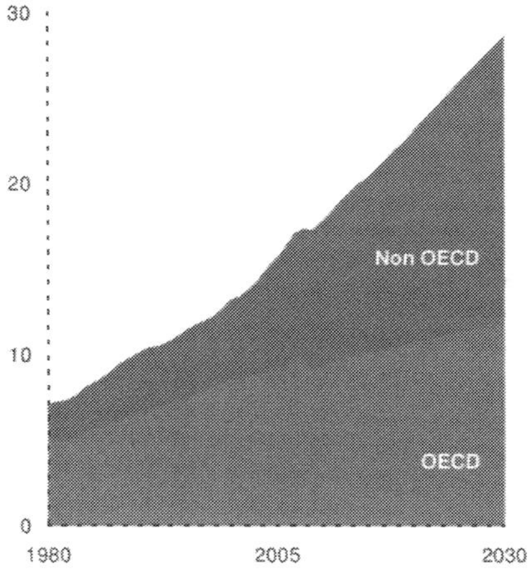

ExxonMobil 2010 Energy Outlook

12

ExxonMobil
Taking on the world's toughest energy challenges.

ExxonMobil 2010 Energy Outlook

13

ExxonMobil

Taking on the world's toughest energy challenges.

Economic Choices for U.S. Power

Baseload, Startup 2025

2010 cents/kWh

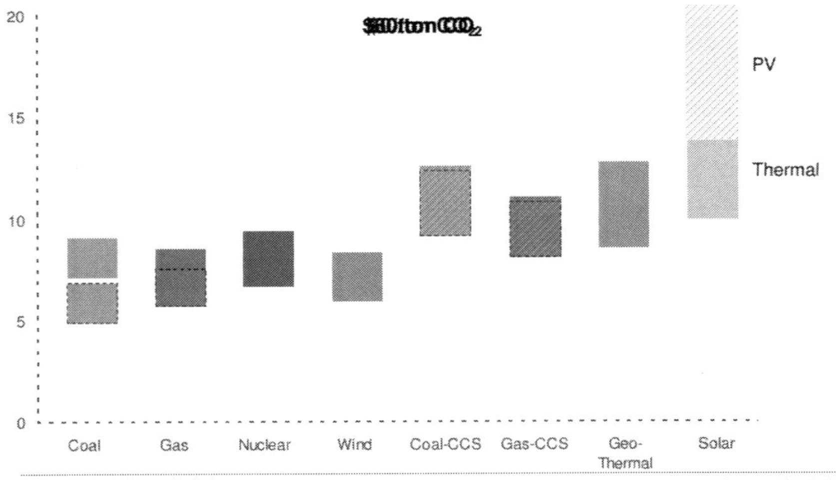

ExxonMobil 2010 Energy Outlook

14

ExxonMobil

Taking on the world's toughest energy challenges.

ExxonMobil 2010 Energy Outlook

15

ExxonMobil

Taking on the world's toughest energy challenges.

Regional Needs Drive Fuels for Power

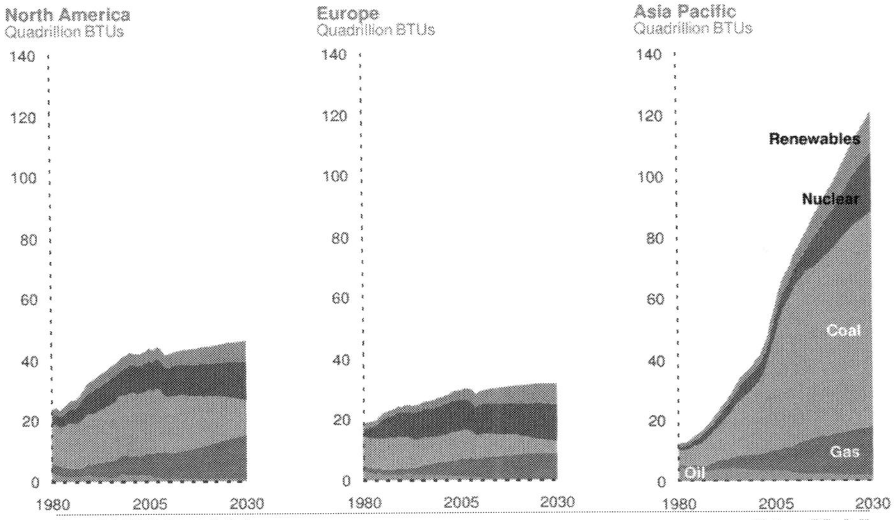

ExxonMobil 2010 Energy Outlook

16

ExxonMobil

Taking on the world's toughest energy challenges.

ExxonMobil 2010 Energy Outlook

17

ExxonMobil
Taking on the world's toughest energy challenges.

Energy Mix Continues to Evolve

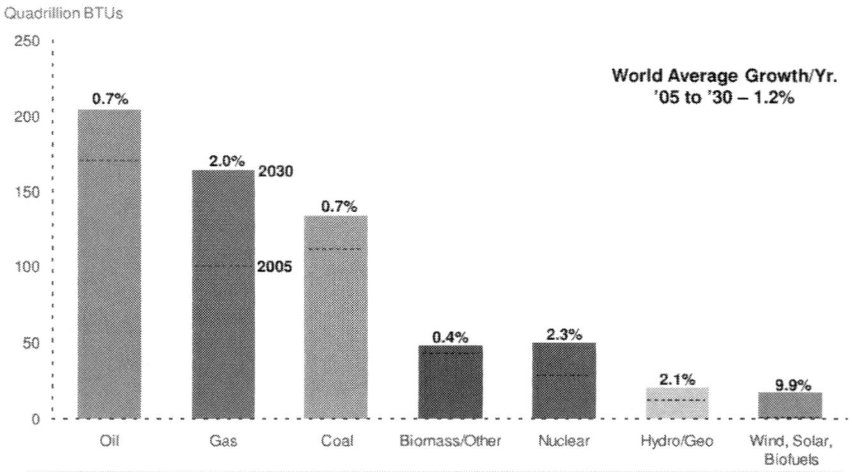

ExxonMobil 2010 Energy Outlook

18

ExxonMobil
Taking on the world's toughest energy challenges.

ExxonMobil 2010 Energy Outlook

19

ExxonMobil

Taking on the world's toughest energy challenges.

Regional CO_2 Emissions Trends Vary

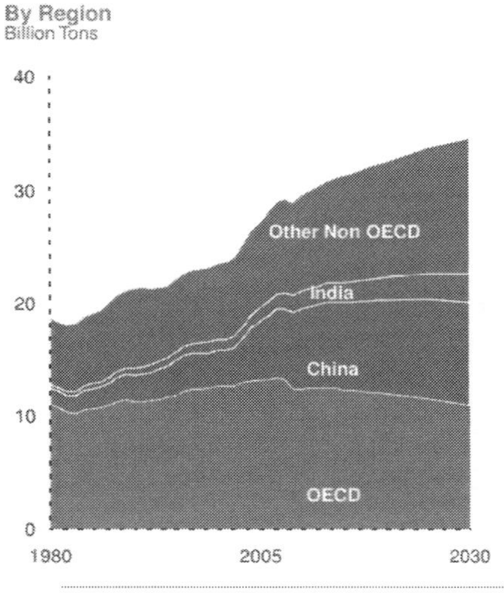

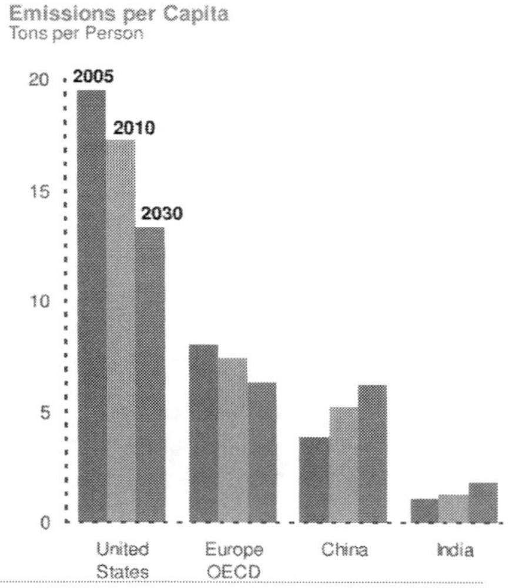

ExxonMobil 2010 Energy Outlook

20

ExxonMobil

Taking on the world's toughest energy challenges.

Development Challenges and Solutions

World development continues, while lives improve and economies grow

Increase Efficiency

Mitigate Emissions

Expand Supplies

ExxonMobil 2010 Energy Outlook

23

ExxonMobil

Taking on the world's toughest energy challenges.

ExxonMobil™

ExxonMobil 2010 Energy Outlook

24

ExxonMobil

Taking on the world's toughest energy challenges.

Insight into Phase Behavior of Complex Fluids

Walter G. Chapman
Chemical and Biomolecular Engineering, Rice University

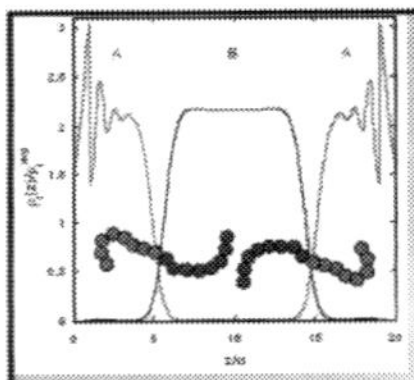

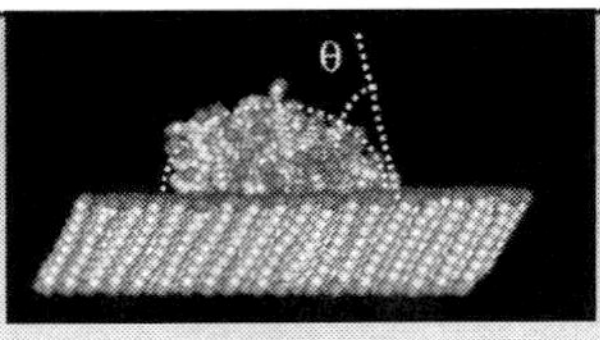

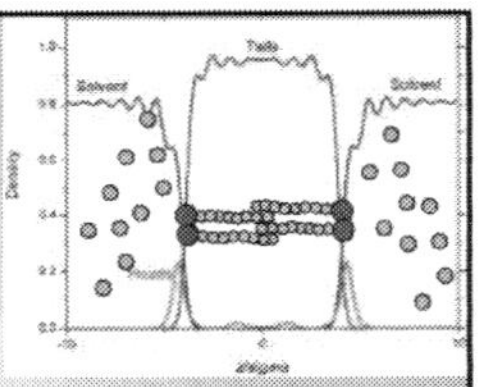

Walter G. Chapman and Kenneth R. Cox

- Molecular Modeling of Complex Fluids and Thin Films
- Flow Assurance (Gas Hydrates & Asphaltenes)
- High Pressure Phase Behavior and NMR
- Subsea Distribution of Gas Hydrates

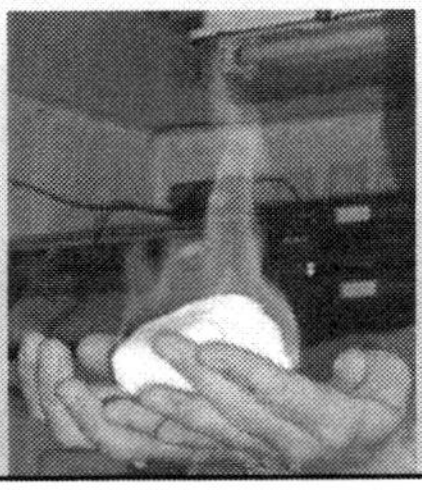

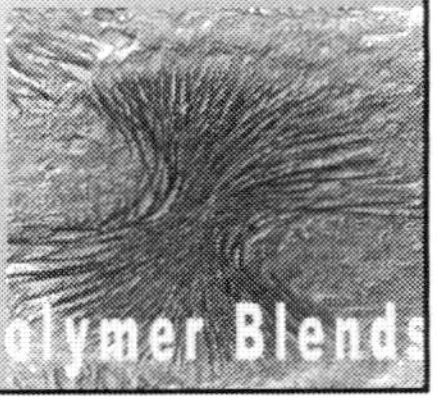

Contributors

Present Students	Former Students	
Guangsheng Gu	Dhananjay Ghonasgi – ConocoPhillips	Sandeep Tripathi – GE
Chris Emborsky	Mario Llano – Univ Valle, Cali, Colombia	Shuqiang Gao – Shell
Zhengzheng Feng	Victor Perez – Chevron	Matthew Yarrison – ExxonMobil
Sayantan Chatterjee	Pallav Jain – Bechtel	Piyush Srivastava – Baker-Petrolite
Aditya Agrawal	Chad Segura – Aspen Tech	Aleksandra Dominik – Shell
Deepti Ballal	Alejandro Garcia – Monterey Tech., Mexico	Doris Gonzalez – Schlumberger
Kai Gong	Prasanna Jog – Dow Chemical	Gaurav Bhatnagar – Shell
Sai Panuganti	Sharon Sauer – Rose-Hulman Institute	Clint Aichele – ConocoPhillips
Ben Marshall	Jie Zhang – Schlumberger	Adam Bymaster – ExxonMobil
	David Ting – Shell Global Solutions	Shekhar Jain – Shell
	Auleen Ghosh – Intel	Francisco Vargas – PI University

Other Collaborators
Dr. Anju Kurup – Rice Univ.
Dr. Kyoo Song – Rice Univ.
Prof. Kenneth Cox – Rice Univ.
Prof. George Hirasaki – Rice Univ.
Prof. Waylon House – Texas Tech. Univ.
Dr. Keshawa Shukla – INTEC Engineering

RICE UNIVERSITY 3

Motivation – Property Prediction for Design

Process Design – Challenges for Conventional Models

- Phase behavior / density / enthalpy
- Polar and Associating components
- Large size asymmetry – polymer solutions
- VLE and LLE over a wide range of T and P
- Viscosity and other transport properties
- Interfacial properties

Properties and Structure for Product Design

- Meso-scale structure determines properties (micro-domains)
- Composites – dispersed nano-particles
- Surfactant formulations and block copolymers

RICE UNIVERSITY 4

Predicting Complex Fluid Properties

Applications

- Polymer solutions and petroleum systems
- Coatings, adhesives, microemulsions, and wettability alteration
- Self-assembly for manufacturing nano-materials (e.g., stabilization in nano-colloid / polymer systems, self-healing materials)
- Chemical sensors and affinity based separations
- Cosmetics, moisturizers, conditioning shampoo

Goal

- Understand how molecular interactions lead to bulk and interfacial properties and to self-assembly

RICE UNIVERSITY

5

Outline

Overview of the SAFT Approach

SAFT Model for Process Design (Bulk Properties)

- Polar Components
- Polydisperse Polymer Solutions
- Asphaltenes
- Transport Properties

SAFT Based Models for Product Design (Microstructured Systems)

- Surfactants
- Nano-fillers / Polymer System

Summary

RICE UNIVERSITY

6

SAFT EQUATION OF STATE

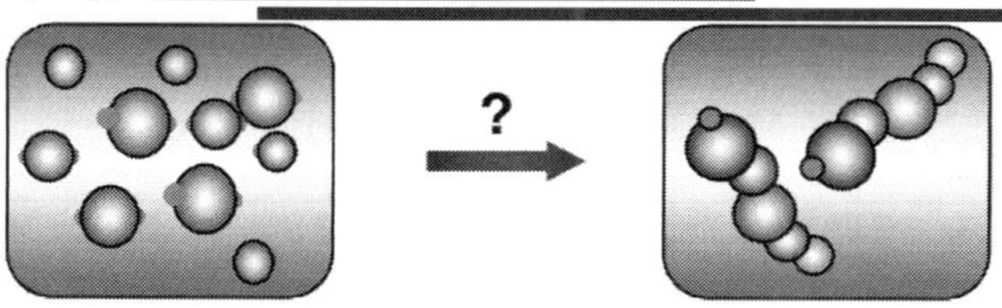

Chapman et al., *Mol. Phys.* (1988)
Wertheim, J. Stat. Phys (1986)

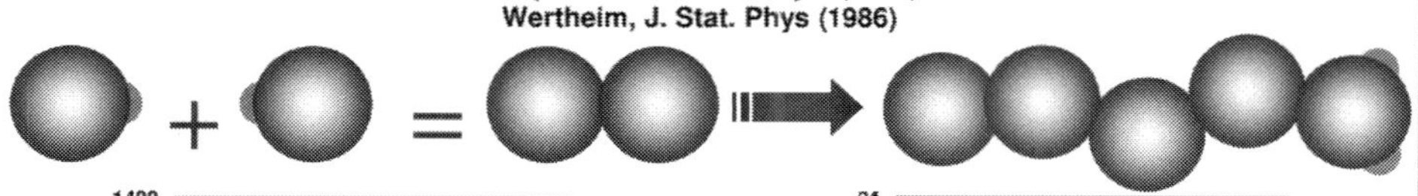

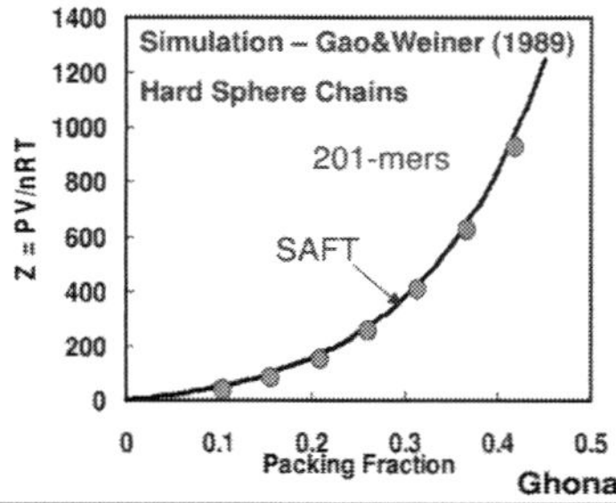

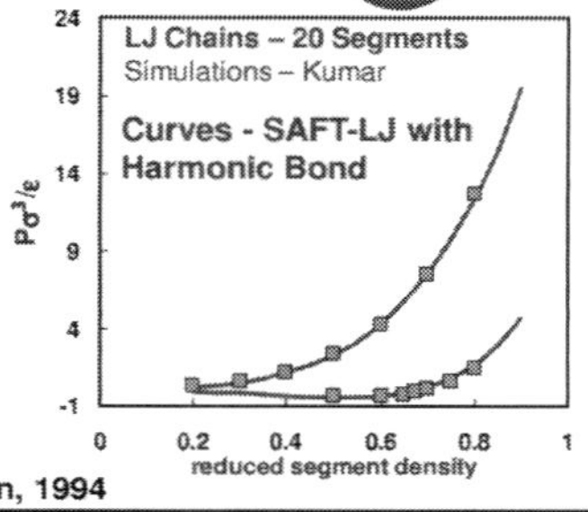

Ghonasgi and Chapman, 1994

The PC-SAFT EoS

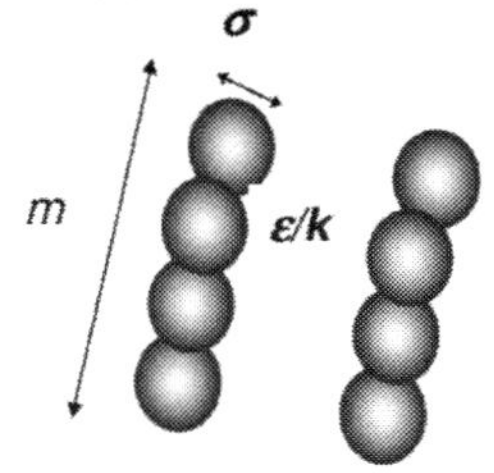

$$\frac{A^{res}}{RT} = \frac{A^{seg}}{RT} + \frac{\Delta A^{chain}}{RT} + \frac{\Delta A^{assoc}}{RT}$$

Chapman, Jackson, and Gubbins, Mol. Phys. 65, 1057 (1988)
Gross & Sadowski, Ind. Eng. Chem. Res., 40, 1244-1260 (2001)
Doris Gonzalez, Doctoral Thesis, Rice Univ. (2008)

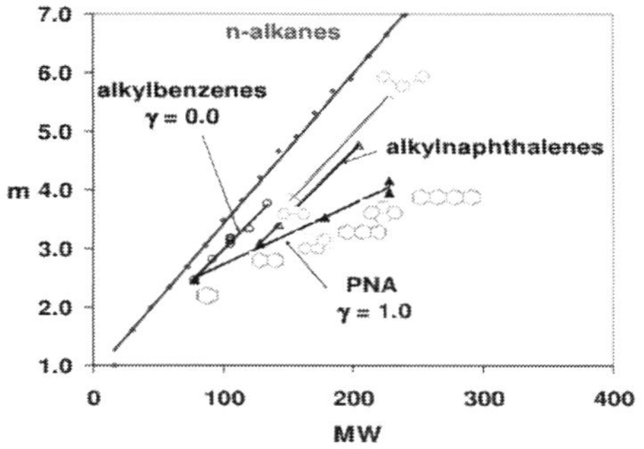

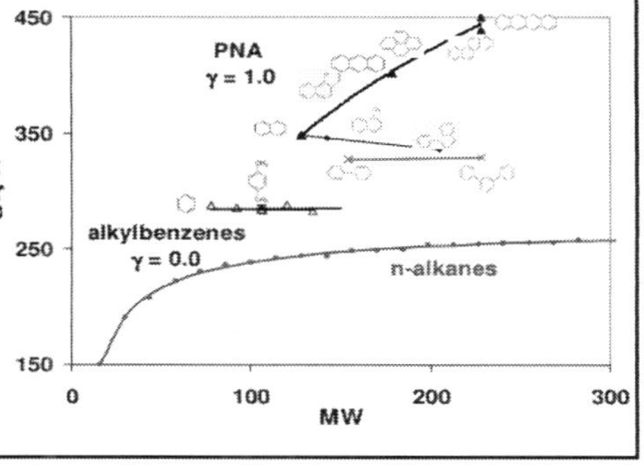

Some Applications of SAFT

SAFT has been Applied to Mixtures Containing

Alcohols	Alkanolamines	Carboxylic Acids
Refrigerants	Amphiphiles	HF
Water	Fluoroalkanes	Polynuclear Aromatics
Electrolytes	Copolymers	Ionic Liquids

Example Corporate Users

- Exxon-Mobil
- Shell Global Soln
- Lyondell
- Dow
- BASF
- Bayer
- Chevron
- DSM
- Nova Chemicals
- BP
- Aspen Tech
- InfoChem / VLXE / Calsep

Comments from Industrial Users

- Gives Qualitatively Correct Phase Behavior before Fitting Mixture Parameters
- Provides a Tool Where None had Previously Existed (e.g., polar polymers)

9

Associating Fluids

Motivation

Hydrate Inhibition

- Methanol Partitioning to Hydrocarbon Phase
- Many models predict false LLE for methanol / alkane systems

Water Content in Gas Processing

Surfactant Manufacture

Non-ideality of phase behavior is caused by differences in intermolecular forces

RICE UNIVERSITY 10

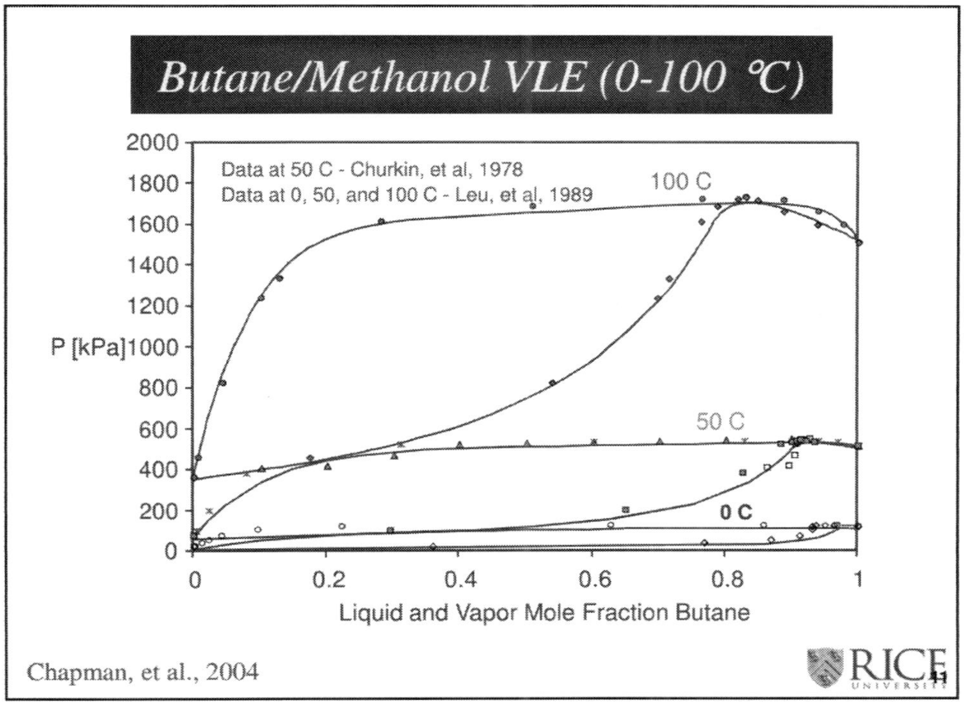
Butane/Methanol VLE (0-100 °C)
Data at 50 C - Churkin, et al, 1978
Data at 0, 50, and 100 C - Leu, et al, 1989
100 C
50 C
0 C
P [kPa]
Liquid and Vapor Mole Fraction Butane
Chapman, et al., 2004
RICE UNIVERSITY
11

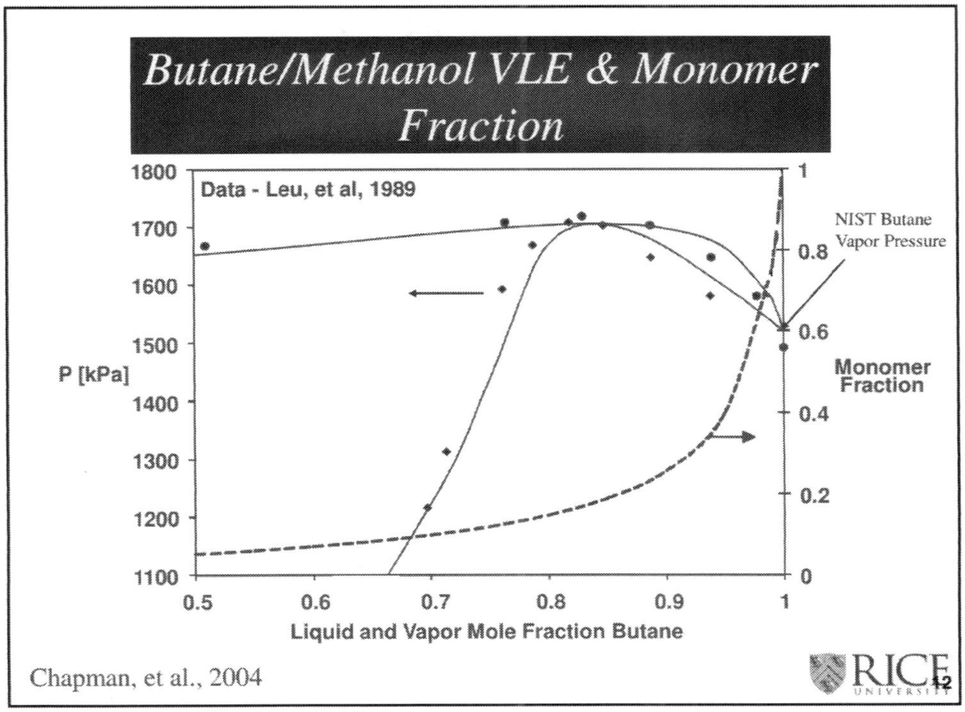
Butane/Methanol VLE & Monomer Fraction
Data - Leu, et al, 1989
NIST Butane Vapor Pressure
Monomer Fraction
P [kPa]
Liquid and Vapor Mole Fraction Butane
Chapman, et al., 2004
RICE UNIVERSITY
12

Butane/Methanol VLE (0-100 ℃)

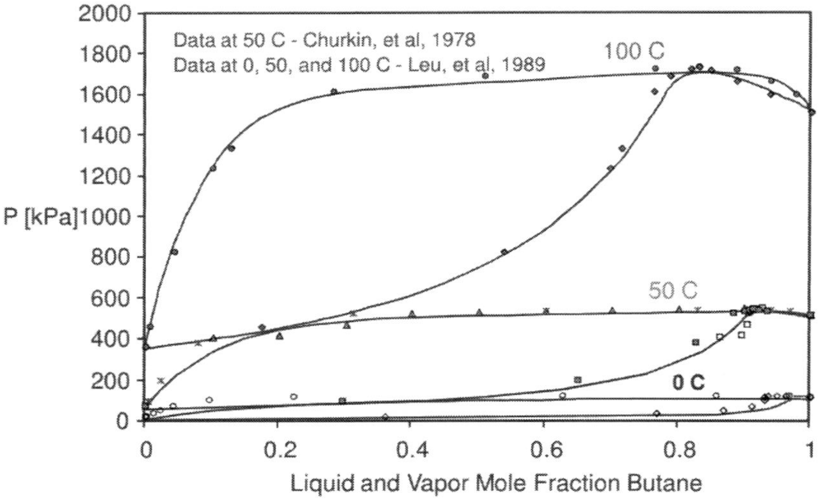

Chapman, et al., 2004

Water Content in Alkanes

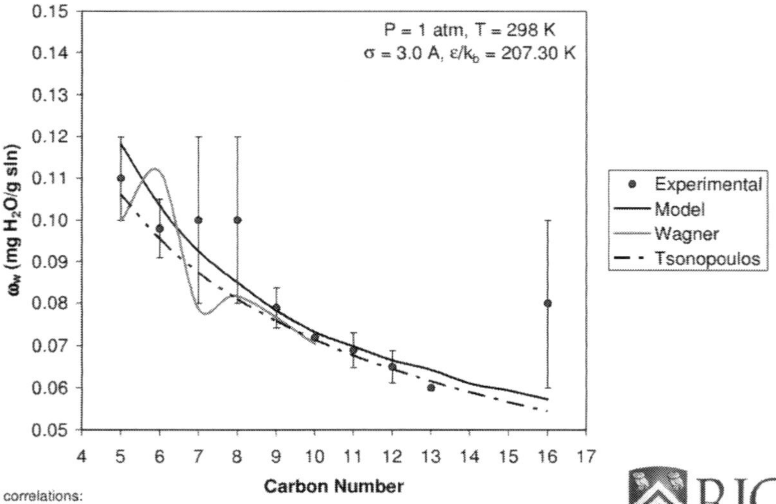

Data & correlations:
[19] *IUPAC/NIST Solubility Database ver. 1.0.* 2003, NIST Measurement Services Division.
[20] Tsonopoulos, C. *Fluid Phase Equil.* **1999**, 156 (1-2), 21-33.
[21] Wagner, J., *GPA Engineering Data Book.* 1999, Gas Processors Association (GPA): Tulsa. p. 1-35.

Water Content in Methane

Emborsky, Cox, and Chapman, IECR, 50, 7791 (2011)

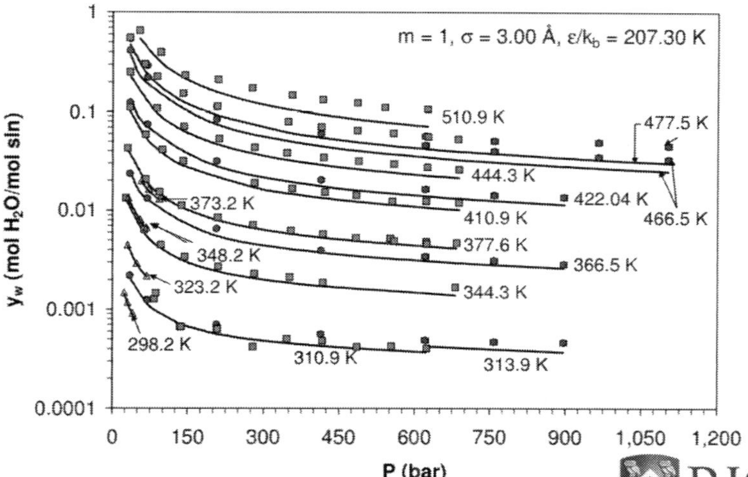

Data:
[22]Yarrison, M., K. R. Cox, and W. G. Chapman. *Ind. Eng. Chem. Res.* **2006**, 45 (20), 6770-7. (circles)
[23]Olds, R. H., B. H. Sage, and W. N. Lacey. *Ind. Eng. Chem.* **1942**, 34 1223-7. (squares)
[24]Rigby, M. and J. M. Prausnitz. *J. Phys. Chem.* **1968**, 72 (1), 330-4. (triangles)

RICE

Water Content in n-Alkanes

Emborsky, Cox, and Chapman, IECR, 50, 7791 (2011)

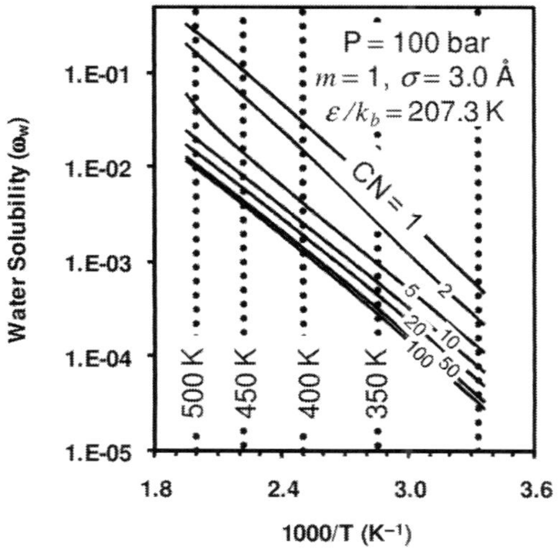

RICE

Polar Mixtures

Applications

- Phase Behavior involving Polar Solvents
- Water Soluble Polymers for Paints and Coatings
- Biochemical Separations

Need to Predict Phase Behavior in Polar Mixtures

- Single polar group per molecule (for example: ketones, ethers, aldehydes, and esters)
- Multiple polar groups (for example, polyesters, polyethers, and polar copolymers)

We have performed studies of ketones, diketones, ethers, diethers, triethers, and esters with alkanes as well as polar copolymer mixtures

RICE UNIVERSITY 17

Theory & Simulation Results

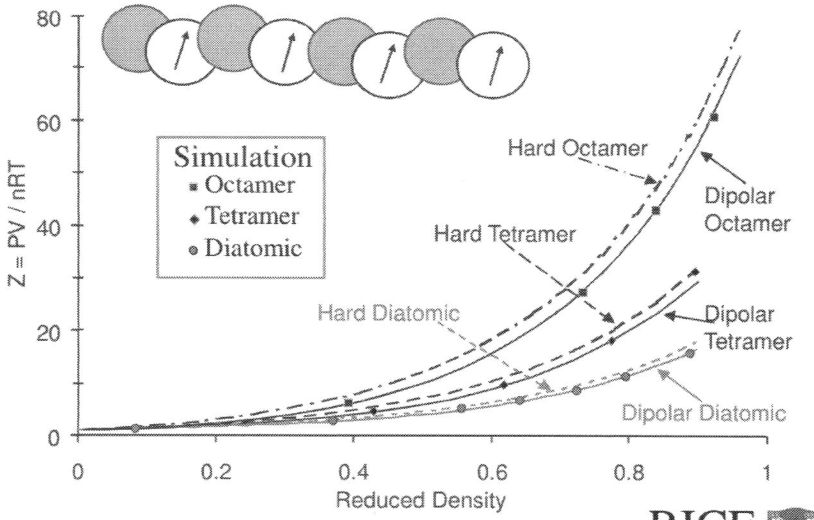

(Jog & Chapman, 1999)

RICE UNIVERSITY 18

Ethyl Acetate / Heptane

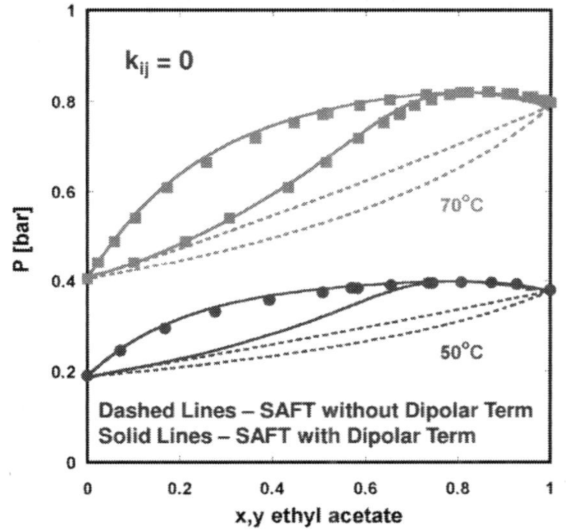

(Dominik & Chapman, 2004)

Ethyl acetate: μ = 1.8D

19

Di-Ketone/Cyclohexane

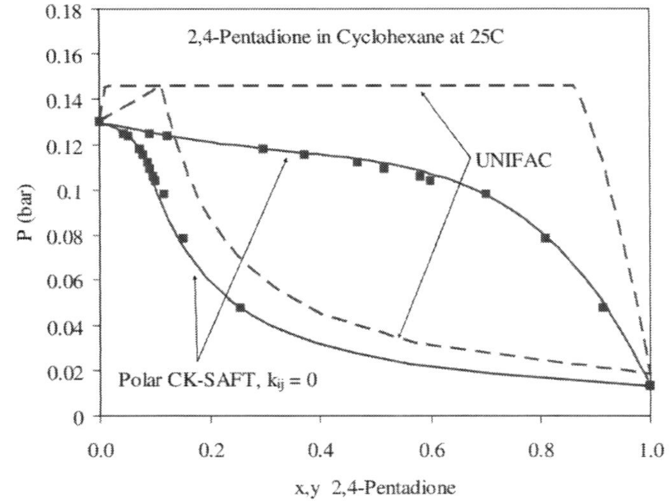

Sharon G. Sauer and Walter G. Chapman, *Ind. Eng. Chem. Res.* **2003,** *42,* 5687-5696

20

H_2S / C1, C2, and C3 Alkanes

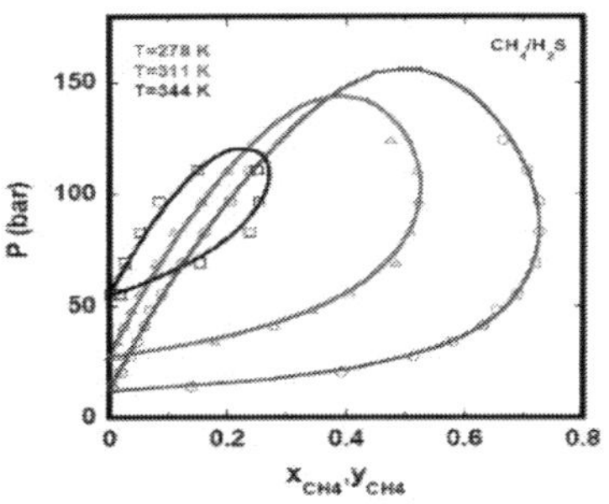

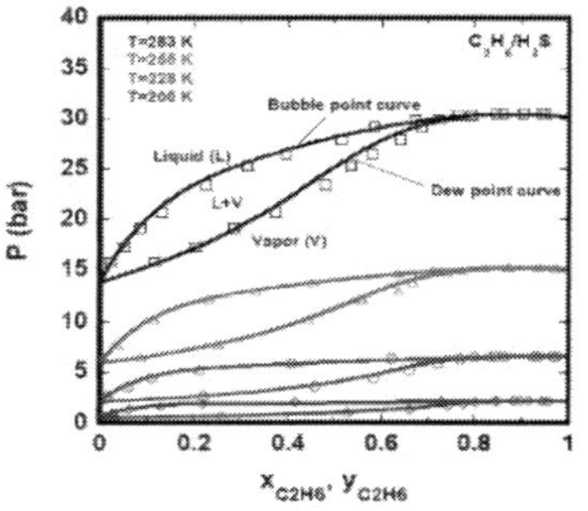

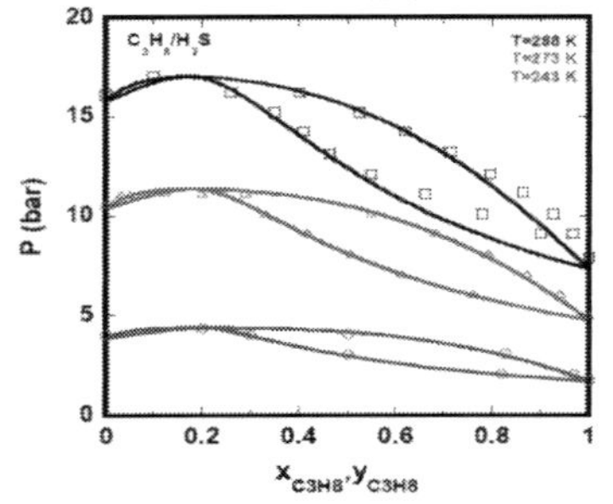

- PC-SAFT results agree well with experimental data
- Model predicts the location of the azeotrope well for C_2H_6 and C_3H_8 at all temperatures

21

Exp. Data from: Kohn, J.P., and F. Kurata, *AIChE J.* **4**, 211 (1958). Kalra et al. *J. Chem. Eng. Data*, **22**, 85 (1977). Carroll et al. *Fluid Phase Equilibria*, **81**, 187 (1992)

POLYMER SOLUTIONS

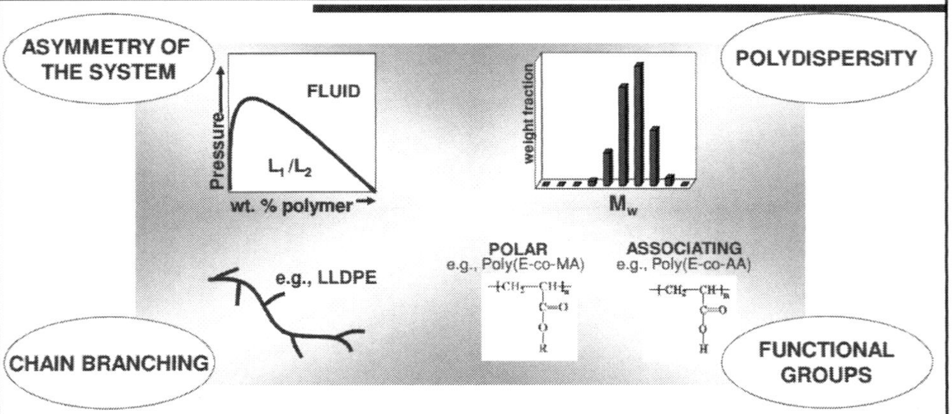

REQUIREMENTS FOR THERMODYNAMIC MODEL

BRANCH LENGTH AND DENSITY — SPECIFIC INTERACTIONS

VALIDITY OVER WHOLE RANGE OF CO-MONOMER CONTENT

POLYMER PVT PROPERTIES — TRANSFERABLE PARAMETERS

Polydisperse Polymers

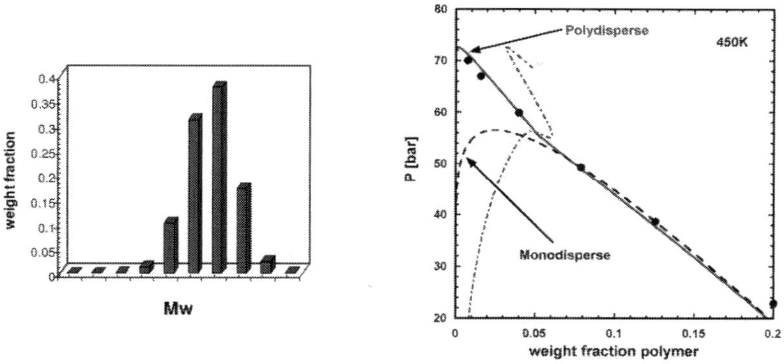

Exp. data: de Loos et al. *Fluid Phase Equilibria* 1996

Algorithms for polydisperse polymer systems nearly independent of the number of pseudo-components

P.K. Jog and W.G. Chapman *Macromolecules* **2002**, *35*, 1002-1011
N. Koak and R.A. Heidemann *AIChE Journal* **2001**, *47*, 1219-1225
S. Behme, G. Sadowski, Y.H. Song and C.C. Chen *AIChE Journal* **2003**, *49*, 258-268
A. Ghosh, P.D. Ting and W.G. Chapman *Ind. Eng. Chem. Res.* **2004**, *43*, 6222-6230

23

Consortium for Complex Fluids

Branched Polyolefins - Concept

Experimental observation
branches on the chain lower the cloud point pressure

Model behavior
Position of cloud point pressure controlled by ε_{ij} between segments

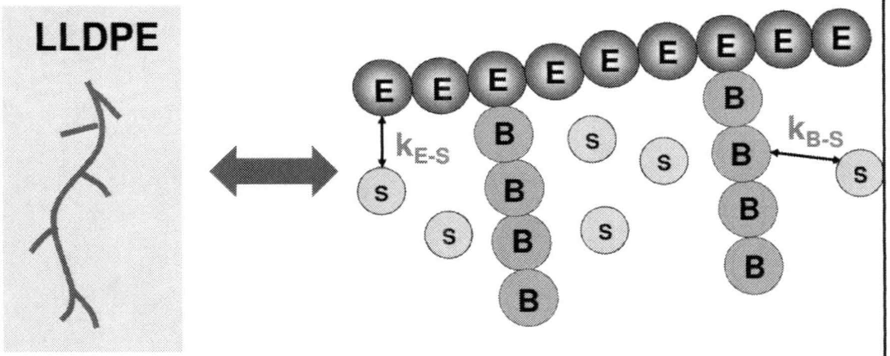

- **Two segment types: Chain and Branch**
- **Same σ and m/Mw (from HDPE data)**
- **Two energy parameters: ε_C and ε_B**

Only one pure component parameter per copolymer type

A. Dominik and W.G. Chapman, *Macromolecules* **2005**, *38*, 10836-10843

24

POLY(ETHYLENE-CO-OLEFINS)

RICE UNIVERSITY

POLY(E-CO-1-OCTENE) / ETHYLENE

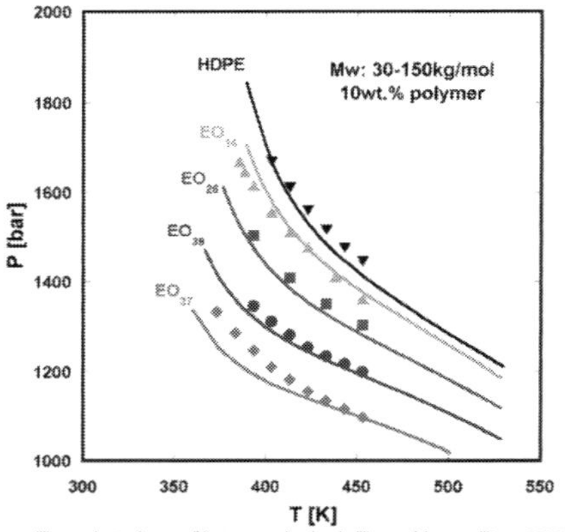

Exp. data from Chan et al., *Ind. Eng. Chem. Res.* 2000

POLY(E-CO-1-BUTENE) / PROPANE

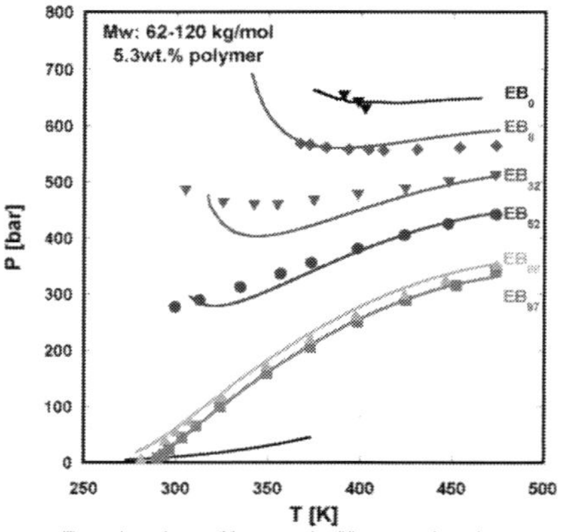

Exp. data from Chen et al., *Macromolecules* 1992

Accurate PVT properties **Valid for extrapolations**

A. Dominik and W.G. Chapman, *Macromolecules* 2005, 38, 10836-10843

Polystyrene / Cyclohexane / CO_2

RICE UNIVERSITY

Exp. data: de Loos (1994), Bungert et al. (1997)

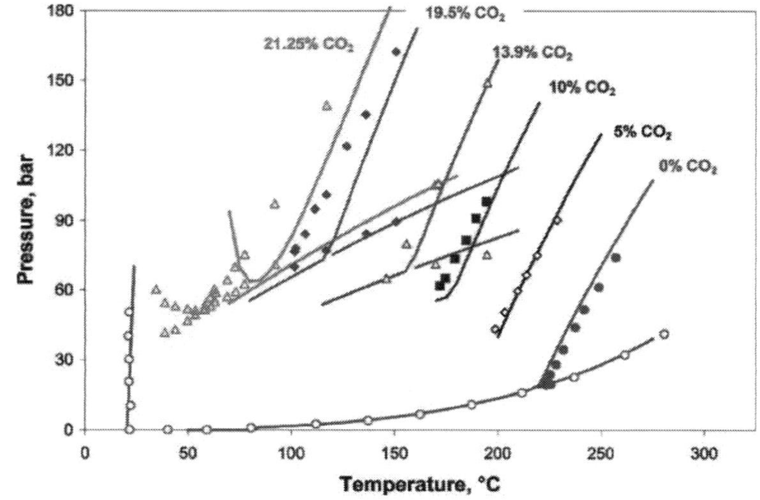

Vargas (2008) – parameters from Gross and Sadowski

Asphaltene Stability in Oil

Comparison of the experimental and predicted recombined oil asphaltene stability at reservoir temperature (71.1°C)

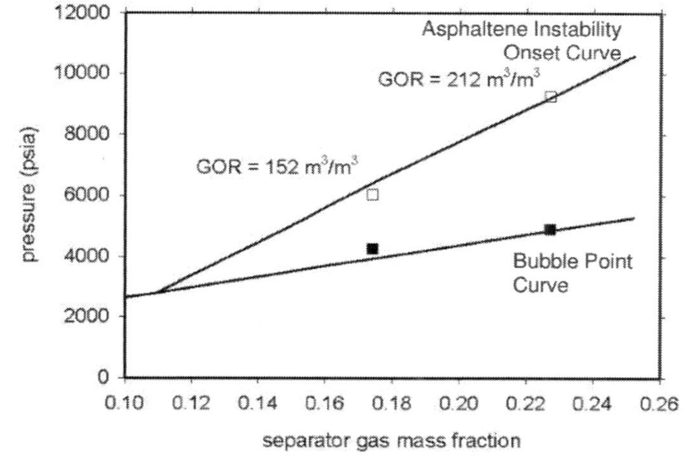

P.D. Ting, G.J. Hirasaki, and W.G. Chapman, *Petroleum Science & Technology. 21(3-4):647-661, 2003.*

Positive Feedback from Industry

RICE UNIVERSITY

- Prasanna Jog, Dow – Documented millions of dollars in cost savings from using SAFT related to process and product development
- Oliver Pfohl, Bayer – SAFT enabled polymer process modeling and development
- Eric Cheluget, Nova – Using SAFT to model polydispersity and phase boundary in the reactor
- Mac Radosz, formerly Exxon – SAFT enabled Metallocene based Polymer Plants for
  - ExxonMobil, Dow, Chevron-Phillips, Mitsubishi Chemicals, Quattor (Brazil), and Borealis

S^E Scaling of Transport Properties

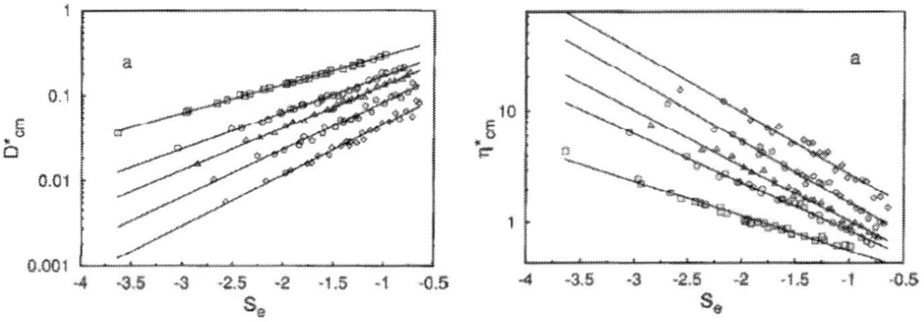

Simulation results for center of mass diffusivity and viscosity scale with SE from SAFT for LJ Chains of 1, 2, 3, 5, and 10 segments (squares through diamonds, respectively).

Goel, et al., J. Chem. Phys., 129, 164904 2008
Rosenfeld, Phys. Rev. A 15, 2545 1977; Chem. Phys. Lett. 48, 467 1977; J. Phys.; Condens. Matter 11, 5415 1999.

RICE UNIVERSITY 29

The Friction Theory (f-theory)

S.E. Quiñones-Cisneros et al, Energy & Fuels 2005, *19*, 1314-1318

$$\tau = \tau_0 + \tau_f \Rightarrow \eta = \eta_0 + \eta_f$$

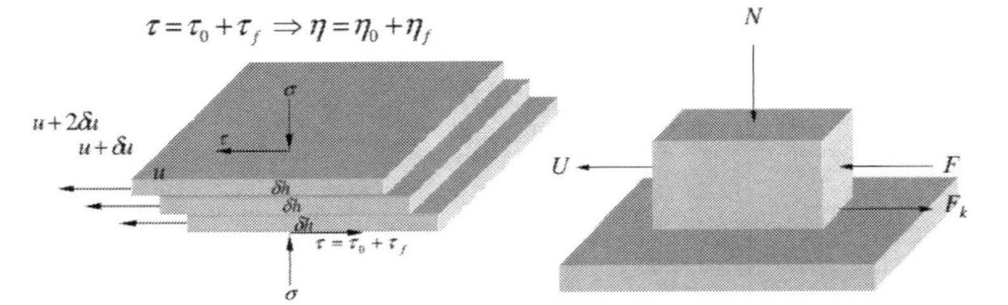

Van der Waals: $\sigma = p = p_r + p_a$ *Amontons-Coulomb:* $F_k = \mu_k N$

$$\tau_f = \tau_r + \tau_a$$

$$\tau_r = \sum_{i=1} \mu_{r,i}\, p_r^i \qquad \tau_a = \sum_{i=1} \mu_{a,i}\, p_a^i$$

 30

F theory + COR

S.E. Quiñones-Cisneros et al, Energy & Fuels 2005, *19*, 1314-1318

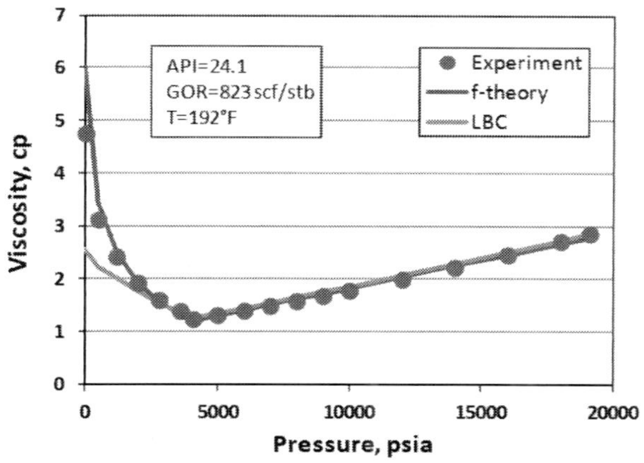

Example Applications

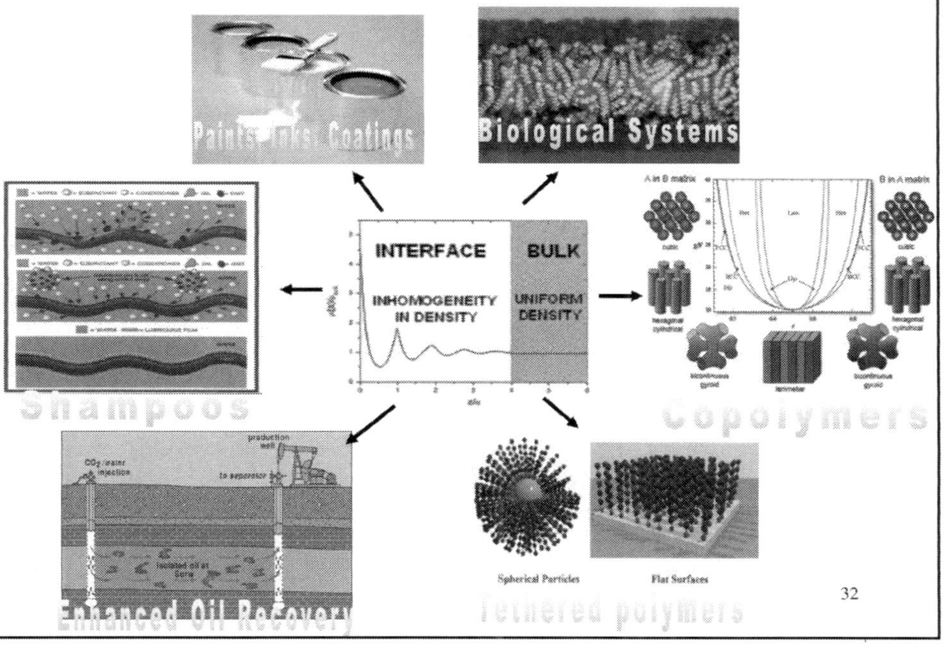

Double-tail Surfactants in Water

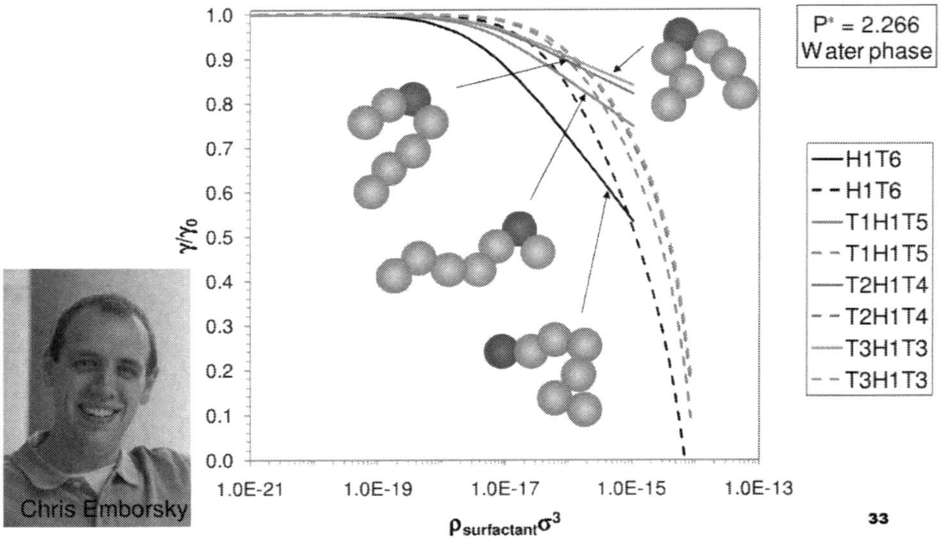

33

Particles with Grafted Chains in Polymer Melt or Solutions

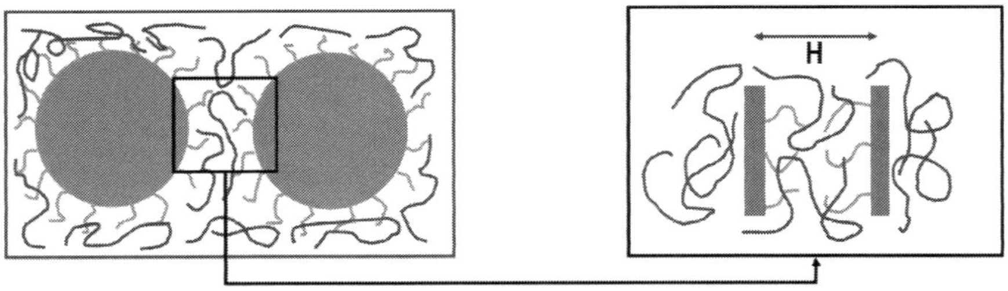

- Particles dispersed in polymer melt or solution can experience polymer-induced attraction ("depletion attraction")
  - Examples include nanoclays in polymer melts, latex particles in polymer solutions, crystallites in rubbery polymers, etc.
- Grafted chains are used to provide steric stabilization against particle aggregation
- We want to study how the effective force between the particles depend on the grafting density, grafted chain length, and free polymer chain length

Page 34

Results – Phase Diagram

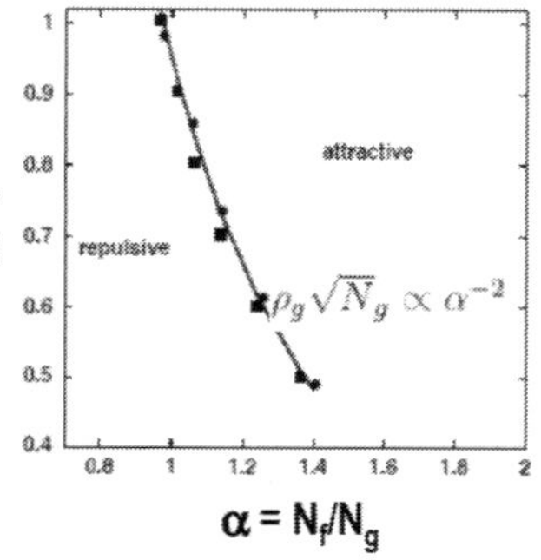

Two sets of iSAFT results: N_g = 101 (squares) and N_g = 151 (circles). Various grafting densities and free polymer chain lengths correspond to various points. Free polymer bulk density φ^b=0.6.

- When free chains are much longer than grafted chains, there is an attractive free energy minimum corresponding to the equilibrium distance between the two particles ("attractive" region)
- The boundary between "attractive" and "repulsive" region is consistent with the scaling law proposed by Leibler *et al.* for the case of particles in polymer solutions (athermal or good solvent)

Summary

SAFT accounts for association, polar interactions, polydispersity, and size/shape effects on bulk phase behavior

Groups are using SAFT to predict the best solvent / carrier for a process

- Slurry polymerization – Ghosh and Chapman, IECR 41, 5529 (2002).
- Solvent design for CO2 capture – Pereira, et al, Comp & Chem Eng., 35, 474 (2011) and Oyarzun, et al., Energy Proc., 4, 282 (2011).

Transport properties are being calculated from SAFT

Interfacial properties and molecular structure can be calculated – Effects of molecular size, hydrophobic, and hydrophilic effects

iSAFT – Offers a platform to simultaneously model bulk and interfacial properties (as well as self-assembly) of polyatomic fluids

Sandia National Lab is implementing iSAFT in the open source Tramonto DFT package

Acknowledgments

The Robert A. Welch Foundation

DeepStar

NSF – CBET-0756166

DOW

Chevron

THE PETROLEUM INSTITUTE

VLXE

POLYMER & COMPLEX FLUIDS

Schlumberger

RESEARCH & DEVELOPMENT

OIL SUB-COMMITTEE

Infinite Dilution Revisited: Filling the Predictive Gaps

Kenneth R. Cox

Department of Chemical and Biomolecular Engineering

Rice University, Houston, Texas USA

7 October 2011

Student View of Design

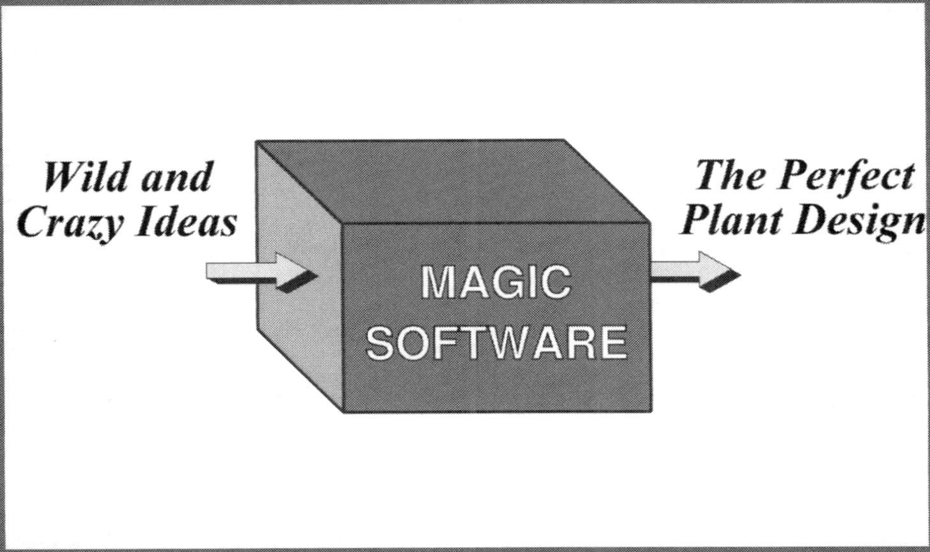

Design Decisions – Heuristics

- Optimum reflux is 1.3-1.5 times minimum reflux
- Column stages is twice minimum stages
- Pressure drop is 0.1 psi per tray
- Column vapor velocity is 2 ft/s at moderate pressure
- 1 pound of steam per gallon of reboiled liquid
- Optimum Kremser absorption factor is 1.25 to 2.0

Where Insight is Needed

- Users are not Thermo Experts – Need Guidelines, not Rigor
- Unreliable Parameters – When to be Skeptical
  - Lack of Relevant Data
  - Inaccurate Data
  - Inappropriate Fitting Algorithm
- Proliferation of Methods – Where do we Start?
- No Suitable Method Available – New Tools may be Needed
- Quick Calculations Outside of Process Simulators

Heuristic View for Phase Equilibria

- Scatchard-Hildebrand Theory
  - Cohesive Energy Density
  - Mixture Properties from Pure Component Properties
  - Solubility Parameters Useful for Solvent Selection
- Solution of Groups
  - Homologous Series Correlations
  - Pierotti → ASOG → UNIFAC
  - Leveraging available data by functional group additivity
- Infinite Dilution Activity Coefficients
  - Best Application of Molecular Insight
  - Most Critical Limit for Distillation Design

Student Design Challenge

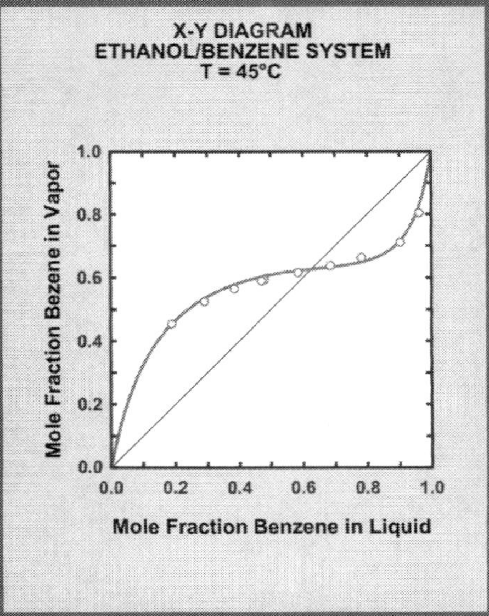

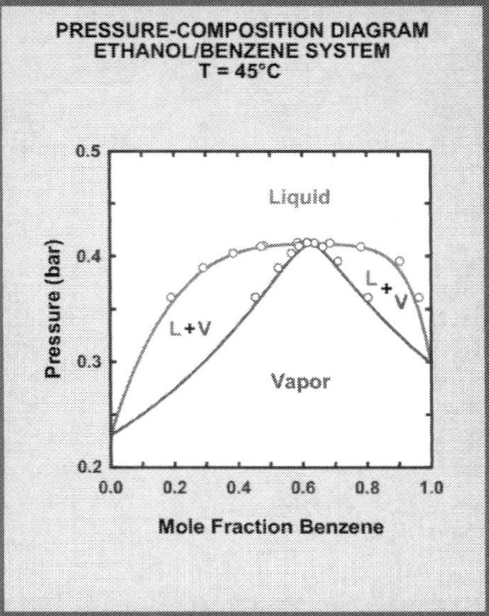

Exploitation of Hydrogen Bonding

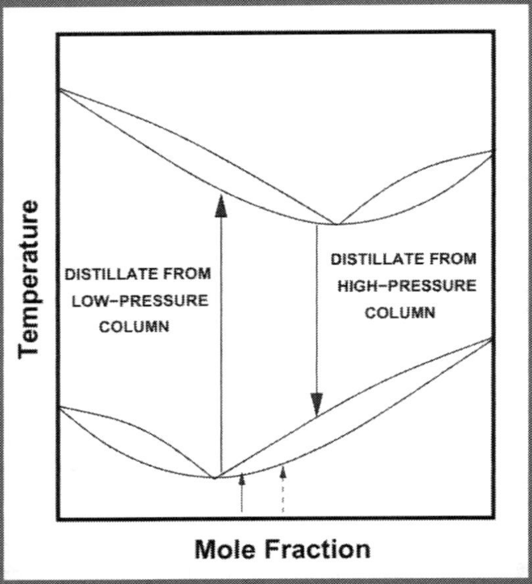

Pressure-Swing Distillation

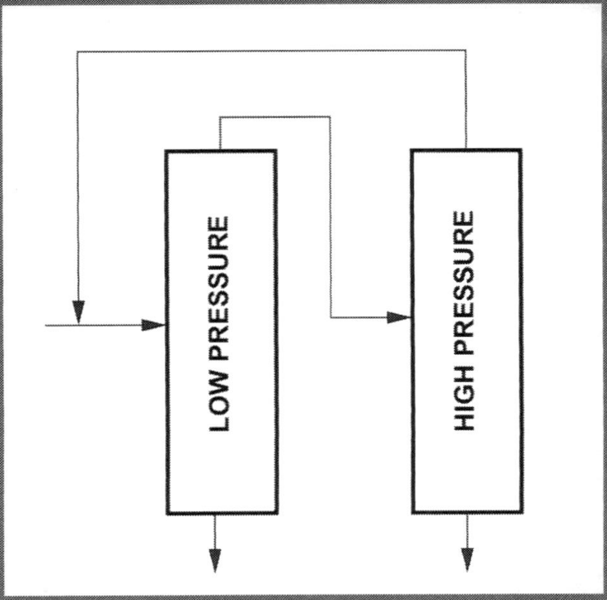

The Need for Heuristics

**TOUGH DECISIONS VS.
DEPTH OF UNDERSTANDING
1980 - PRESENT**

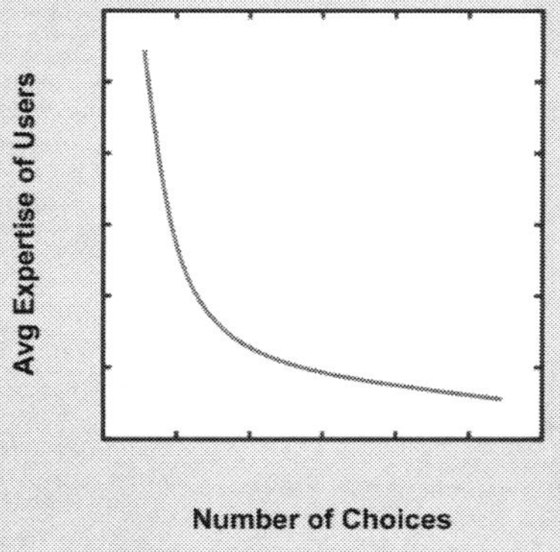

EOS MODELS	GE MODELS
Peng Robinson	NRTL
SRK	Margules
Kabadi Danner	UNIQUAC
Lee Kesler	van Laar
Zukevitch-Joffee	Wilson
BWR-Lee Starling	Bromley-Pitzer
PR-MHV2	Electrolyte NRTL
PR-Wong Sandler	Redlich-Kister
Nothnagel	UNIFAC
Ideal Gas	Ideal Liquid
Hayden-O'Connell	SH-RST
NBS/NRC Steam	Chien-Null
Boston-Matthias	Pitzer

Many Confusing Choices!

Activity Coefficient Models

- Workhorse for "chemical" systems including polar interactions
- Capable of fitting associating sytems where data is available
- Parameters can be predicted using solution of groups approach
- Temperature range very Limited by range of data
- Different LLE/VLE parameter sets
- Not valid for compressible phases
- Vapor phase correction by EOS
- Difficult to use for systems with supercritical components

Model View of Associated Fluid

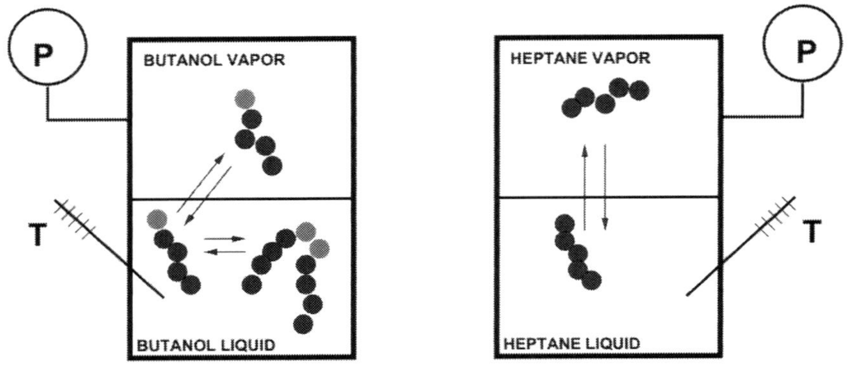

In the associating pure fluid, only the monomer is considered volatile

Basis of Relative Volatilities

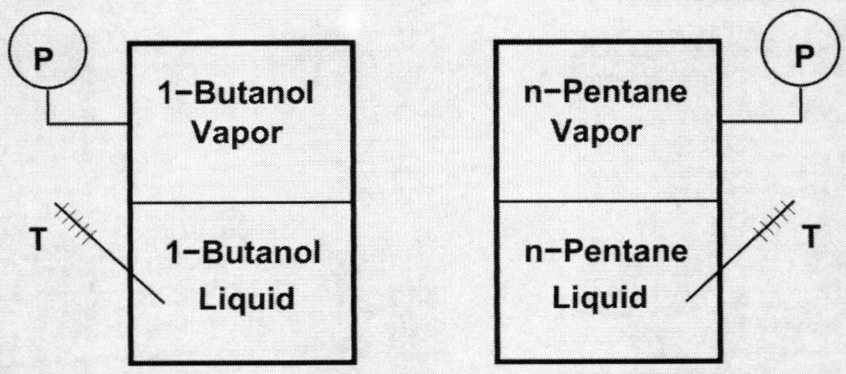

T = 25 C	T = 25 C
P = 6.5097 torr	P = 512.6 torr
N/V = 0.3501 mol/m^3	N/V = 28.53 mol/m^3

Relative Volatilities in Mixtures

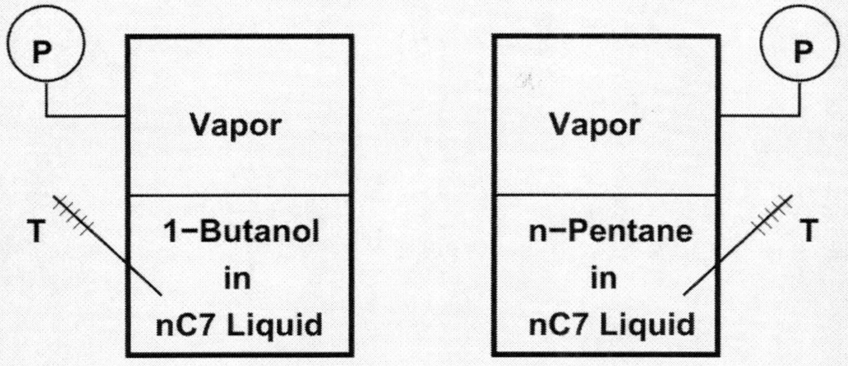

P = 760 torr
X = 0.0001
P BuOH = 0.236 torr
N/V = 0.0102 mol/m^3

P = 760 torr
X = 0.0001
P C5 = 0.286 torr
N/V = 0.0129 mol/m^3

Equations for Real Solutions

$$f_i^l = x_i \gamma_i P_i^0$$

$$f_i^v = y_i \phi_i P$$

$$k_i = \frac{y_i}{x_i} = \frac{\gamma_i P_i^0}{\phi_i P}$$

The above equations apply only to systems near atmospheric pressure. At higher pressures, other corrections need to be applied to describe the liquid phase fugacity.

Estimations based on Analog Volatilities

PURE C4OH

OH---OH---OH---
R R R

APPROXIMATION

$$P_{C4OH} = x_{C4OH}\gamma^{\infty}_{C4OH}P^0_{C4OH}$$

$$\approx x_{C4OH}P^0_{nC5}$$

$$\gamma^{\infty}_{C4OH} \approx \frac{P^0_{nC5}}{P^0_{C4OH}}$$

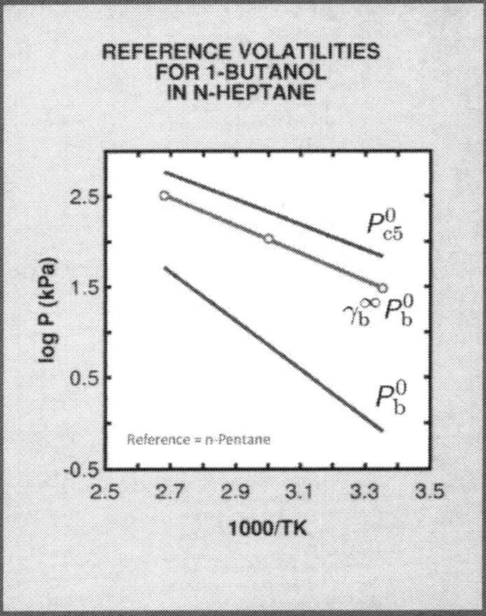

n-Pentane as Analog

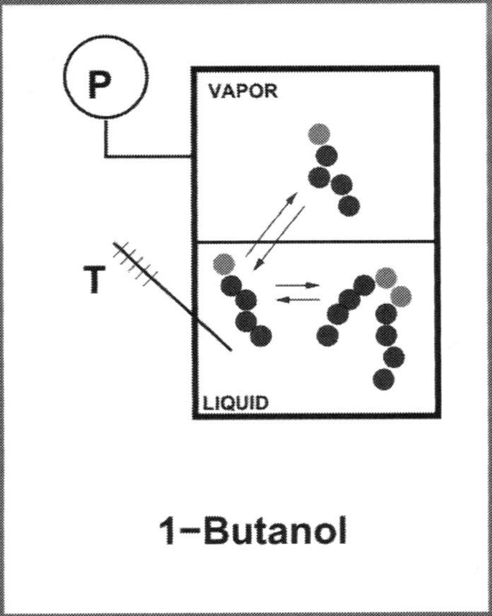

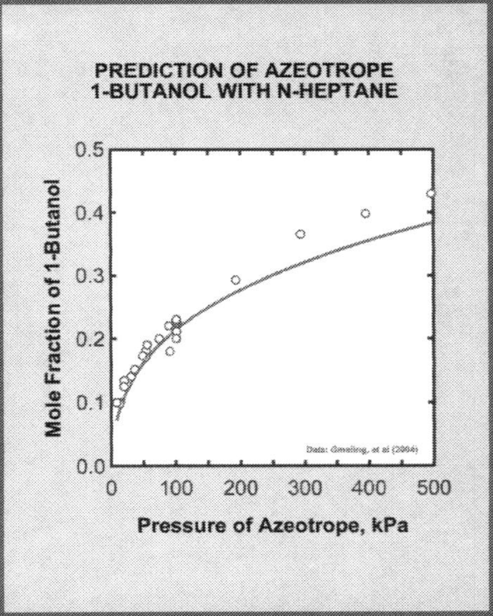

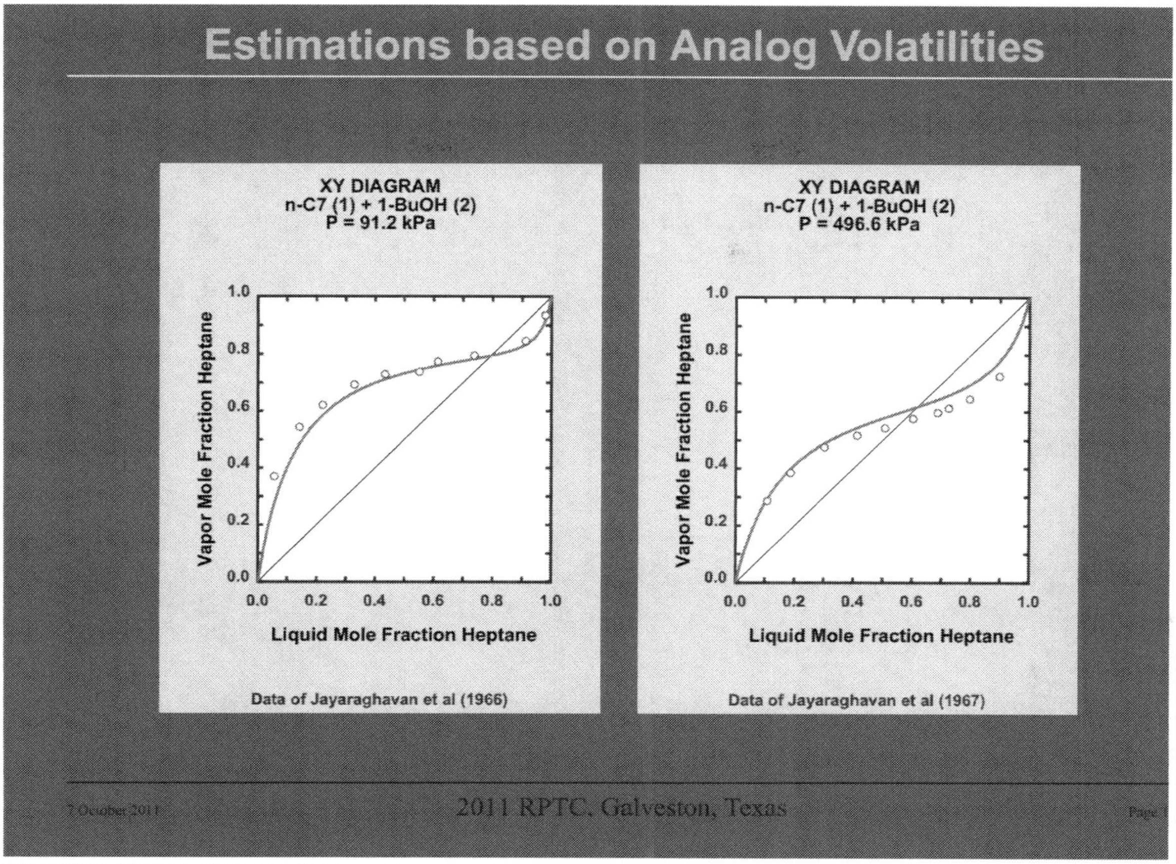

Estimations based on Analog Volatilities
XY DIAGRAM
n-C7 (1) + 1-BuOH (2)
P = 91.2 kPa
Vapor Mole Fraction Heptane
Liquid Mole Fraction Heptane
0.0
0.2
0.4
0.6
0.8
1.0
Data of Jayaraghavan et al (1966)
XY DIAGRAM
n-C7 (1) + 1-BuOH (2)
P = 496.6 kPa
Vapor Mole Fraction Heptane
Liquid Mole Fraction Heptane
Data of Jayaraghavan et al (1967)
7 October 2011
2011 RPTC, Galveston, Texas

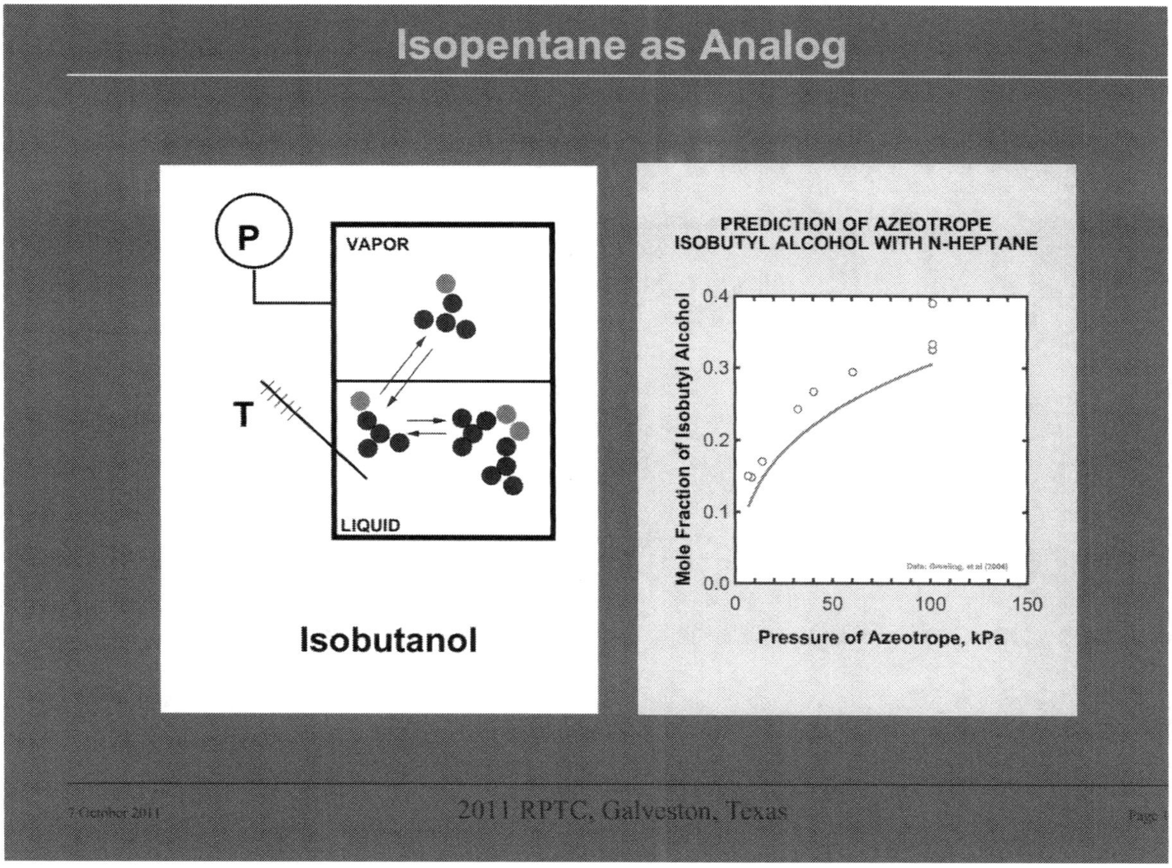

Isopentane as Analog
P
VAPOR
T
LIQUID
Isobutanol
PREDICTION OF AZEOTROPE
ISOBUTYL ALCOHOL WITH N-HEPTANE
Mole Fraction of Isobutyl Alcohol
0.0
0.1
0.2
0.3
0.4
0
50
100
150
Pressure of Azeotrope, kPa
7 October 2011
2011 RPTC, Galveston, Texas

Isopentane as Analog

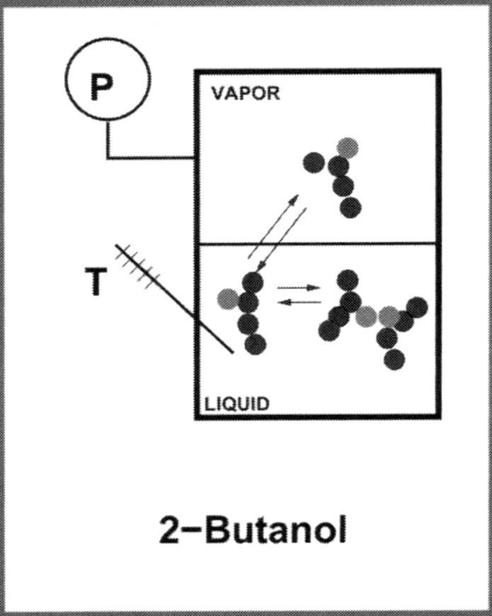

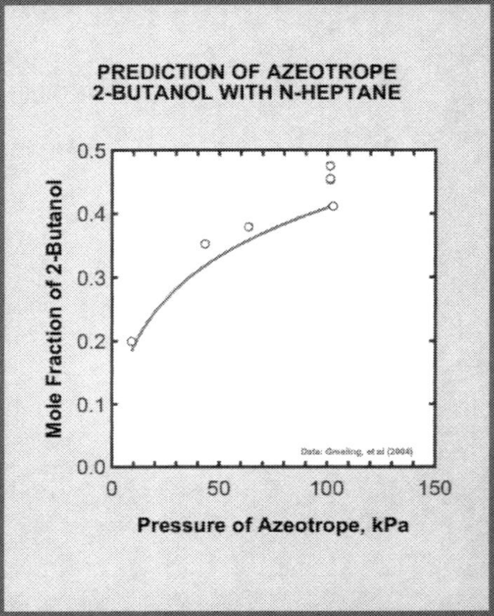

Neopentane as Analog

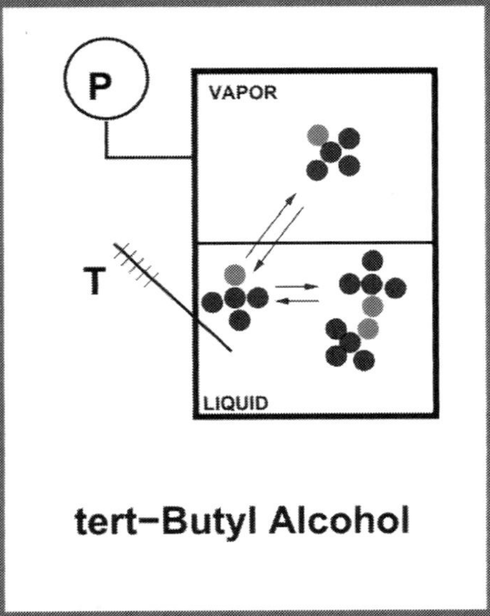

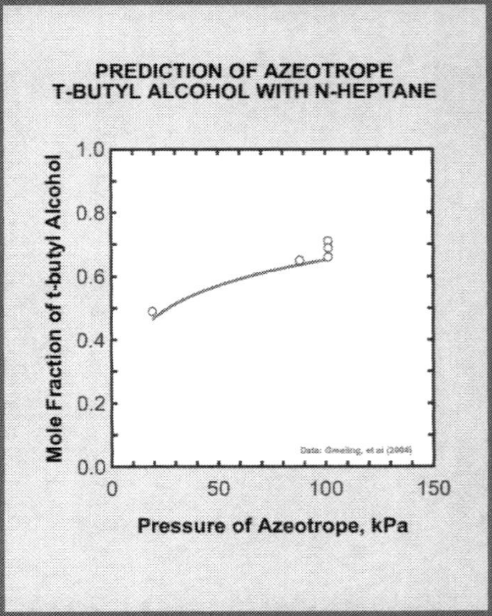

Comparison with Simulator Correlation

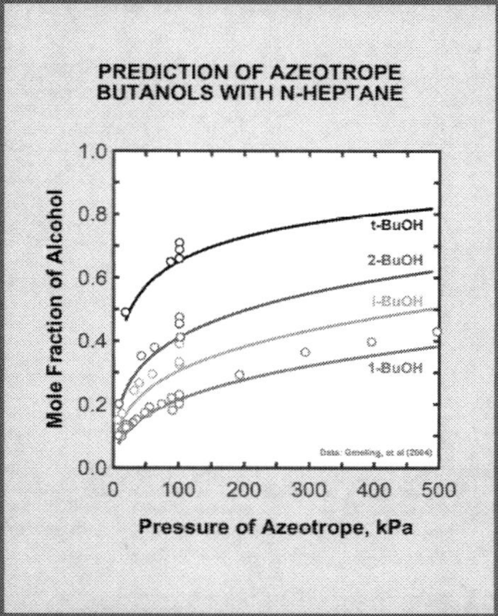

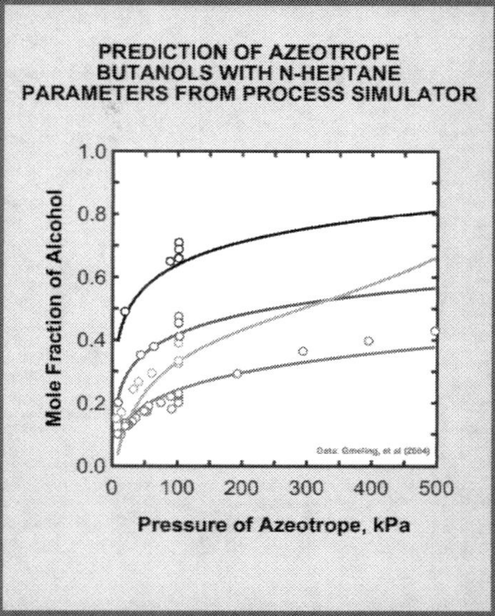

n-Hexane as Analog

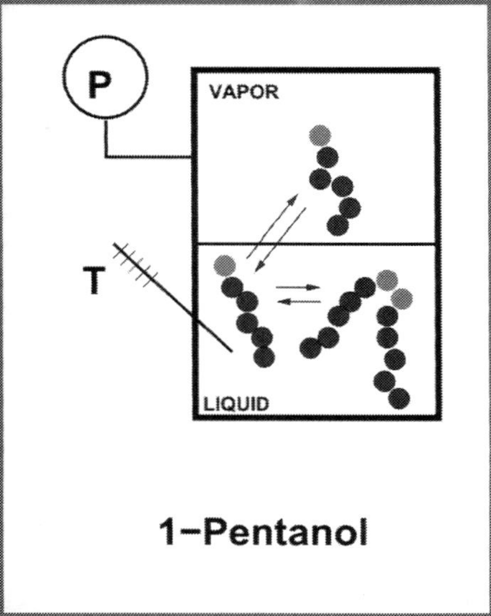

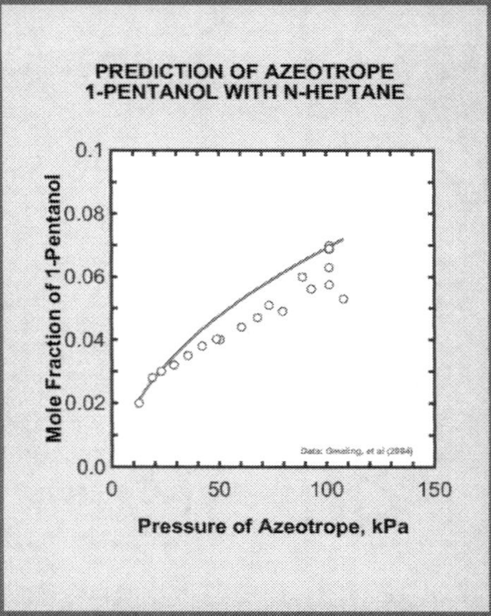

Neopentane as Analog

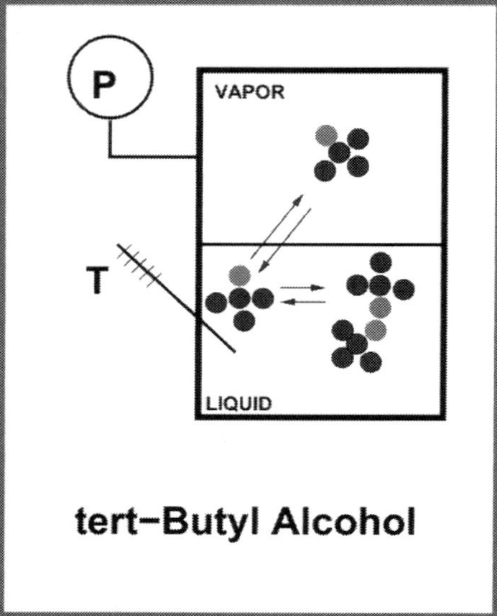

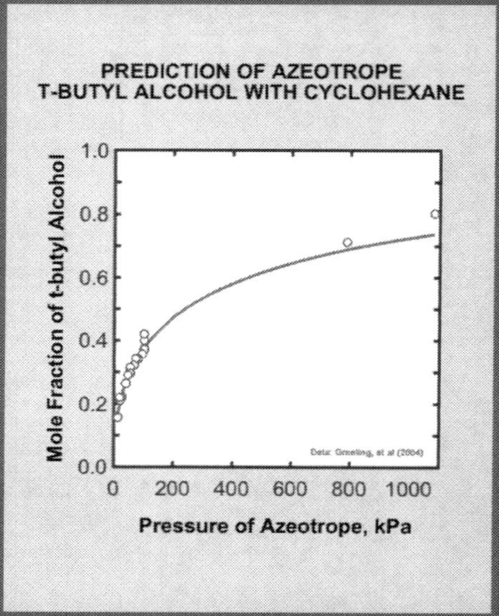

A Distillation Mystery

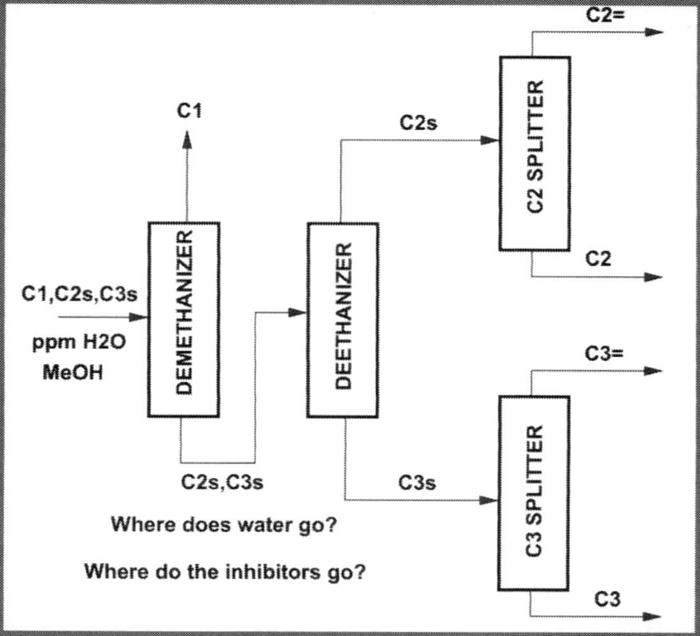

Component Properties

Compound	T_b/K	T_c/K	P_b/bar
Methane	111.66	190.56	45.99
Ethane	184.55	305.32	48.72
Propane	231.02	369.83	42.48
Methanol	337.11	470.00	72.30
Water	373.15	647.14	220.64

A Distillation Mystery

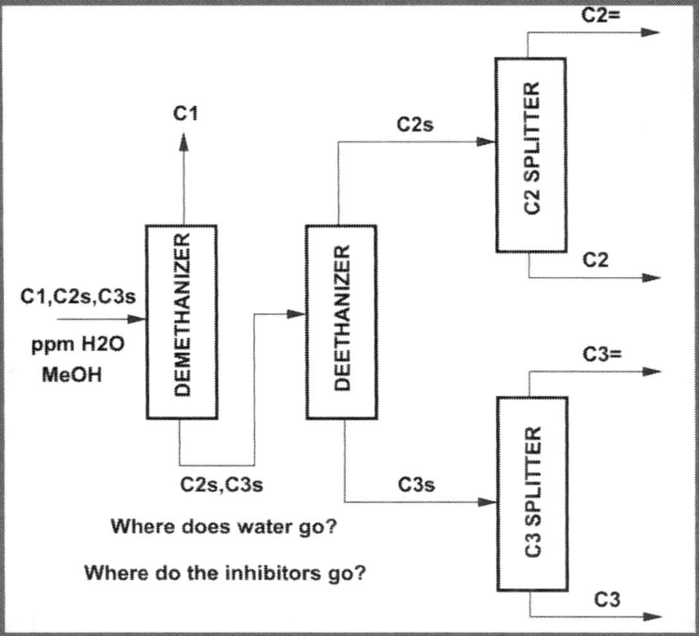

Reason for Interest

- Concerns about Water Entrapment
  - Corrosion
  - Slugs of Fluid – "Hiccups"
  - Hydrates Formation
- Unexpected Methanol Volatility – Product Contamination
- Observations/Causes
  - Both Methanol and Water Much More Volatile than Expected
  - Water and Methanol "Repelled" by Hydrocarbon Phase
  - "Oil and Water Don't Mix" – Hydrophobic Effect
  - Strong Temperature Effect of K-factor Reported in Literature – Temperature Gradient in Column
  - Should Use Ice Reference for Calculations with Water

These causes are very useful as mnemonics, but they can be misleading.

7 October 2011 2011 RPTC, Galveston, Texas Page 2

Analog for Methanol?

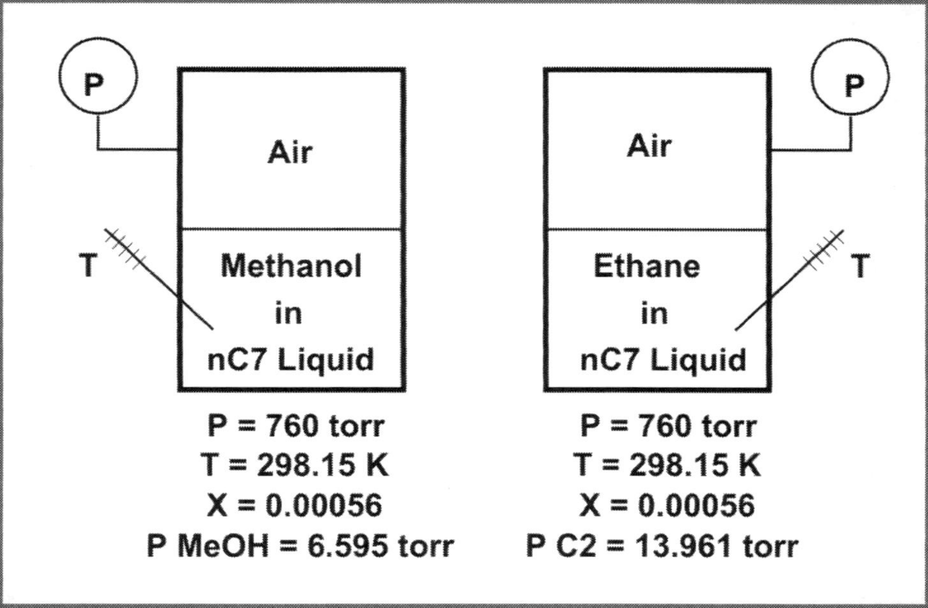

7 October 2011 2011 RPTC, Galveston, Texas Page 2

Analog for Methanol?

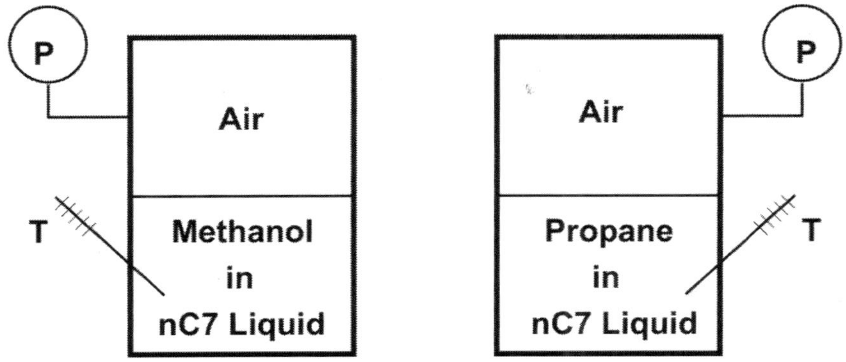

P = 760 torr	P = 760 torr
T = 298.15 K	T = 298.15 K
X = 0.00056	X = 0.00056
P MeOH = 6.595 torr	P C3 = 3.751 torr

What Reference for Water?

$$\frac{f^0_{water}}{P_{c,analog}} = \text{function of } \frac{T}{T_{c,analog}}$$

$$f_{water} = x_w f^0_w \exp\left[\frac{v_1(\delta_2 - \delta_1)^2}{RT}\right]$$

"1" refers to the analog

"2" refers to the solvent

$$\frac{1}{x_w} = \frac{f^0_w \exp\left[\frac{v_1(\delta_2-\delta_1)^2}{RT}\right]}{P^0_{water}}$$

REFERENCE FUGACITIES
OF LIGHT COMPONENTS
METHOD OF SHAIR AND PRAUSNITZ

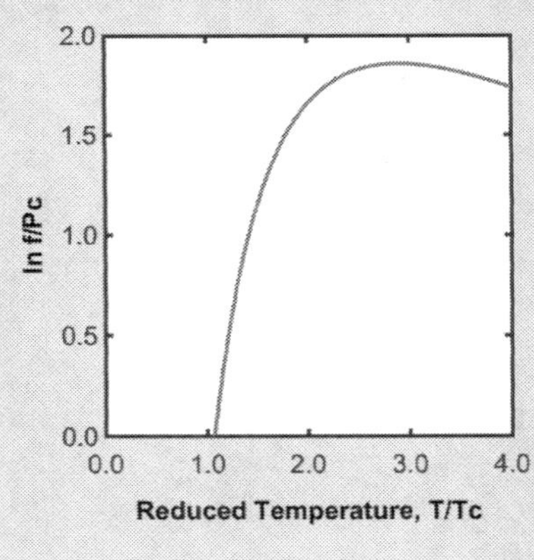

Analog Behavior for Water

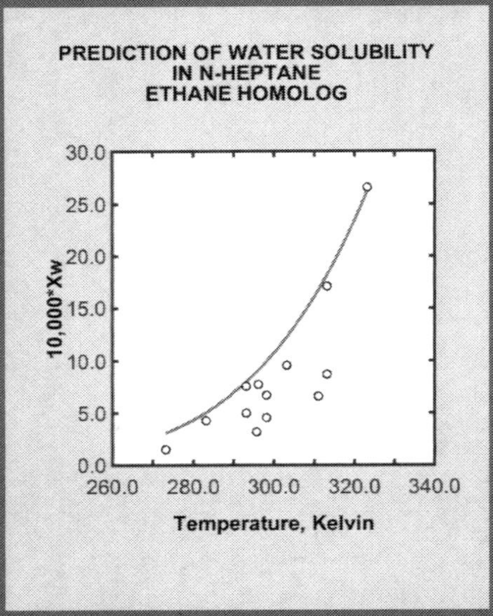

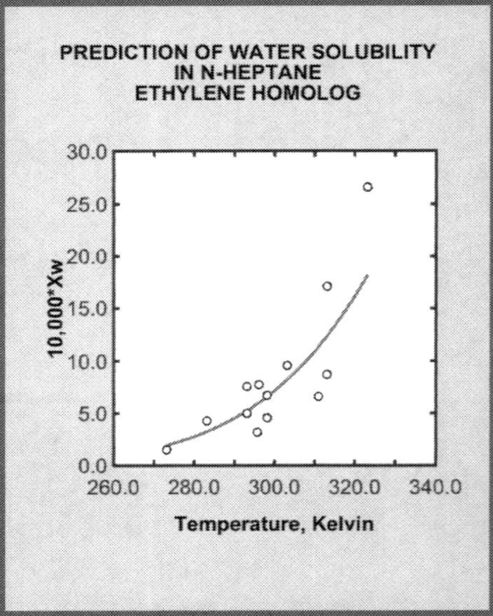

Observations

- At infinite dilution in hydrocarbons, "hydrophobic" interaction description is misleading
- In n-heptane at 25°C, water behaves like a hydrocarbon with two carbon atoms
- In n-heptane at 25°C, methanol behaves like an alkane with 2.57 carbon atoms
- Can we apply these results at temperatures other than 25°C?

Observations Reported in Literature

- Gleich 1975 – US Patent – C2 Splitter operation to prevent hydrate formtion
- Kister and Hower 1987 – Water slugs from refluxed gas plant deethanizer
- Saletan 1994 – Creative Troubleshooting Book – Explanation of water entrainment in C2 splitters
- Kister 2006 – Distillation Troubleshooting Book – Methanol overhead in C3 Splitter

Typical Gamma Infinity Behavior

Compound	γ_{hc}^{∞}	γ_{water}^{∞}	Ratio
n-Decane	5.26×10^8	1.67×10^3	3.16×10^5
n-Octane	9.09×10^6	1.67×10^3	5.45×10^3
n-Heptane	2.33×10^6	1.79×10^3	1.30×10^3
n-Hexane	4.35×10^5	2.13×10^3	2.04×10^2

Calculated from mutual solubility data (IUPAC)

Hydrogen Bonding and Distillation

- Responsible for surprising trace component volatility and entrapment behavior
- Difficult to model infinite dilution behavior can lead to mis-sized columns
- Most models incorrectly handle temperature dependence
- Opportunity to take advantage of temperature dependence of relative volatilities

Water in Methane

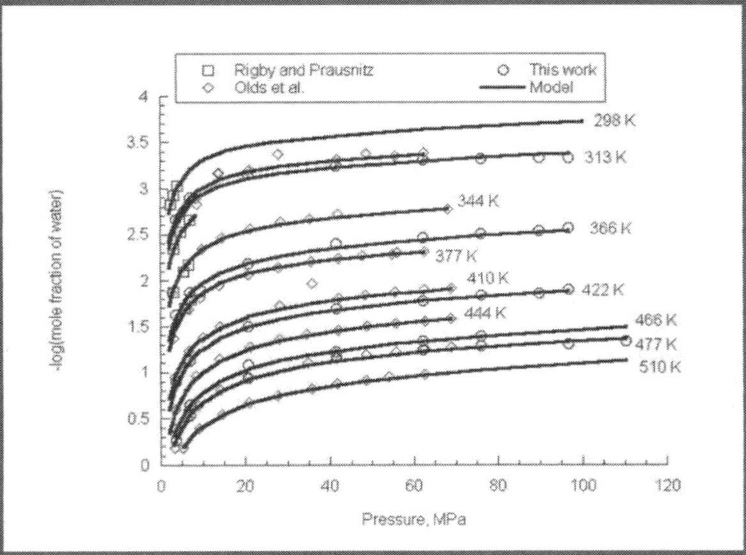

Yarrison, Cox, and Chapman 2006

Water in Ethane

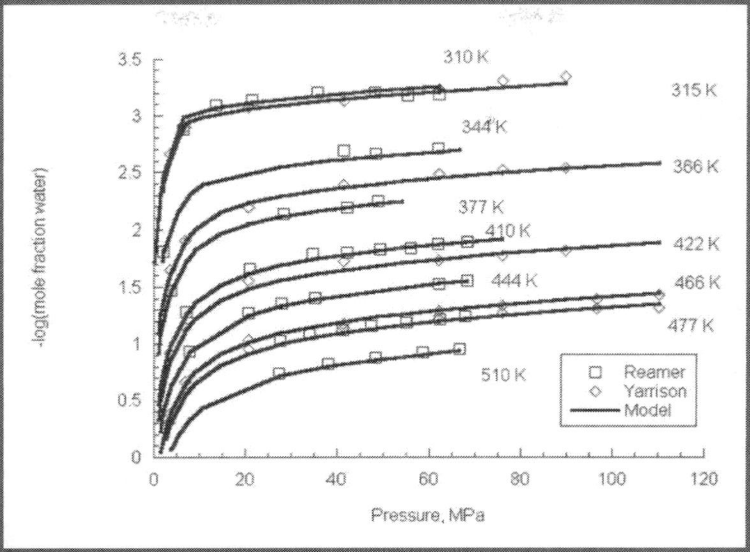

Yarrison, Cox, and Chapman 2006

Water in Methane Comparisons

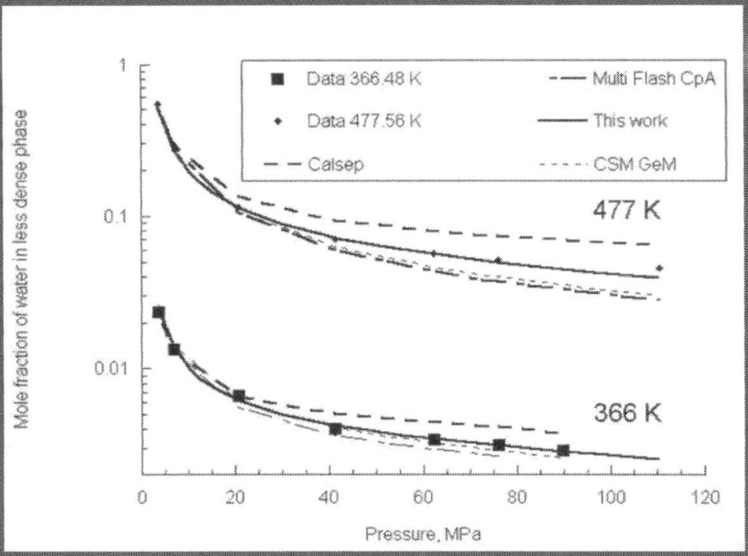

Yarrison, Cox, and Chapman 2006

Water in Methane Comparisons

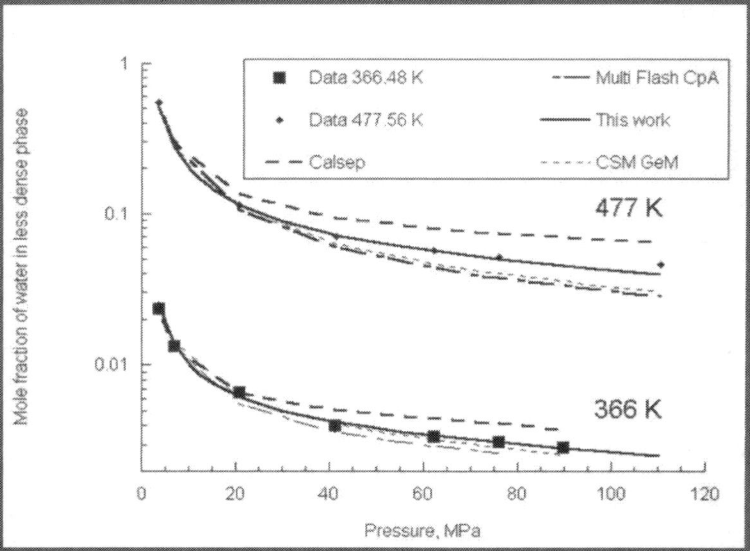

Yarrison, Cox, and Chapman 2006

Observations from Experimental Studies

- The model parameters used for fitting data confirms alkane-like behavior of water
- A lack of data for water solubility in olefins limits the accuracy of the models
- The water content in olefins is signicantly higher than the corresponding alkanes
- The volatility of water in olefins is lower than the corresponding alkanes

Important Lessons Learned

- Hydrocarbon-like behavior of hydrogen-bonding molecules at infinite dilution
- Discussions of “hydrophobic” behavior can be misleading
- Insufficient data available for olefins to develop accurate parameters
- Asymmetry in infinite dilution activities makes modeling difficult and leads to errors in parameter fitting
- More development and better parameters needed for associated fluid models in the process simulators

7 October 2011 2011 RPTC, Galveston, Texas Page 4

Prediction of Mineral Scaling Using a Comprehensive Thermodynamic Model

A. Anderko, P. Wang, M.M. Lencka, R.D. Springer and J.J. Kosinski

OLI Systems Inc.

aanderko@olisystems.com

Objective

- Develop an accurate tool for predicting scale formation as a function of
  - Production conditions
    - Composition of reservoir water
    - Temperature and pressure conditions
    - Presence of sour gases
    - Solids introduced during drilling, corrosion products, etc.
  - Presence of additional components (methanol, glycols)
- Develop methodology for predicting the interactions of minerals with CO_2 in environments related to CO_2 sequestration

2

Scope

- Outline of the thermodynamic model
- Modeling the behavior of Na – K – Mg – Ca – Ba – Fe – Cl – SO_4 – CO_3 – S – H_2O – methanol – glycol systems
- Verification of the model for scale formation in aqueous solutions
- Application of the model to predict mineral transformations in CO_2 - rich systems

Thermophysical Framework

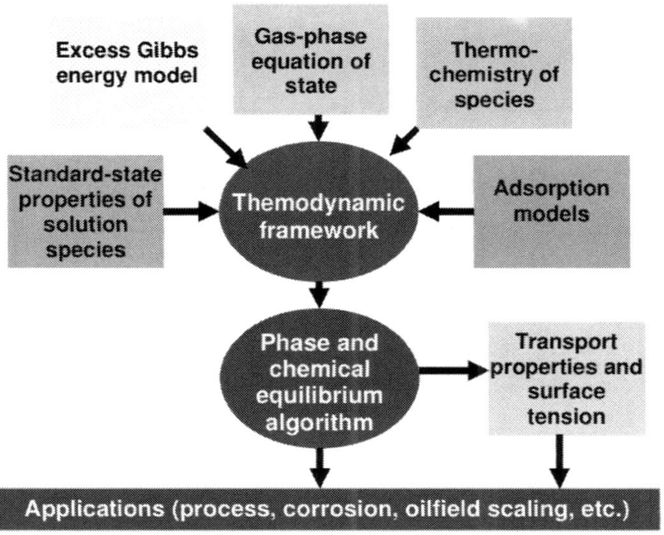

Mixed-solvent electrolyte model

- **Excess Gibbs energy** $$\frac{G^{ex}}{RT} = \frac{G^{ex}_{LR}}{RT} + \frac{G^{ex}_{LC}}{RT} + \frac{G^{ex}_{II}}{RT}$$

LR Long-range electrostatic interactions

LC Local composition term for neutral molecule interactions

II Ionic interaction term for specific ion-ion and ion-molecule interactions

- **Standard-state properties**
  - Helgeson-Kirkham-Flowers equation for ionic and neutral aqueous species
  - Standard thermochemistry for solid and gas species

Mixed-solvent electrolyte model
Chemical equilibrium calculations

Simultaneous treatment of phase and chemical equilibria (acid-base reactions, hydrolysis of cations, complexation, etc.)

$$\sum \nu_i A_i = 0$$

Standard-state chemical potential of i

At equilibrium

$$-\frac{\Delta G^0}{RT} = \ln \prod (x_i \gamma_i)^{\nu_i} \quad \text{with} \quad \Delta G^0 = \sum_i \nu_i \mu_i^0$$

- Applicable to electrolytes from infinite dilution to the fused-salt limit

Strategy for developing model parameters

- Step 1: Establishing parameters for binary subsystems
  - Binary combinations of cations: Na^+, K^+, Mg^{2+}, Ca^{2+}, Ba^{2+}, Fe^{2+} and anions: Cl^-, SO_4^{2-}, CO_3^{2-}, S^{2-}
  - Obtained parameters:
    - Interaction parameters between cations and anions
    - Thermochemical properties of hydrated salts
    - Thermochemical properties of selected ion pairs
- Step 2: Establishing parameters for available ternary subsystems
  - Obtained parameters:
    - Thermochemical properties of double salts
    - Cation-cation and anion-anion interaction parameters
    - Interactions between ions and some neutral molecules
- Step 3: Verification for multicomponent systems

Binary systems: $CaSO_4 - H_2O$

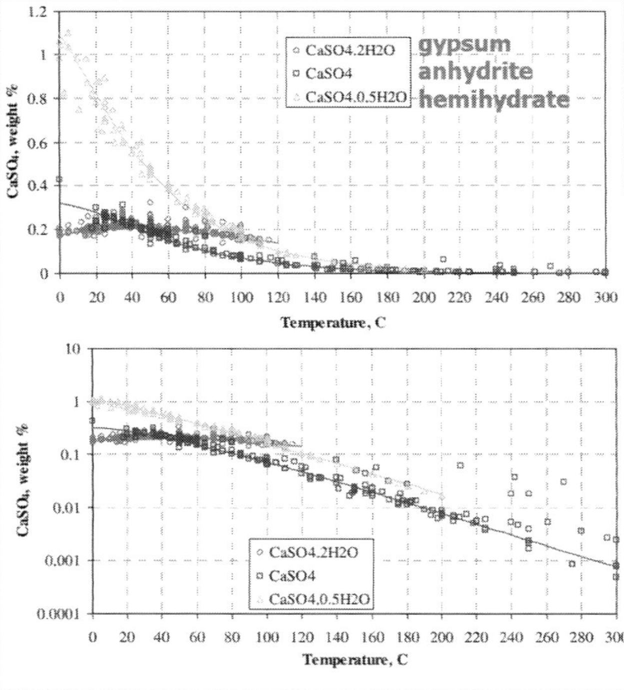

- Metastable phases are important
- The lowest solubility corresponds to the stable phase
- Solubilities of metastable phases can be obtained over wide ranges of temperature
- Metastable phases persist in multicomponent systems

Binary systems: $CaSO_4 - H_2O$

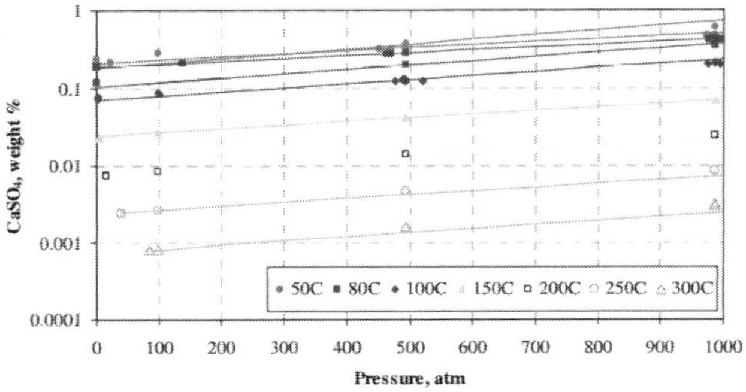

- Pressure effect on solubility is not negligible but rather small
- It is handled entirely by the pressure dependence of standard-state properties of ions – it is not a chemical effect

Ternaries: $CaSO_4$ in $CaCl_2$ and $MgSO_4$ solutions

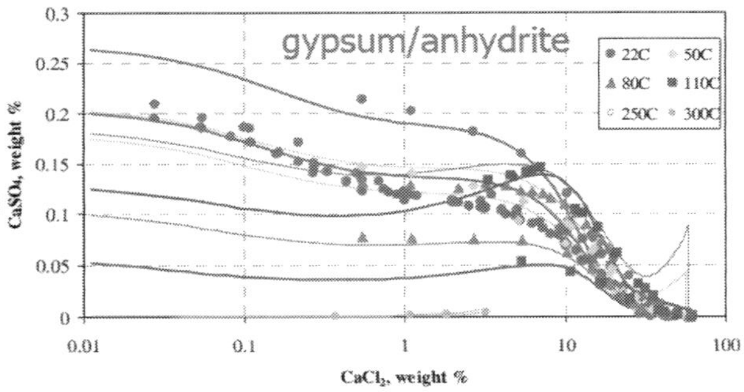

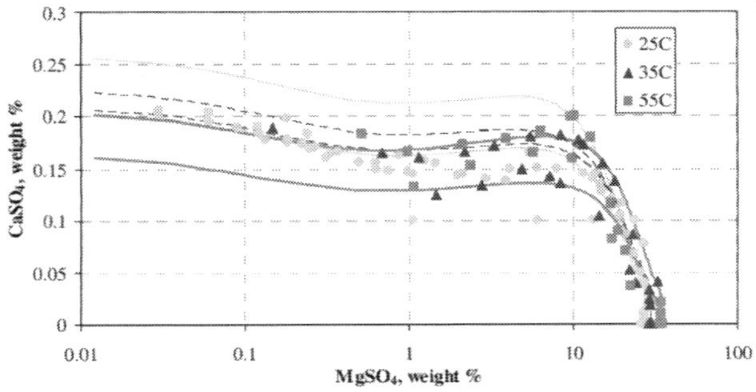

- Common ion effect (Ca^{2+} or SO_4^{2-}) causes a drop in solubility
- Metastability persists in ternary solutions – stable and metastable solubilities are nearly parallel

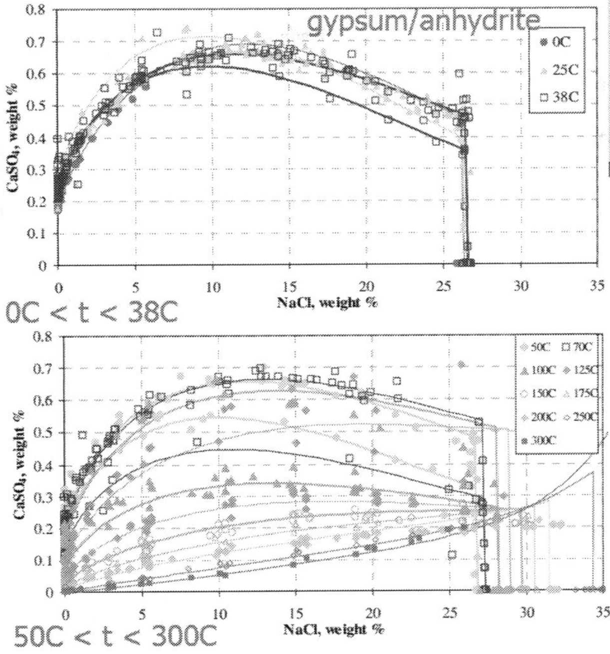

Ternary systems: $CaSO_4$ – NaCl

- No common ion effects – solubility is determined by ion interaction effects
- Solubility of both gypsum ($CaSO_4.2H_2O$) and anhydrite ($CaSO_4$) can be obtained
- Metastability persists in ternary systems

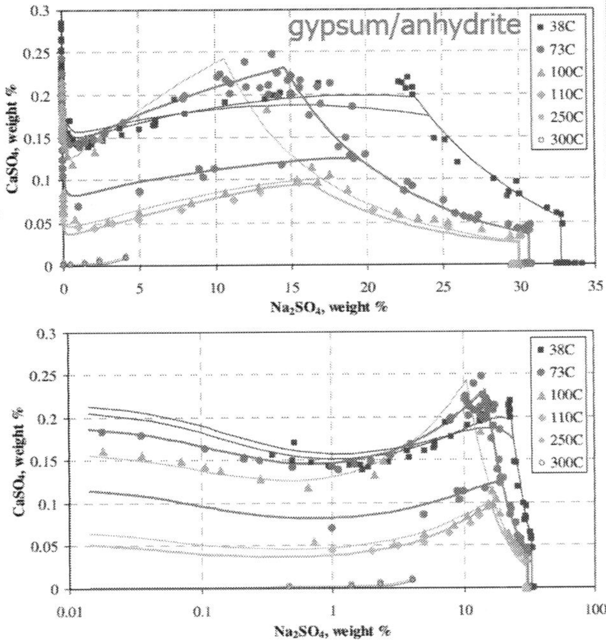

Ternary systems: $CaSO_4$ – Na_2SO_4

- Metastability, common ion effect, and double salt formation cause very complex solubility behavior

Ternary systems: Effect of various salts on $BaSO_4$ solubility

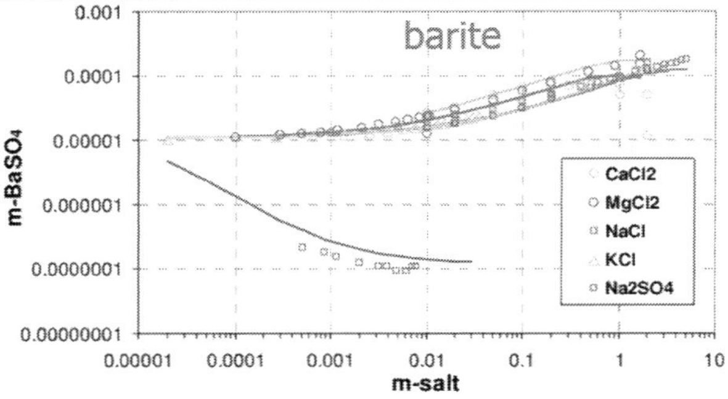

- Solubility patterns are similar for salts that do not have a common ion with $BaSO_4$

CO_2 solubility up to 4000 atm

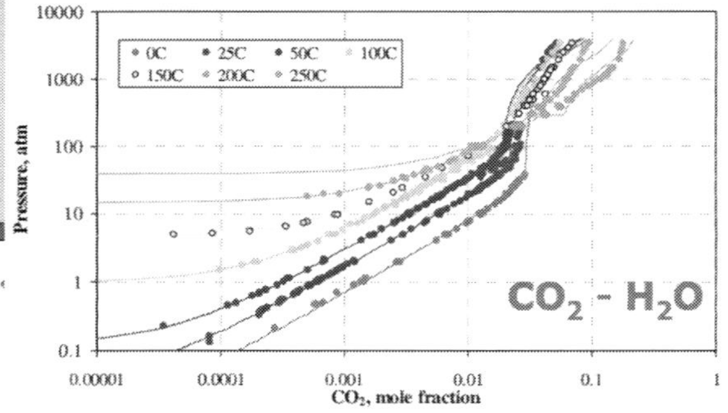

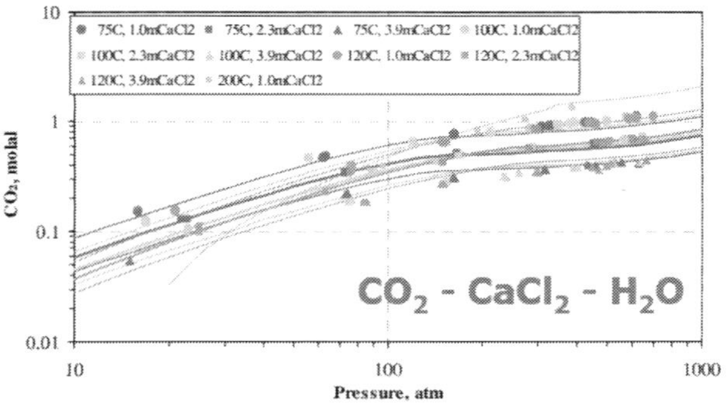

- Valid up to high pressures
- Model transitions from vapor-liquid to liquid-liquid-like behavior
- Reproduces experimental data for CO_2 with various salts

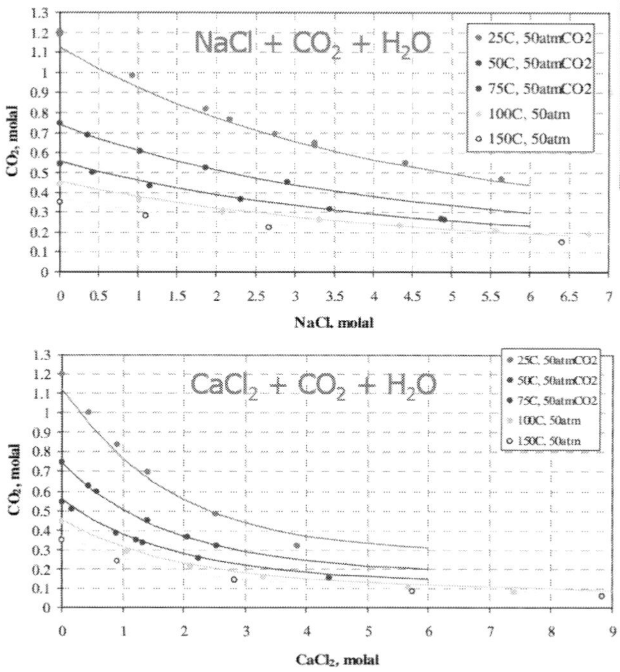

CO_2 - salt systems

- Solubility of CO_2 in NaCl and $CaCl_2$ solutions at elevated CO_2 partial pressure and various temperatures

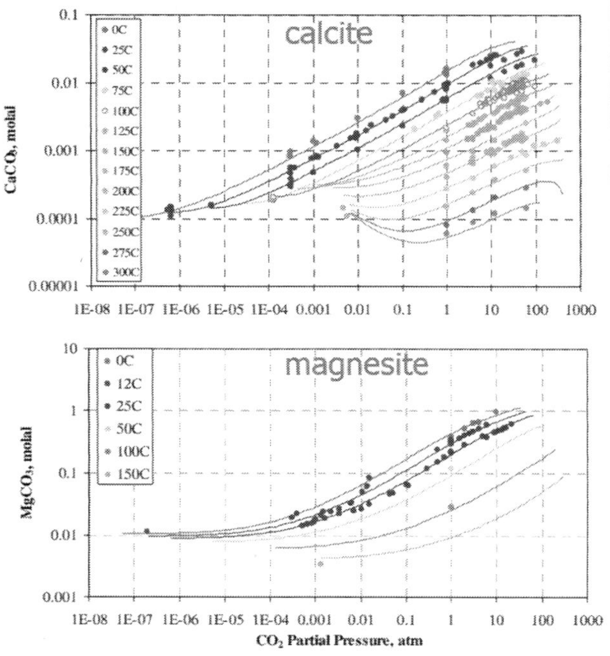

Ca and Mg carbonates

- Strong dependence on CO_2 partial pressure as well as temperature
- This is due to the aqueous reactions between $CO_{2(aq)}$, HCO_3^-, CO_3^{2-}, and their effects on solution pH – dissolved CO_2 promotes the dissolution of carbonates

$CaCO_3$ solubility as a function of NaCl and CO_2

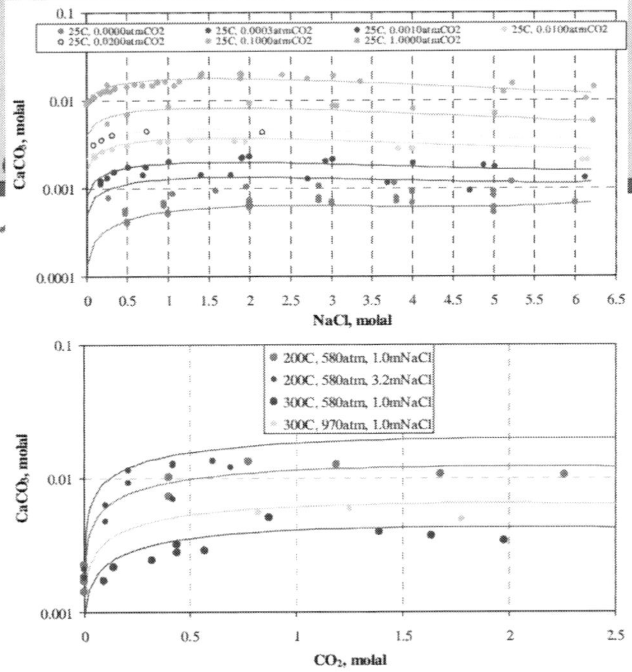

- Effect of NaCl and CO_2 partial pressure in a 3-phase VLS system
- Effect of dissolved CO_2 content and NaCl in a 2-phase LS system

Salt – methanol systems:

Solubility as a function of solvent composition

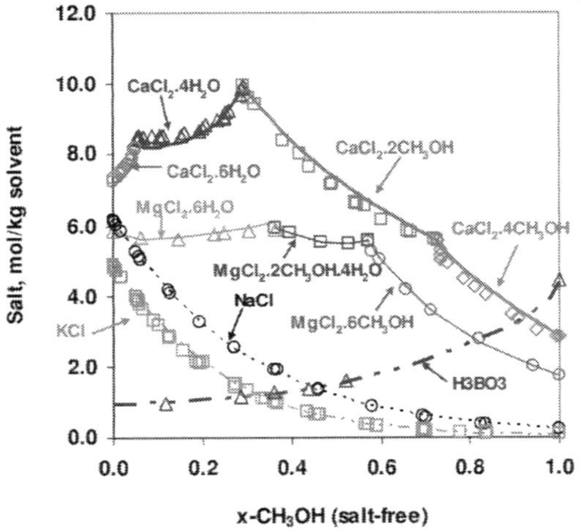

- Solubility patterns are very specific to the salt; generalizations in terms of solvent properties (e.g. dielectric constant) are not possible
- SLE complexity depends on the formation of hydrated or solvated phases

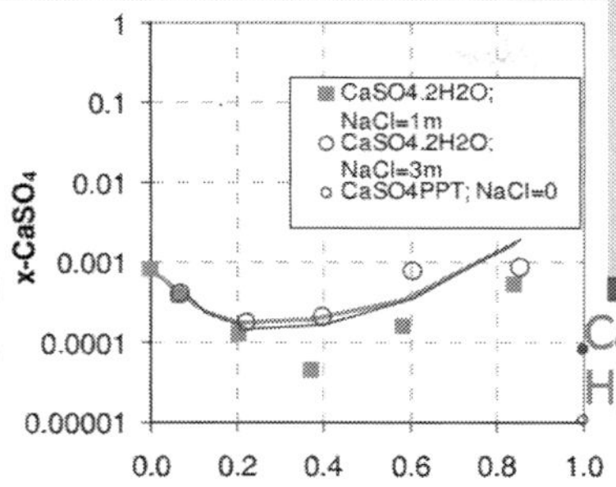

$CaSO_4$ in H_2O + methanol

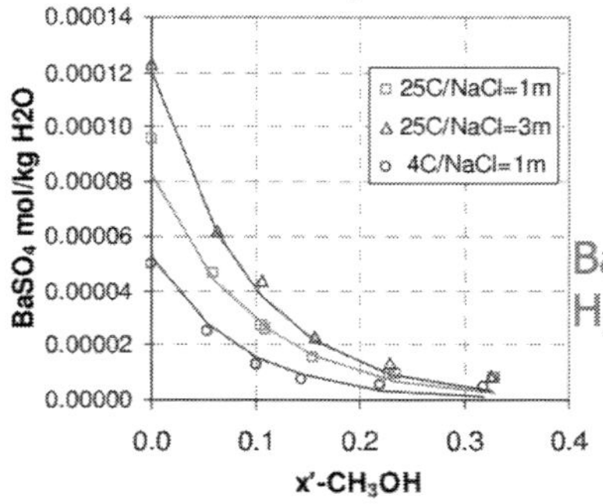

$BaSO_4$ in H_2O + methanol

Effect of methanol on scale formation

- Methanol effect is controlled by specific ion – methanol interactions
- The effect depends on the particular scale

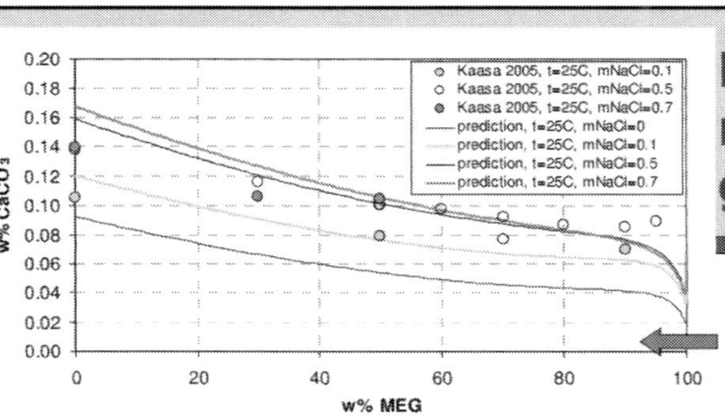

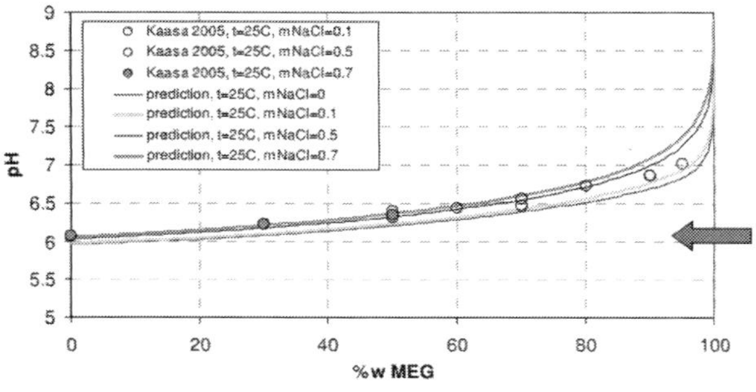

NaCl – $CaCO_3$ – monoethylene glycol – H_2O

Solubility of $CaCO_3$ as a function of glycol content

pH of $CaCO_3$ - saturated solutions: calculated simultaneously with phase equilibria

Thermodynamics of CO_2 sequestration systems: Interactions with minerals

- CO_2 can become water-saturated after disposal
- "Wet" CO_2 is likely to interact with rocks
  - Very reactive
  - High effective partial pressure of CO_2
- Almost unstudied
- Reaction mechanisms are likely to be different from those in aqueous solutions
  - Mineral transformation rather than dissolution – precipitation

21

Interactions of CO_2 with minerals

- Availability of experimental data
  - Plentiful data are available for CO_2 - water - salt systems dominated by water
  - No data are available for salt solubilities in CO_2-rich phases
  - Even for the binary CO_2 - H_2O system, there are large discrepancies in experimental data on the CO_2-rich side
- Approach:
  - Calibrate the MSE model using available CO_2 - salt - water data
    - CO_2 - HCO_3 - CO_3 - Cl - Na - K - Mg - Ca
  - Predict the properties of CO_2-rich phases

Predicting the water content in the CO_2-rich phase

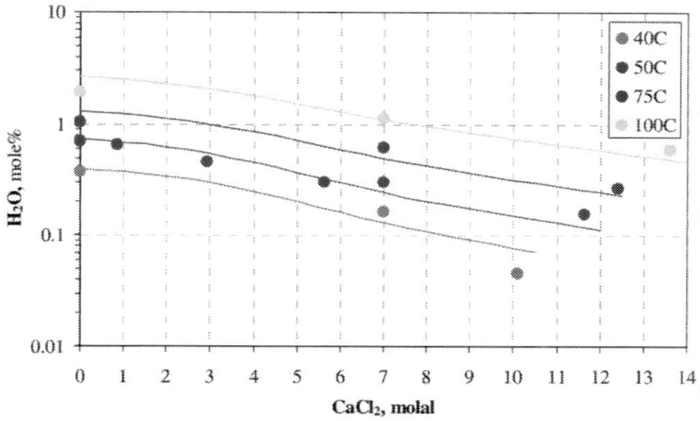

Experimental data: A.R. Felmy et al., PNNL

- These data were not used to fit the parameters
- Filling the missing "half of the story"
- Implication: Water activity in the CO_2 phase will control mineral reactivity
- It also controls corrosivity of CO_2 phases

Mineral reactivity in CO_2 as a function of water content

- Divalent orthosilicates (Mg_2SiO_4, Ca_2SiO_4, ...) as model minerals
  - Important in reservoir environments (basalts)
  - Transform to carbonates on a measurable time scale
- Conversion of forsterite (Mg_2SiO_4) investigated at PNNL as a function of time and water content

Mg_2SiO_4 conversion in supercritical CO_2

% Initial water saturation	Phase equilibria (MSE)	Solid Phases (exp)	Solid Phases Predicted - final stable	Solid Phases (MSE) – quartz, talc suppressed
14,900	Aq, V, S	SiO2(am), MgCO3	MgCO3, Quartz	MgCO3, SiO2(am)
7430	Aq V, S	SiO2(am), MgCO3	MgCO3, Quartz	MgCO3, SiO2(am)
1410	Aq, V, S	SiO2(am), MgCO3	MgCO3, Quartz	MgCO3, SiO2(am)
743	Aq, V, S	SiO2(am), MgCO3, Q3 species	MgCO3, Quartz	(V,S) MgCO3, SiO2(am), Mg5(OH)2(CO3)4.4H2O
149	V, S	SiO2(am), MgCO3, Mg5(OH)2.4/5H2O, Q3 species	MgCO3, Quartz	(V,S) MgCO3, SiO2(am)
74	V,S	Carbonate (am), low coordinated Si	MgCO3, Quartz	(V,S) MgCO3, SiO2(am)

Experimental data: A.R. Felmy et al., PNNL

25

Summary

- A comprehensive model has been developed for predicting mineral scaling as a function of
  - Brine composition
  - Temperature
  - Pressure
  - Presence of methanol and glycols
- The model can be also used for predicting the behavior of CO_2-rich, water-depleted phases
  - Effect of salts on water content in supercritical CO_2
  - Mineral transformations in CO_2-rich phases

Acknowledgements

- Andy Felmy (PNNL)
- Peiming Wang, Margaret Lencka, Ronald Springer, Jerzy Kosinski, Honggang Zhao, Robert Young (OLI Systems)

Use of API Standards for Phase Equilibrium and Thermophysical
Properties in the Simulation of Refinery Processes
For the Improved Design & Operation of Refining Facilities
Calvin Spencer
Chairman, API TDC
Todd Willman
Director of Development
Arjun Gujar
Process Engineer
energy API
TECHNICAL DATA BOOK 9th Edition
EPCONSOFTWARE
It's time to upgrade to the latest technology

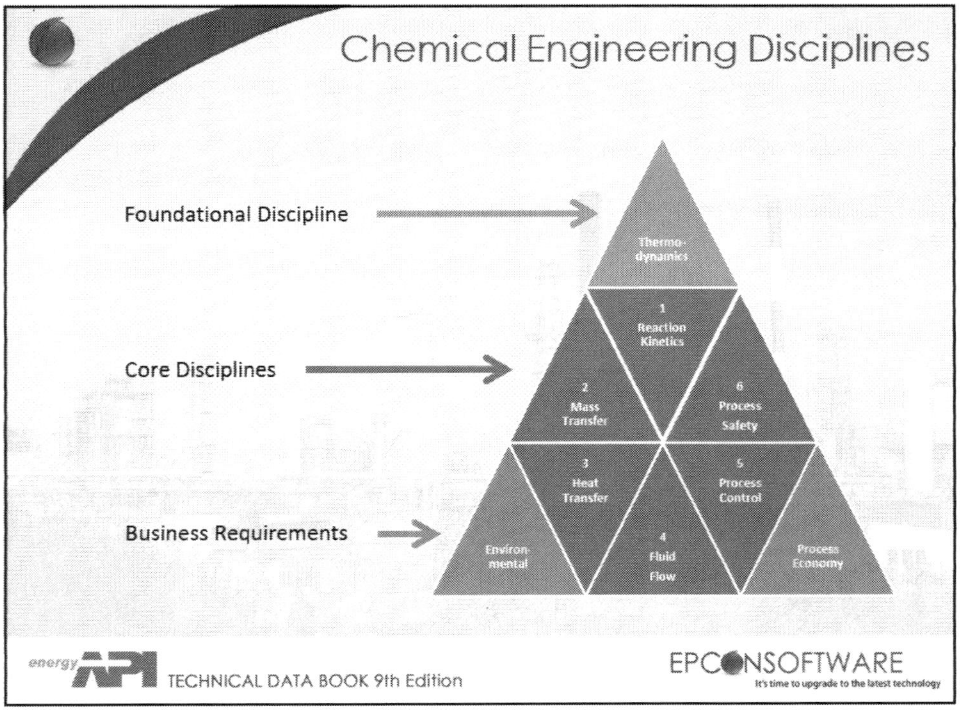

Chemical Engineering Disciplines
Foundational Discipline
Core Disciplines
Business Requirements
Thermo-dynamics
1 Reaction Kinetics
2 Mass Transfer
6 Process Safety
3 Heat Transfer
5 Process Control
4 Fluid Flow
Environ-mental
Process Economy
energy API
TECHNICAL DATA BOOK 9th Edition
EPCONSOFTWARE
It's time to upgrade to the latest technology

API Technical Data Book Version 9
Technical Overview

- ***An API Standard*** - Developed under an *API Sponsored Industrial Consortium* over a 50 year period (1961-2011)
  - *Mission Statement: provide a critically reviewed compilation of the physical and thermodynamic data and correlations that are of most interest to petroleum refiners for process evaluation and equipment design*
- Provides the technical expert's recommendations on what data and methods should be employed to accurately predict the phase equilibrium and thermo-physical properties of chemical and refining streams.
- This information is necessary for the cost effective design of and for the safe and efficient operation of process units.
- Over 10,000 copies of the printed Data Book have been sold in the US and abroad
- Over $6MM in Investments made by API since 1961
- Over $1MM in Software Development funded by EPCON since 1998
- An educational tool in thermodynamics for the practicing Chemical Engineer

energy API TECHNICAL DATA BOOK 9th Edition

EPCONSOFTWARE
It's time to upgrade to the latest technology

API Technical Data Book Version 9
Technical Coverage

Volume I - General Properties and Characterization

- ***Chapter 1 - General Data (2009)***
- ***Chapter 2 - Characterization of Hydrocarbons (2011)***
- ***Chapter 3 - Petroleum Fraction Distillation Interconversions (2011)***
- ***Chapter 4 - Critical Properties (2006)***
- ***Chapter 5 - Vapor Pressure (2006)***
- ***Chapter 6 – Density (2011)***

Volume II - Thermal Properties and Phase Equilibria

- ***Chapter 7 - Thermal Properties (2006)***
- ***Chapter 8 - Vapor-Liquid Equilibrium K-Values (2006)***
- ***Chapter 9 - Phase Equilibria in Systems Containing Water (2009)***

Volume III - Surface & Transport Properties and Combustion

- ***Chapter 10 - Surface and Interfacial Tension (2006)***
- ***Chapter 11 – Viscosity (2006)***
- ***Chapter 12 - Thermal Conductivity (2006)***
- ***Chapter 13 – Diffusivity (2005)***
- ***Chapter 14 – Combustion (2005)***
- ***Chapter 15 - Adsorption Equilibria (2005)***
- ***Chapter 16 – Software Applications (2011)***

energy API TECHNICAL DATA BOOK 9th Edition

EPCONSOFTWARE
It's time to upgrade to the latest technology

API Technical Data Book Version 9
Committee Members

- Stewardship of the Electronic API Technical Data Book – Petroleum Refining – An API Standard
- Member Companies
  - BP (Joel Glassett)
  - CB&I/Lummus (Homy Maerefat)
  - Chevron (Henry Ren)
  - ConocoPhillips (Howard Wilson)
  - Shell Global Solutions (Patrick Malone)
- Member Companies
  - Fluor (Pending)
  - Invensys (David Bluck)
  - KBR (Cal Spencer – Chairman)
  - Marathon Oil (Alex Velikoff)
  - ExxonMobil (Jack Heidman)
- Contractor – EPCON International
  - Todd Willman, Director of Development
  - Chaitanya Belwal, Product Manager
  - Arjun Gujar, Data Book Coordinator

energy API TECHNICAL DATA BOOK 9th Edition

EPCON SOFTWARE
It's time to upgrade to the latest technology

API Technical Data Book Version 9
Common Uses

- Independent cross-check of process simulator results
- Research and development tool for conceptual design and feasibility studies
- User-friendly design tool for engineers who infrequently use simulation software
- Can be used as a basis for guarantees, commodity transfer, etc. as a commonly accepted industry standard
- Useful in companies without readily available/timely technical support through a central engineering group
- Good estimation tool for refineries and operating facilities (allows consistency with the central office)
- Education/Training tool for young engineers and undergraduates in chemical engineering programs
- Low-cost alternative for the consulting engineer to provide highly accurate flash and mixture properties of chemical and petroleum streams

energy API TECHNICAL DATA BOOK 9th Edition

EPCON SOFTWARE
It's time to upgrade to the latest technology

API Technical Data Book Version 9
Example of Strengths

- Phase Equilibrium of Water-Hydrocarbon Systems
- Thermal Conductivity
- Kinematic Viscosity of Petroleum Fractions
- Hydrogen Solubility
- Physical Properties of Sulfur Containing Components
- Petroleum Fraction Characterization
- Petroleum Fraction Properties
- Pure Component Properties

energy API TECHNICAL DATA BOOK 9th Edition

EPCONSOFTWARE
It's time to upgrade to the latest technology

API Technical Data Book Version 9
Example of Strengths

Phase Equilibrium of Water-Hydrocarbon Systems

- Can be Calculated on 3 Platforms in the API Tech Data Book 9
  - Direct Solubility Equations (semi-empirical, temperature dependent correlation)
  - NRTL (Flash/Mixture Calculations)
  - SRK-Kabadi-Danner (Flash/Mixture Calculations)
- The Solubility Equation provides the best fit of the experimental data
- NRTL parameters for 100+ hydrocarbons and water have been highly tuned to fit the experimental data and essentially provide the same results of the Solubility Equation plus extremely accurate solubility for:
  - Mixtures as well as pure components
  - Pressures other than Atmospheric
- SRK-KD provides reasonable estimates for Water-Hydrocarbon Systems when accurate NRTL parameters are not available for all of the components in the given system

energy API TECHNICAL DATA BOOK 9th Edition

EPCONSOFTWARE
It's time to upgrade to the latest technology

Water – Propane Solubilities

(140 Deg.F)

Data Type	*Water in Propane (mole%)*	*Propane in Water (mole%)*
Experimental Data	*.184*	*.0202*
API Solubility Equation	*.186*	*.0200*
API-NRTL	*.186*	*.0200*
API-SRK-KD	*.250*	*.0270*

energy API TECHNICAL DATA BOOK 9th Edition

EPCONSOFTWARE It's time to upgrade to the latest technology

Mutual Solubilities of Water

Hydrocarbon Systems

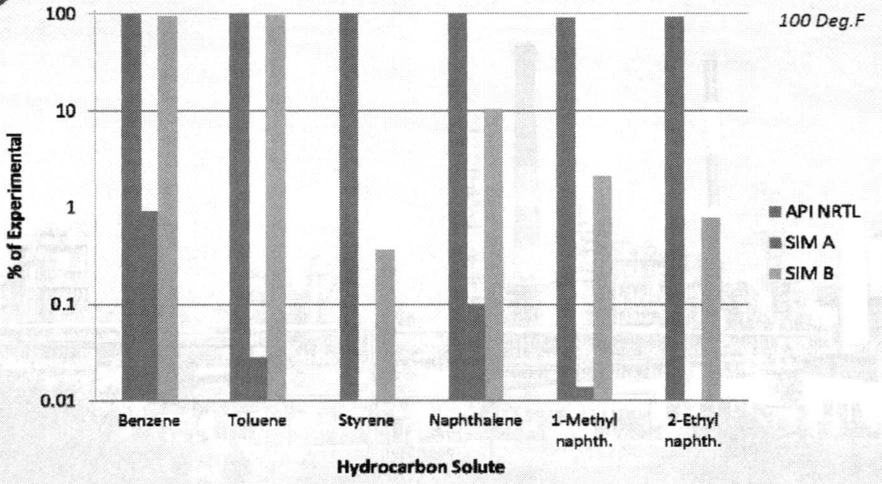

energy API TECHNICAL DATA BOOK 9th Edition

EPCONSOFTWARE It's time to upgrade to the latest technology

Version 9 Enhancements

Major Revisions

- ASTM D-2887 -> ASTM D-86
- ASTM D-2887 -> Mean Avg. Boiling Point (MeABP)
- Completely revised Chapter 6
- Flash Point of Petroleum Fractions
- Henry's Constants (in water)
- DIPPR 2009 Public Release Database

energy API TECHNICAL DATA BOOK 9th Edition

EPCONSOFTWARE
It's time to upgrade to the latest technology

API Methodology

- General Approach

$$\Theta_p = f\,(\text{Boiling Point, SG } 60^\circ F/60^\circ F)$$

- Watson Characterization Factor

$$K = \frac{\sqrt[3]{MeABP, {}^\circ R}}{SG\ 60^\circ F/60^\circ F}$$

Where,
MeABP – Mean Average Boiling Point

Typically Computed from D-86 Distillation Curve Data

energy API TECHNICAL DATA BOOK 9th Edition

EPCONSOFTWARE
It's time to upgrade to the latest technology

Overview of Current API TDB Methods

For Which Distillation Curve/Boiling Point Data Are Required

Procedure	Petroleum Fraction Property	Boiling Point Type
2B1.1	Boiling Point Characterization	VABP, Slope
2B2.1	Molecular Weight	MeABP
2B3.1	Acentric Factor	MeABP
2B5.1	Refractive Index	MeABP
2B6.1	Winn Chart	MeABP
2B7.1	Flash Point	IBP, 5%
2B8.1	Pour Point	MeABP
2B9.1	Aniline Point	MeABP
2B10.1	Smoke Point	MeABP
2B11.1	Freezing Point	MeABP
2B12.1	Cloud Point	MeABP
2B13.1	Cetane Index	MeABP
4D1.1	True Critical Temperature	VABP
4D2.1	True Critical Pressure	VABP, Slope

energy API TECHNICAL DATA BOOK 9th Edition

EPCONSOFTWARE It's time to upgrade to the latest technology

Overview of Current API TDB Methods

For Which Distillation Curve/Boiling Point Data Are Required

Procedure	Petroleum Fraction Property	Boiling Point Type
4D3.1	Pseudocritical Temperature	MeABP
4D4.1	Pseudocritical Pressure	MeABP
5A1.19	Vapor Pressure (MB)	MeABP
5B1.1	True Vapor Pressure	Slope (10%)
6A3.5	Liquid Density	MeABP
7B4.2, (Figures 7B4.3 - 7B4.6)	Enthalpy	MeABP
7B4.7	Enthalpy (Alternate Computer Method)	MeABP
7D2.2	Isobaric Heat Capacity	MeABP
7D4.2	Isobaric Heat Capacity (Vapor)	MeABP
8D1.5	Vapor-Liquid Equilibrium	MeABP
10A3.1	Surface Tension	MeABP
10A3.2	Surface Tension (Alternate Computer Method)	MeABP
11A4.2	Liquid Viscosity	MeABP
12A3.2	Liquid Thermal Conductivity	MeABP

energy API TECHNICAL DATA BOOK 9th Edition

EPCONSOFTWARE It's time to upgrade to the latest technology

Key Issues

Using the foregoing framework, the MeABP is derived from ASTM Method D-86 Distillation Data; whereas today's standard is generally ASTM Method D-2887, Simulated Distillation by Gas Chromatography.

energy API TECHNICAL DATA BOOK 9th Edition

EPCONSOFTWARE It's time to upgrade to the latest technology

ASTM D-2887 Distillation Basis

Gasoline, Naphtha Range Cuts

3A3.2 – Current Data book method (ASTM D-2887 -> ASTM D-86)
D-2887 50 % temperature < 440 ^{0}F

Kerosene, Jet Fuels, Diesel

3A3.3 – New Procedure (ASTM D-2887 -> ASTM D-86)
440 ^{0}F ≤ D-2887 50 % temperature ≤ 650 ^{0}F

Heavy Fractions and Whole Crudes

2B1.3 – New Procedure (ASTM D-2887 -> MeABP)
D-2887 50 % temperature > 650 ^{0}F

energy API TECHNICAL DATA BOOK 9th Edition

EPCONSOFTWARE It's time to upgrade to the latest technology

Procedure 2B1.3

ASTM D-2887→MeABP

$$MeABP = WABP_{D\text{-}2887} - 1.675 \cdot (S)^{1.83}$$

where,

$$WABP_{D\text{-}2887} = 0.015 \cdot (T_{IBP} + T_{FBP})\, 0.051 \cdot (T_5 + T_{95}) + 0.116 \cdot (T_{10} + T_{90}) + 0.212 \cdot (T_{30} + T_{50} + T_{70})$$

$$S = (T_{90} - T_{10}) / 80$$

where,

T = D-2887 Distillation Temperature, ° F

energy API TECHNICAL DATA BOOK 9th Edition

EPCONSOFTWARE It's time to upgrade to the latest technology

Alaska North Slope (K = 11.74)

Comparison of MeABP Methodologies

Fraction	SG 60F/60F	MeABP, °F				
		D-86	D-2887→D-86	Dev	D-2887*	Dev
320-450	0.8095	361.6	369.5	7.8	372.6	11.0
450-500	0.8458	460.0	462.4	2.3	466.2	6.2
500-550	0.8633	508.5	516.0	7.5	521.7	13.1
550-600	0.8697	557.4	567.3	9.9	574.2	16.8
600-650	0.8866	602.2	618.2	16.1	626.5	24.3
450-650	0.8665	533.0	535.5	2.5	539.8	6.8
650-750	0.9018	670.3	690.1	19.8	698.2	27.9

* Direct from Procedure 2B1.3

energy API TECHNICAL DATA BOOK 9th Edition

EPCONSOFTWARE It's time to upgrade to the latest technology

Alaska North Slope (K = 11.74)

Comparison of MeABP Methodologies

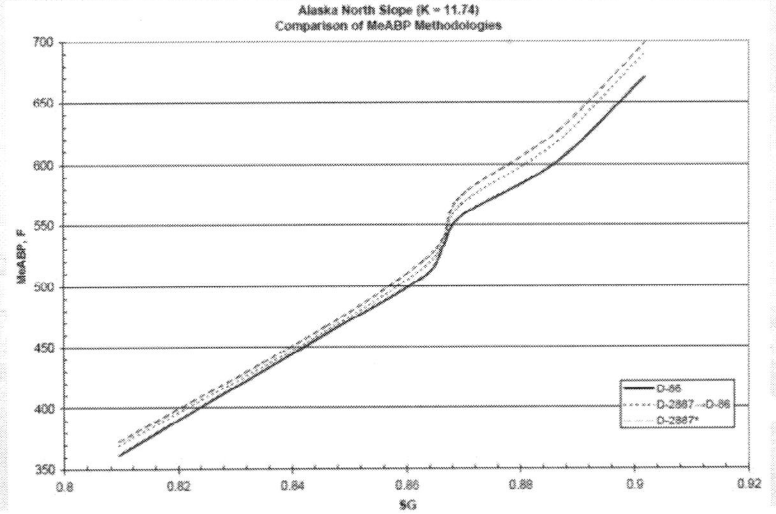

energy API TECHNICAL DATA BOOK 9th Edition

EPCONSOFTWARE
It's time to upgrade to the latest technology

San Joaquin Valley (K = 11.16)

Impact of MeABP on Predicted Physical Constants Watson K

Fraction	SG 60F/60F	MeABP, °F			Watson K				
		D-86	D-2887→D-86	D-2887*	D-86	D-2887→D-86	Dev.	D-2887*	Dev.
320-450	0.8532	415.4	411.85	414.91	11.21	11.20	-0.02	11.21	0.00
450-500	0.8745	468.97	466.69	475.33	11.16	11.15	-0.01	11.18	0.03
500-550	0.8914	518.51	517.2	525.2	11.14	11.13	0.00	11.16	0.03
550-600	0.9124	581.61	587.03	597.68	11.11	11.13	0.02	11.17	0.06
600-650	0.9317	633.4	637.12	646.67	11.06	11.07	0.01	11.10	0.04
450-650	0.9079	566.85	545.24	548.79	11.11	11.03	-0.08	11.05	-0.07
650-750	0.9483	-	682.89	690.61	-	11.02	-	11.05	-

* Direct from Procedure 2B1.3

energy API TECHNICAL DATA BOOK 9th Edition

EPCONSOFTWARE
It's time to upgrade to the latest technology

Utah Altamont (K = 11.16)

Impact of MeABP on Predicted Physical Constants

Viscosity

Fraction	SG 60F/60F	MeABP, °F			Viscosity, cSt (210 °F)				
		D-86	D-2887→D-86	D-2887*	D-86	D-2887→D-86	% Dev.	D-2887*	% Dev.
320-450	0.7649	363.4	367.8	370.6	0.603	0.613	1.7	0.619	2.7
450-500	0.7853	454.5	464.9	468.3	0.878	0.915	4.2	0.928	5.6
500-550	0.7945	513.5	522.4	528.5	1.13	1.17	3.6	1.20	6.2
550-600	0.8003	565.0	574.8	583.3	1.40	1.46	4.0	1.51	7.5
600-650	0.8044	612.6	627.5	637.1	1.71	1.81	5.9	1.88	9.9
450-650	0.7964	534.9	538.3	542.1	1.23	1.25	1.4	1.27	2.9
650-750	0.8208	669.7	677.8	686.4	2.24	2.31	3.2	2.38	6.6

* Direct from Procedure 2B1.3

energy API TECHNICAL DATA BOOK 9th Edition

EPCONSOFTWARE
It's time to upgrade to the latest technology

API Technical Data Book Version 9

Chapter 6 – Liquid Density

- Completely revised Chapter
- Modified Rackett Equation - Z_{RA} regressed from DIPPR 2009 Public Primarily Accepted Density Data for approximately 240 compounds
- New procedure for Volumetric Shrinkage

energy API TECHNICAL DATA BOOK 9th Edition

EPCONSOFTWARE
It's time to upgrade to the latest technology

API Technical Data Book Version 9
Modified Rackett Equation - Z_{RA}

Equation

$$\frac{1}{\rho_s} = \left(\frac{RT_c}{p_c}\right)(Z_{RA})^{\left[1.0+(1.0-T_r)^{2/7}\right]}$$

Z_{RA} – Empirically Derived Constant

Z_{RA} Regressed Using

- AIChE DIPPR 2009 Public Release - Primarily Accepted Data
- AIChE DIPPR 2009 Public Release - Critical Constants

energy API TECHNICAL DATA BOOK 9th Edition

EPCONSOFTWARE It's time to upgrade to the latest technology

API Technical Data Book Version 9
Revised Modified Rackett Z_{RA}

Component Name	Number of Points	Z_{RA}	Avg. Abs. % Deviation
Ethene	34	0.2814	0.32
Propene	96	0.2776	0.39
1,3-Butadiene	87	0.2710	0.10
Benzene	121	0.2699	0.17
Naphthalene	16	0.2611	0.07

energy API TECHNICAL DATA BOOK 9th Edition

EPCONSOFTWARE It's time to upgrade to the latest technology

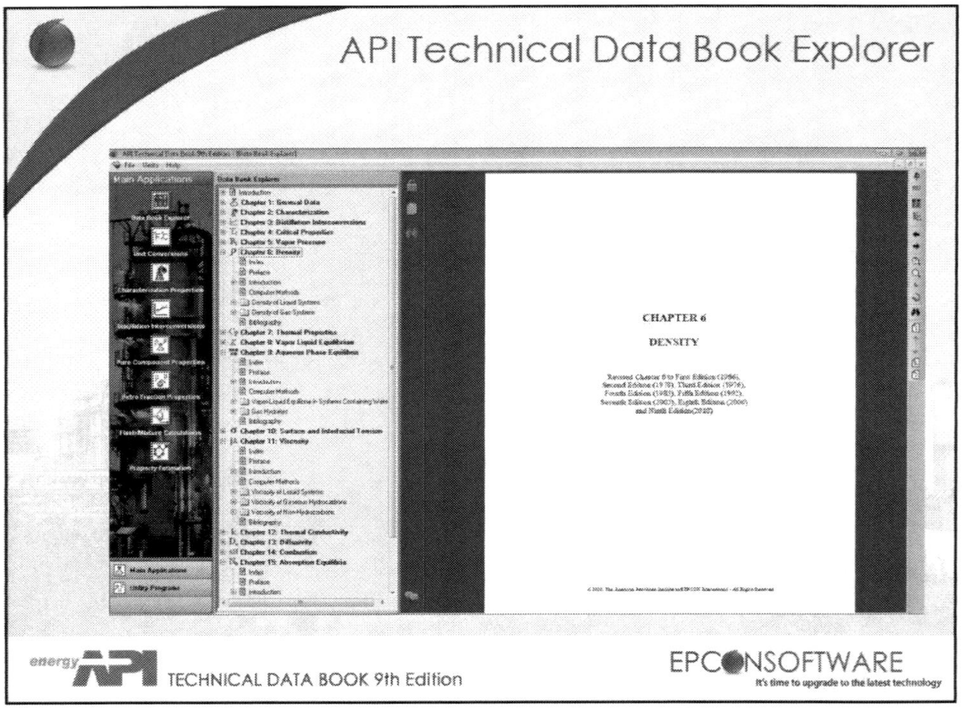
API Technical Data Book Explorer
CHAPTER 6
DENSITY
energy API
TECHNICAL DATA BOOK 9th Edition
EPCONSOFTWARE
It's time to upgrade to the latest technology

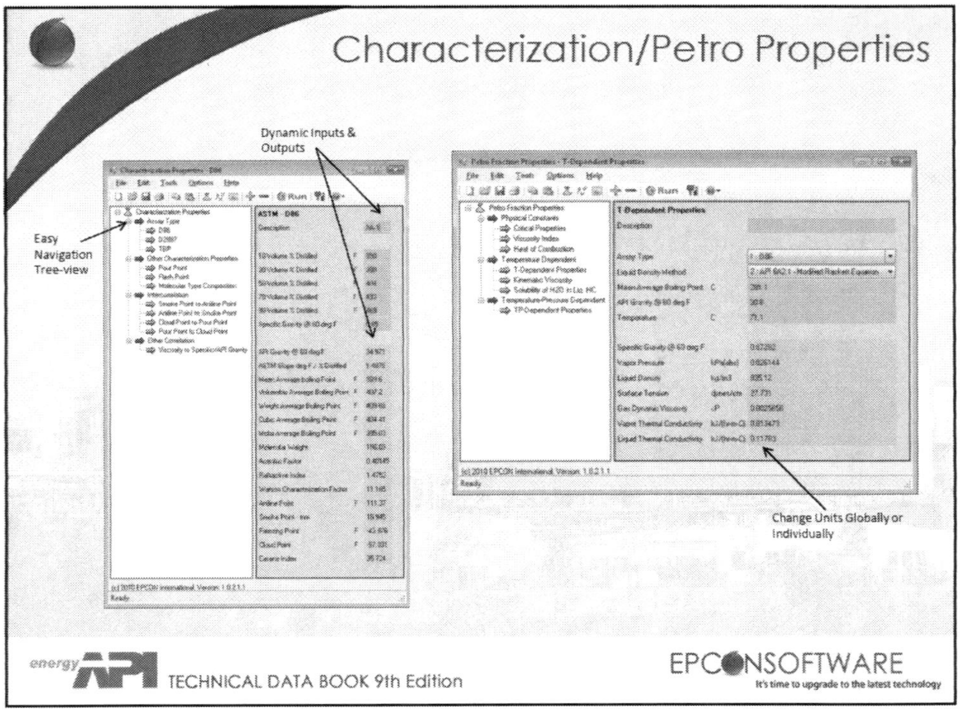
Characterization/Petro Properties
Dynamic Inputs & Outputs
Easy Navigation Tree-view
Change Units Globally or Individually
energy API
TECHNICAL DATA BOOK 9th Edition
EPCONSOFTWARE
It's time to upgrade to the latest technology

Pure Component Properties

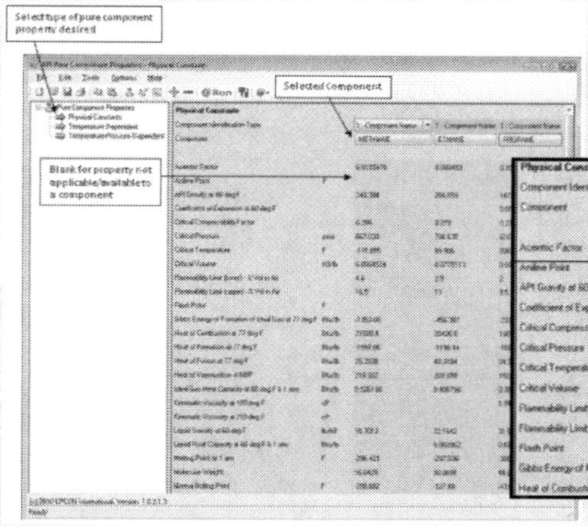

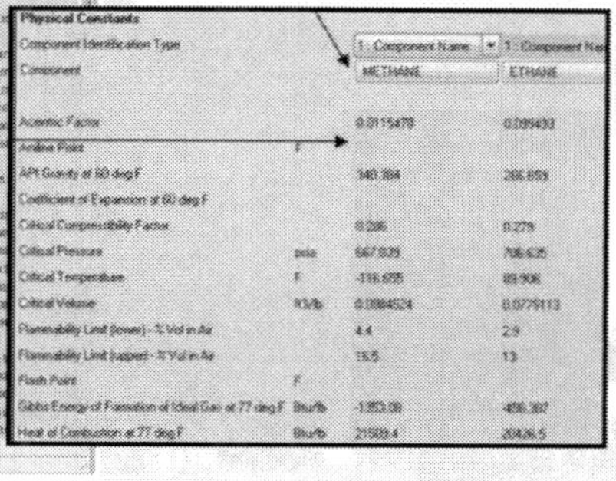

energy API TECHNICAL DATA BOOK 9th Edition

EPCONSOFTWARE
It's time to upgrade to the latest technology

Mixture Properties

The 3-phase flash has been proven in the API Technical Data Bank for over 10 years to provide a superior foundation for your process engineering work.

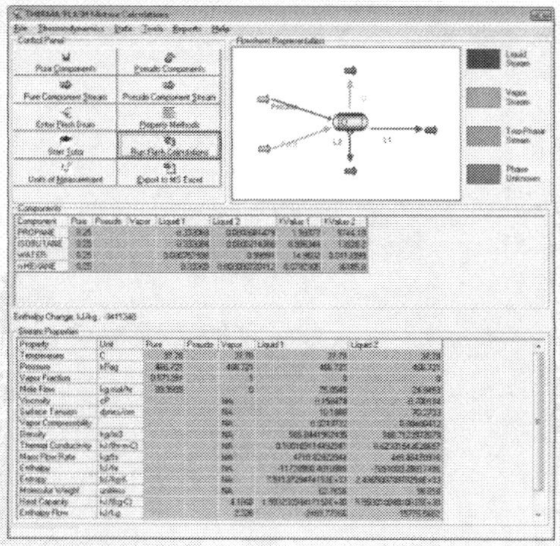

The API Technical Data Committee identified several areas where this flash engine outperforms some of the other simulators including prediction of the 2nd liquid (aqueous) phase.

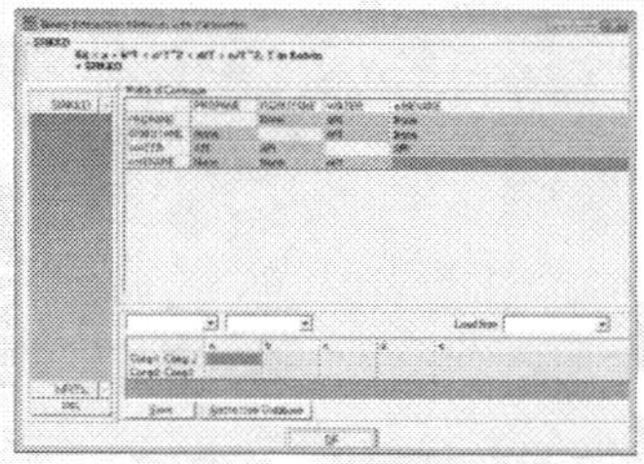

energy API TECHNICAL DATA BOOK 9th Edition

EPCONSOFTWARE
It's time to upgrade to the latest technology

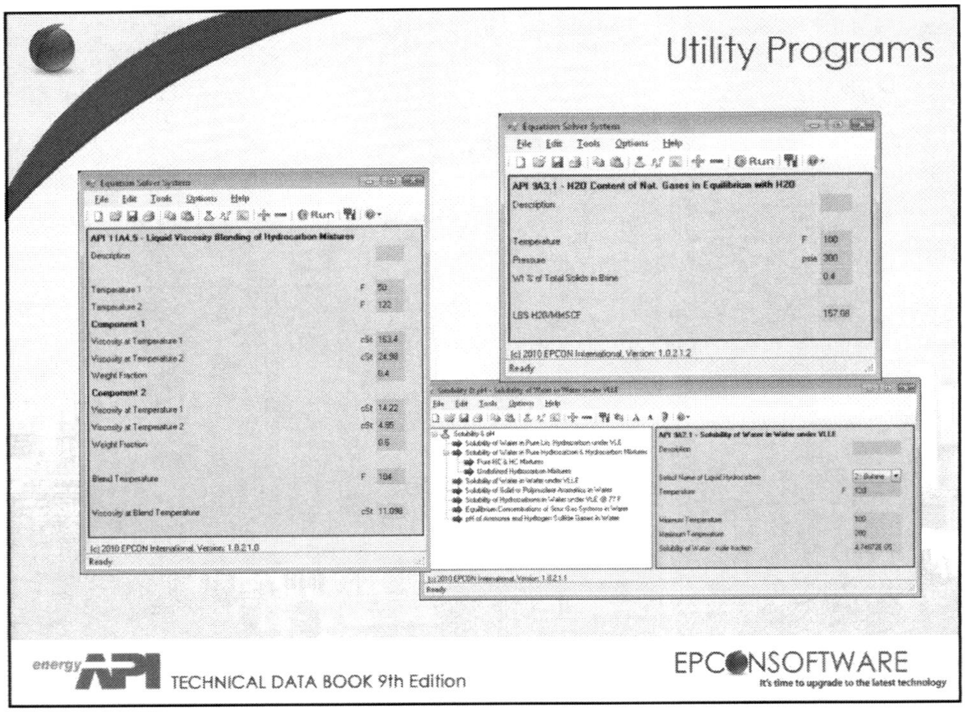
Utility Programs
Equation Solver System
File Edit Tools Options Help
API 11A4.5 - Liquid Viscosity Blending of Hydrocarbon Mixtures
API 9A3.1 - H2O Content of Nat. Gases in Equilibrium with H2O
(c) 2010 EPCON International, Version: 1.0.2.1.0
Ready
energy API TECHNICAL DATA BOOK 9th Edition
EPCONSOFTWARE
It's time to upgrade to the latest technology

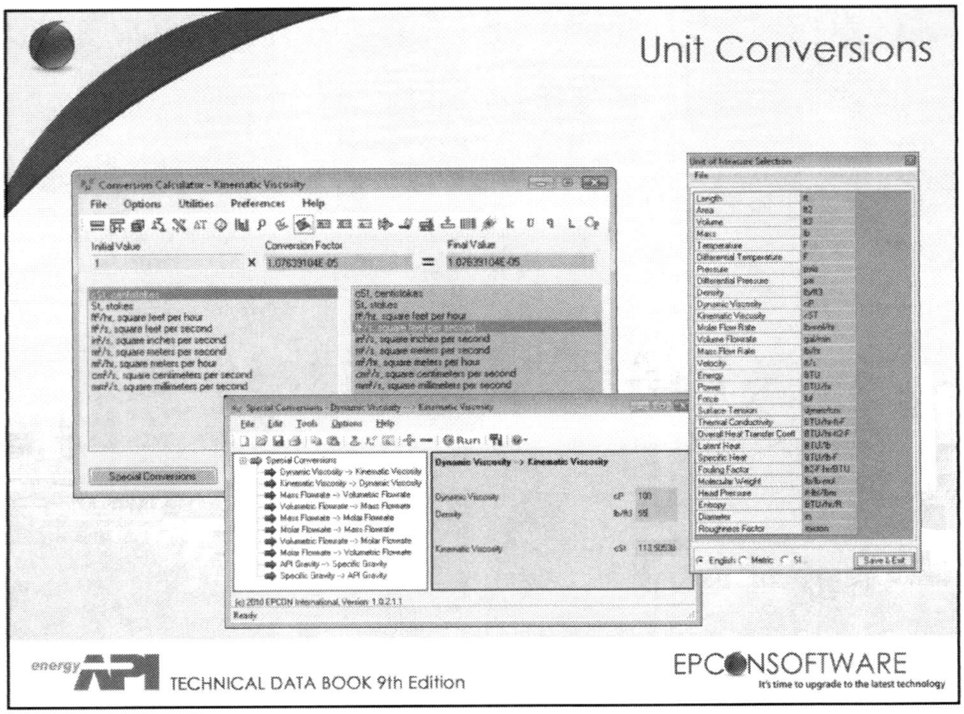
Unit Conversions
Conversion Calculator - Kinematic Viscosity
File Options Utilities Preferences Help
Initial Value
Conversion Factor
Final Value
1
X 1.07639104E-05
= 1.07639104E-05
Special Conversions
Dynamic Viscosity -> Kinematic Viscosity
Unit of Measure Selection
energy API TECHNICAL DATA BOOK 9th Edition
EPCONSOFTWARE
It's time to upgrade to the latest technology

Future Work

- Revised Chapter 7 – Thermal Properties
- Crude Assay Database
- New Equation for Cetane Index
- New Equation for Smoke Point
- SRK-KD expanded to systems containing Mercury
- Revised Method for estimating the Liquid Viscosity of Heavy Petroleum Fractions Including Bitumen's
- Expanded Henry's Constants coverage
- Continued evaluation of correlations

energy API TECHNICAL DATA BOOK 9th Edition

EPCONSOFTWARE
It's time to upgrade to the latest technology

Thermodynamic Modeling of Composition Dependence of Relative Volatility

Kai Fischer
Eliane Kuhne
Eric Hendriks

3rd AIChE Regional Process Technology Conference, October 6-7, 2011, Moody Gardens Hotel, Galveston, TX

Separation by Distillation

- Total world's energy consumption was 15 TW (2008)
- 5 % of world's (3 % of US's) energy consumption is used for distillation

Separation by Distillation

- Total world's energy consumption was 15 TW (2008)
- 5 % of world's (3 % of US's) energy consumption is used for distillation

- Preferred separation technology
- Pressure controlled fluids processing
- Easy phase separation

Separation by Distillation

- Total world's energy consumption was 15 TW (2008)
- 5 % of world's (3 % of US's) energy consumption is used for distillation

- Preferred separation technology
- Pressure controlled fluids processing
- Easy phase separation

- Accurate simulation has strong motivation (CAPEX & OPEX)

Shell Deutschland Oil GmbH,
Rheinland Raffinerie Werk Wesseling,
Xylene Plant

Shell Deutschland Oil GmbH,
Rheinland Raffinerie Werk Wesseling,
Xylene Plant
Xylene Splitter,
130 3-Pass Trays,
ID = 4,5 m
> 70 t of Tray
Material

One of many Trays in a Column...

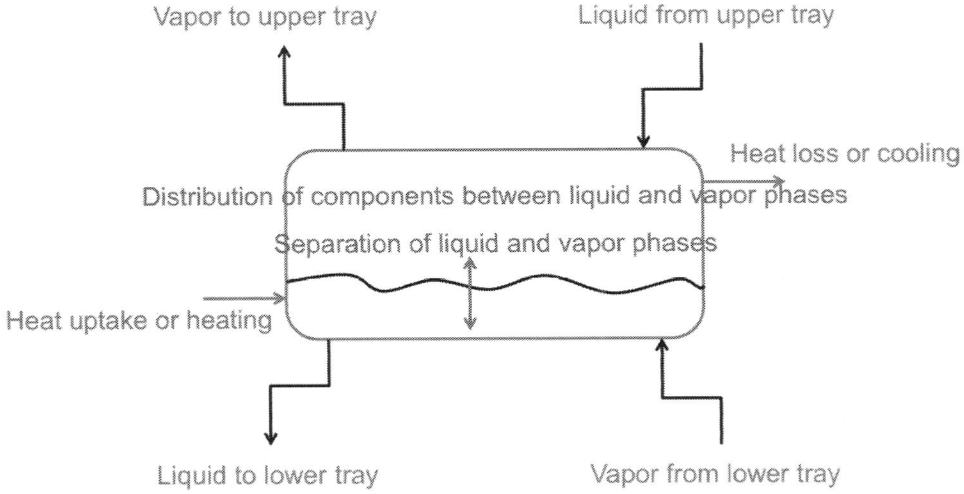

October 2011 6

...the Driving Forces...

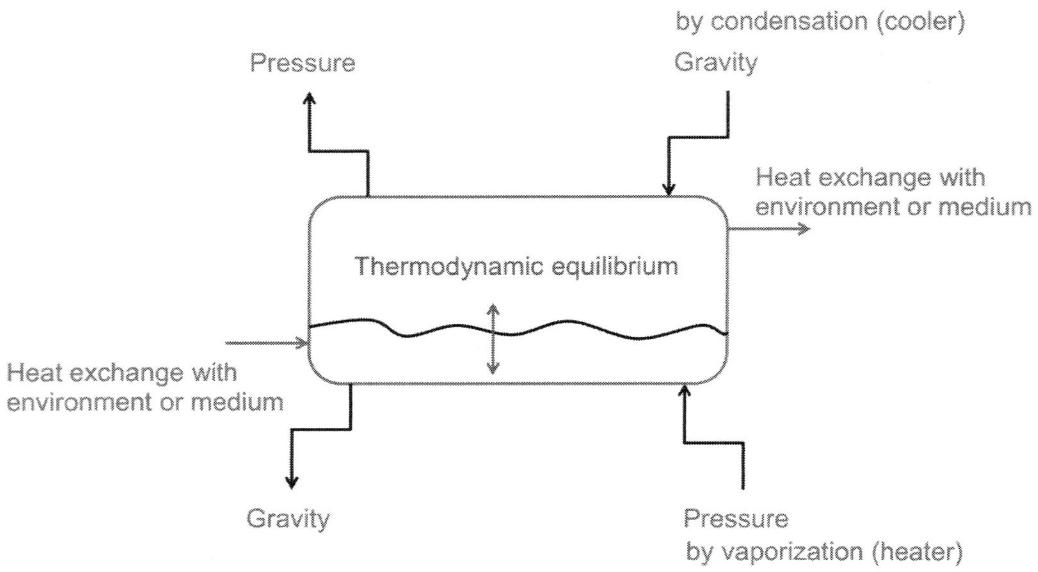

October 2011 9

...and the Resistances and Limitations

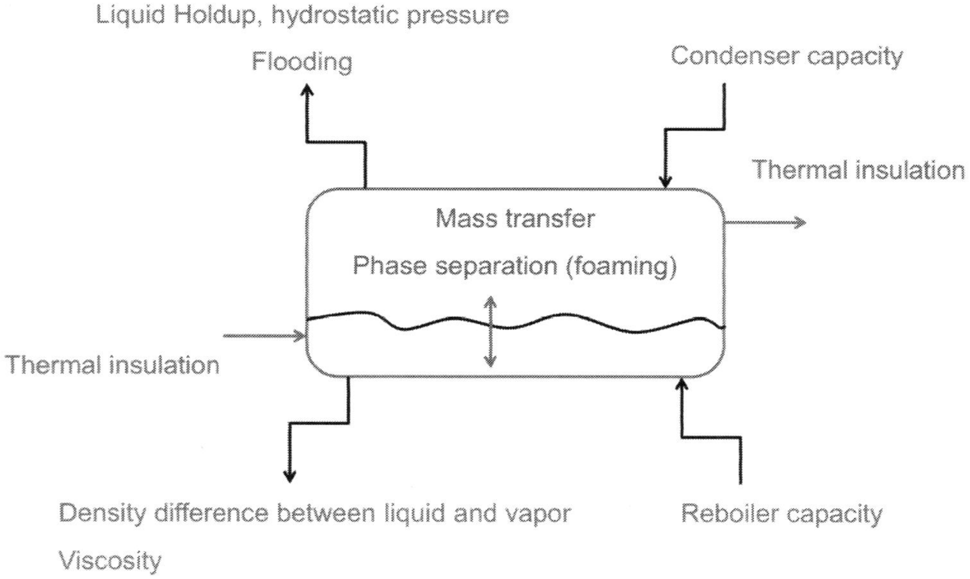

October 2011 10

Core Principle: Phase Equilibrium (Vapor – Liquid)

- High accuracy is decisive for equipment design
- Heat balance: energy consumption
- Mass balance: feasibility
- Vapor mole fraction $y_i(x_i,P)$ as function of liquid composition and pressure required

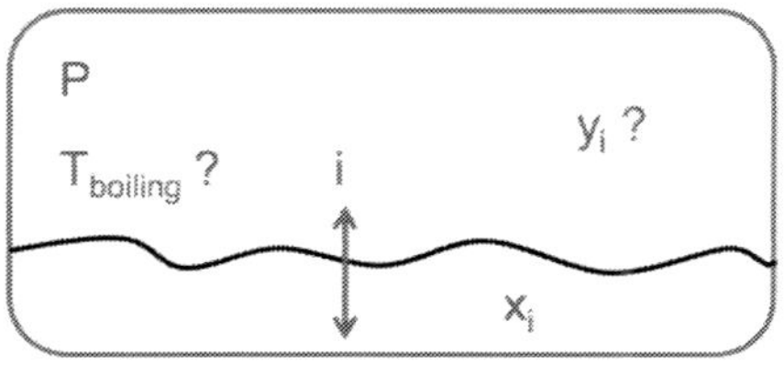

K-values

$$K_i = \frac{y_i}{x_i}$$

relative volatility

$$\alpha_{ij} = \frac{K_i}{K_j}$$

October 2011 11

Core Principle: Phase Equilibrium (Vapor – Liquid)

- Classical approaches: γ-φ (Departure from ideal mixing of real compounds)

$$y_i \cdot \varphi_i^V \cdot P = x_i \cdot \gamma_i \cdot \varphi_i^L \cdot P_i^s \cdot Poy_i \qquad K_i = \frac{\gamma_i \cdot P_i^s}{P}$$

simplified

$$y_i \cdot P = x_i \cdot \gamma_i \cdot P_i^s \qquad \alpha_{ij} = \frac{\gamma_i \cdot P_i^s}{\gamma_j \cdot P_j^s}$$

With vapor pressure correlation for each component $P_i^s(T)$

($\geq$ 3 parameters)

And activity coefficient correlation for each binary pair $\gamma_i(x_i,T)$

($\geq$ 2 parameters)

- Fitted to experimental data (vapor pressures and vapor-liquid equilibrium)
- Supported by process simulation software vendors through parameter databanks
- Limitations of γ-φ approach: supercritical components, higher pressure

October 2011 12

Core Principle: Phase Equilibrium (Vapor – Liquid)

- Classical approaches: φ-φ (through Equation of State)
  (Departure from ideal mixing of ideal gas compounds)

$$y_i \cdot \varphi_i^V \cdot P = x_i \cdot \varphi_i^L \cdot P$$

With pure component data T_c, P_c, ω (critical data and acentric factor)
(≥ 3 parameter corresponding states principle)

And binary interaction parameter for each binary pair $k_{ij}(T)$
(≥ 1 parameters)

- Fitted to experimental data (vapor pressures and vapor-liquid equilibrium)
- Supported by process simulation software vendors through parameter databanks
- Limitations of φ-φ approach: polar components

Result of Classical Approaches: Multiplicity of Property Methods

- Depending on process conditions, different approaches have to be selected for different process units.
- Divergence of process model versions with time
- Local fitting leads to "island solutions" and an inconsistent process model

Suggestion: Apply EoS with g^E Mixing Rules

- Combines advantages of γ-φ and φ-φ approaches
- binary interaction parameter for each binary pair $k_{ij}(T)$ is replaced by activity coefficient correlation for each binary pair $\gamma_i(x_j,T)$
- Important improvement of model flexibility and application ranges
- Important savings:
  - Henry-correlations of supercritical components (e.g. volatile traces) become redundant
  - only one property method to be maintained
  - robust extrapolation (global method, less data needed for correlation)

October 2011 15

Options of EoS with g^E Mixing Rules in Process Simulation Tools

- Proll: "SRKH" only option

  = SRK equation of state + α-function + Huron-Vidal mixing rule + modified NRTL
- Aspen Plus: "PSRK" selected option

  = SRK equation of state + α-function + PSRK mixing rule + NRTL
- 2 workhorse models with the potential of global thermodynamic (not only) phase equilibrium correlations
- Exact choice of elements of hybrid model is not decisive
- Potential of these methods is largely underestimated in industry

October 2011 16

Some of the Reservations

- Column convergence critically depends on the method, which should not be more complex.
- Correlation parameters are not available.
- Vapor pressures have to be very accurate. This cannot be achieved with cubic EoS.
- This is special for complex mixtures and not needed for hydrocarbon mixtures.
- Others don't use these methods.

October 2011 17

Example: α-Olefin Separation (Fractionation)

- Catalytic oligomerization of ethylene
- Crude α-olefin product is first separated into four fractions:
  - C_4
  - C_6 through C_{10}
  - C_{12} through C_{20}
  - C_{22+}
- C_6 - C_{10} and C_{12} - C_{20} fractions are then redistilled to produce individual carbon number cuts

October 2011 18

Flowsheet for α-Olefin Fractionation

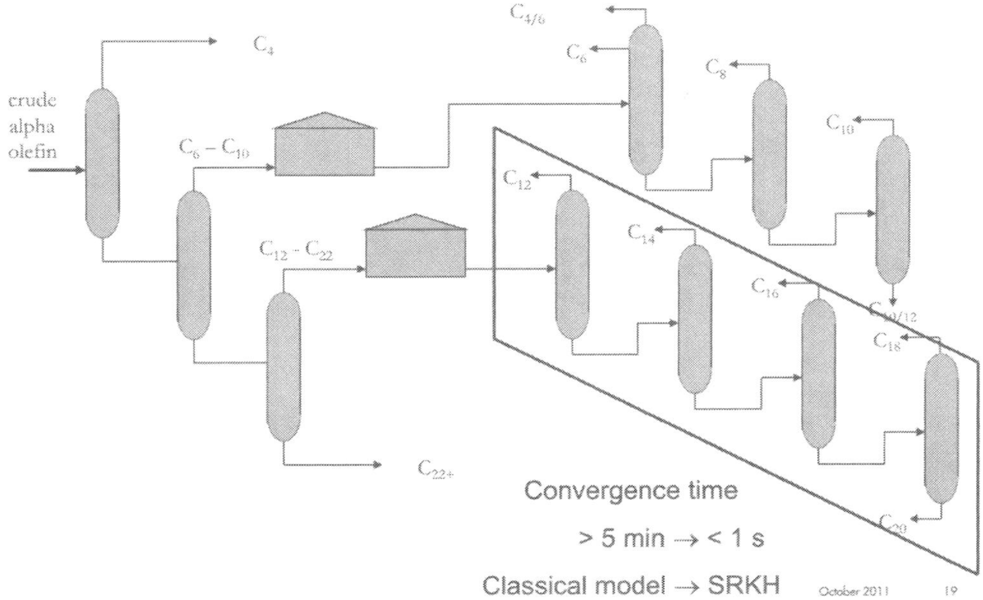

Convergence time

> 5 min → < 1 s

Classical model → SRKH

October 2011 19

SRKH Correlations for Olefin Pairs

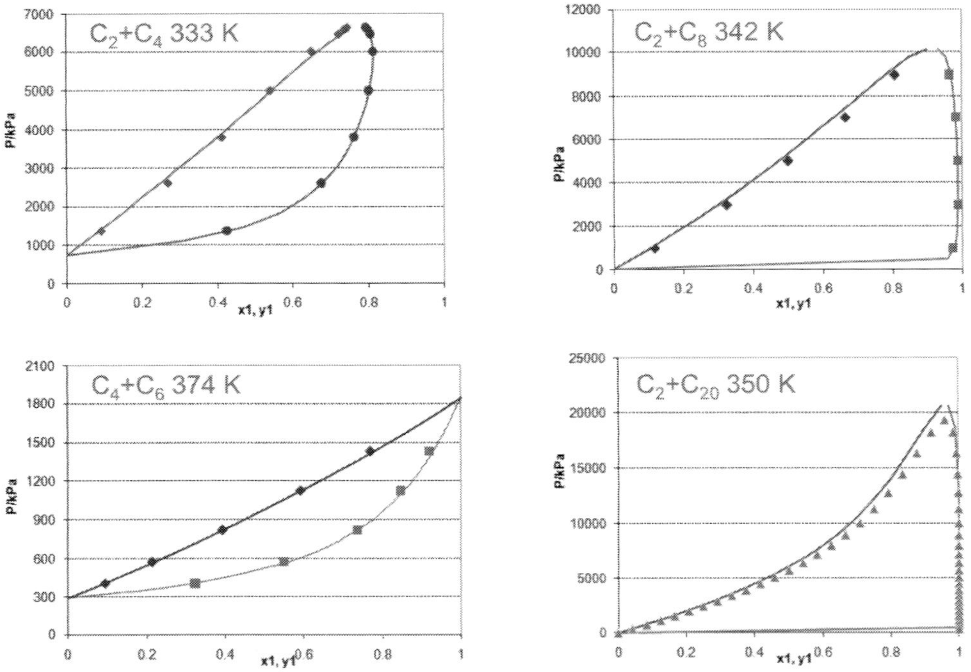

Example of Binary Interaction Parameter Generalization

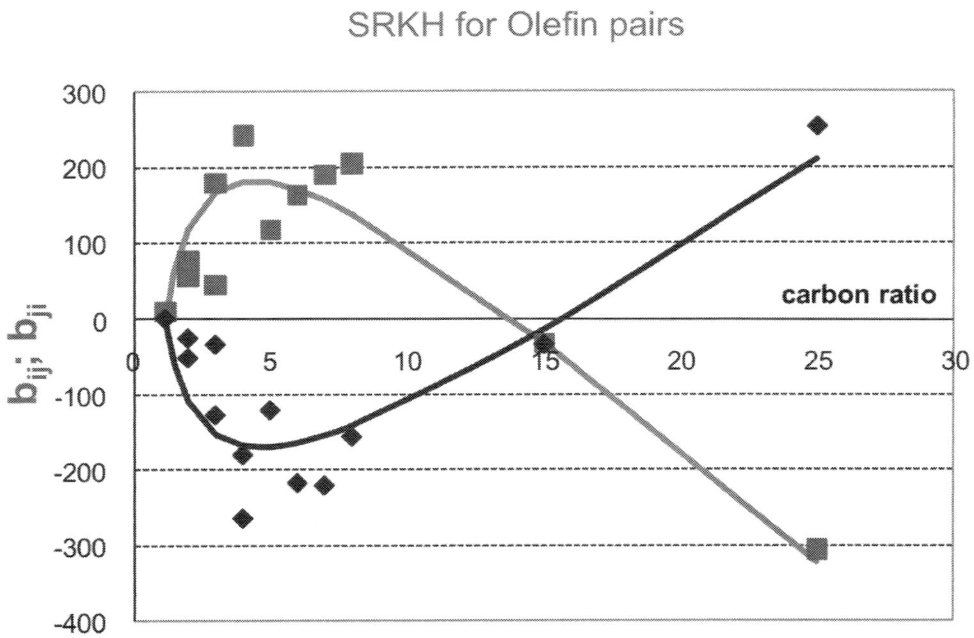

Strategies for Binary Interaction Parameters

- Fitting to experimental data (< 10 %)
- Derivation of parameter generalizations (for homologues series)
- Prediction of data by validated group contribution methods and correlation of those estimated data

- **Use best knowledge, not current feeling of relevance!**

October 2011 22

Example: Xylene Splitter

- Vapor pressures using SRK + Mathias-Copeman α-function

m-Xylene

extended Antoine equation | SRK + MC

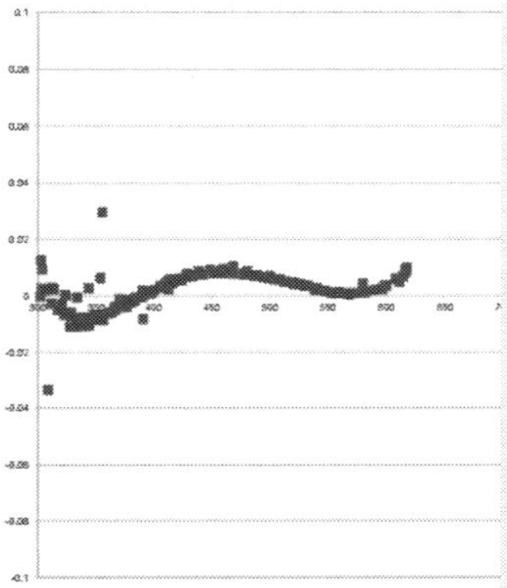

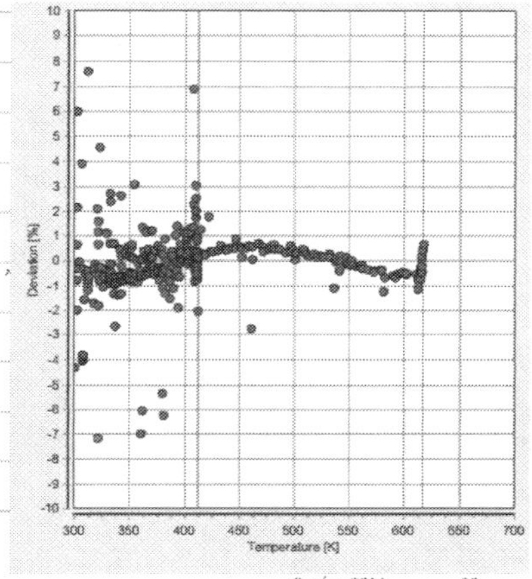

Case Study vs. Plant data

- VLE data are available at max. 1 bar
  Splitter operates at 6 bar
- Reflux is far overestimated

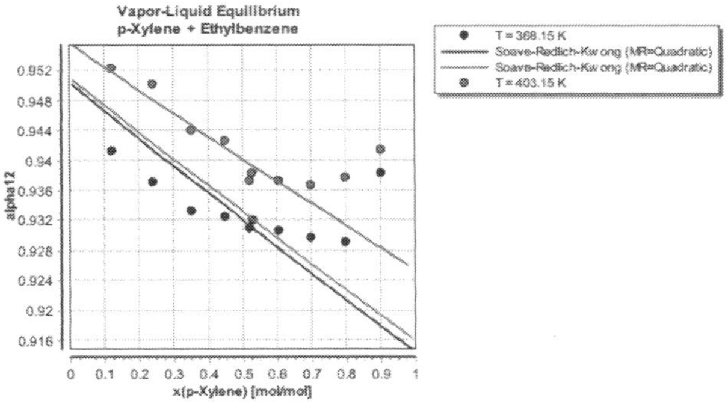

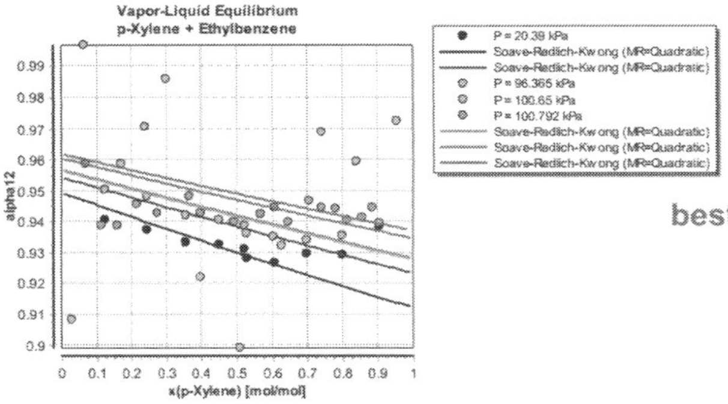

best fit with classical SRK!

Case Study vs. Plant data

- VLE data are available at max. 1 bar Splitter operates at 6 bar
- Reflux is correct

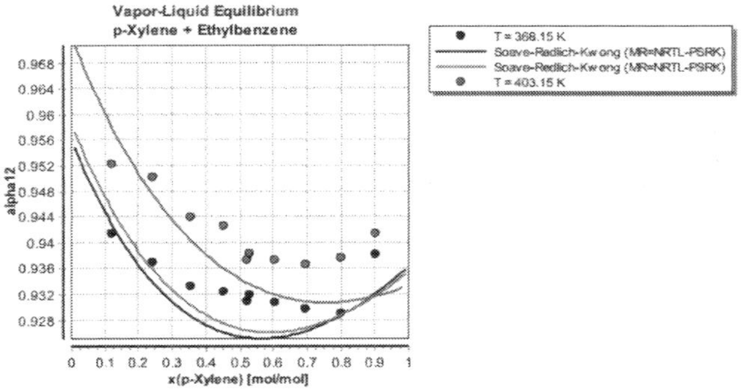

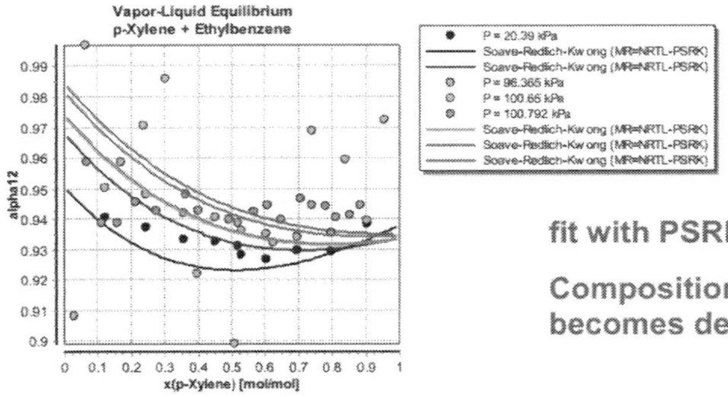

fit with PSRK-NRTL

Composition dependence of relative volatility becomes decisive!

Still Reservations ?

- Column convergence critically depends on the method, which should not be more complex.
- Correlation parameters are not available.
- Vapor pressures have to be very accurate. This cannot be achieved with cubic EoS.
- This is special for complex mixtures and not needed for hydrocarbon mixtures.
- Others don't use these methods.

Still Reservations ?

- Column convergence critically depends on the method, which should not be more complex.
- Correlation parameters are not available.
- Vapor pressures have to be very accurate. This cannot be achieved with cubic EoS.
- This is special for complex mixtures and not needed for hydrocarbon mixtures.
- Others don't use these methods.

October 2011 27

Still Reservations ?

- Column convergence critically depends on the method, which should not be more complex.
- Correlation parameters are not available.
- Vapor pressures have to be very accurate. This cannot be achieved with cubic EoS.
- This is special for complex mixtures and not needed for hydrocarbon mixtures.
- Others don't use these methods. ✓ ?

October 2011 28

Introduction to Process Safety Principles

Dr. Angela E. Summers
President, SIS-TECH Solutions, LP
Julie D. White
Formerly Lloyd's Register

AIChE Regional 2011 Process Technology Conference
Young Professional (YP) Session

Contents

- What is Process Safety?
- Standards and Rules
- Resources for a Young Professional (YP)

What is Process Safety?

Process Safety is:

- Major Accident Prevention
- Fires, Explosions, Toxics
- Accidental chemical releases
- Unintended reactions

Personal Safety is:

- Classic worker health and safety
- 'Slips, Trips, and Falls'
- Protective Equipment
- Hot work permits
- Industrial Hygiene

Process Safety - On-shore

- Federal Agency
  - Occupational Safety and Health Administration (OSHA)
- Promulgated – 1990
  - Clean Air Act Amendments
- Standard
  - OSHA 1910.119 "Process Safety Management (PSM) of Highly Hazardous Chemicals"
- Applies - now
  - Locations with specific chemicals and quantities

Process Safety – Offshore – Outer Continental Shelf

- Federal Agency
  - US Bureau of Ocean Energy Management, Regulation and Enforcement (BOEMRE)
- Ruling - Sept. 2010
  - Workplace Safety Rule on Safety and Environmental Management Systems (SEMS)
- Standards
  - API RP 75
  - 30 CFR Part 250, Subpart S
- Applies - Comply by Nov. 15, 2011

Practical Limitation:

This presentation cannot cover all elements of PSM or SEMS.

Everyday Impact of PSM

- Process Safety Information – OSHA PSM
  - Safety and Environmental Information - API RP 75
- Operating Procedures
- Hazard Analysis
- Management of Change
- Pre-Startup Safety Review
- Mechanical Integrity
  - Quality Assurance/Mechanical Integrity - API RP 75

Process Safety Information (PSI)

- Timing - Written prior to hazard analysis and thereafter part of MOC
- Resource - Process Technology, Equipment, and Chemical Hazard Information

Process Flow Diagram	P&IDs	Materials of Construction
Safe Operating Limits	Relief System	Thermal and Chemical Stability
Consequence of Deviations	Interlocks, ESD	Inadvertent Mixing

- Similar to Safety and Environmental Information within API RP 75

Operating Procedures

- Timing - Initially written prior to PSSR and thereafter reviewed and certified annually (and with each MOC)
- Resource - PSI and operating steps for each phase

Normal, temporary, emergency operations	Start up (Initial, post turnaround, post unplanned shutdown)	Shutdown (Normal and Emergency)

- Tip - Consider all modes of operations during your work.
  - For petrochemical facilities, less than 10% of the time is spent in transient operations; however, 75% of major accidents occur during non-routine operations

Hazard Analysis

- Timing - Every 5 years
  - 10 years for low priority facilities per API RP 75
- Simple Description - Qualitative evaluation of hazards by experienced team

Process Hazards	Previous Incidents	Human factors
Facility siting (potential releases and their potential impact)	Consequence of failure of Monitoring, Controls, Alarms, Interlocks, ESD	Various methods (HAZOP, FMEA, What-If, LOPA, Fault Tree, etc)

- Tip - In practical terms, establishes hazard baseline. After such, managing change is key.

Management of Change (MOC)

- Timing - Prior to non 'replacement in kind' change
- Simple description - Written procedures to manage change to process, technology, equipment, facilities, procedures, and personnel (RP 75).

Technical basis	Potential impact on Safety and Health	Update PSI and procedures
Prior to PSSR and start up	Inform and train employees and contractors	Timing and Authorization requirements

- Tip - Managing change is key to safe design and maintaining safe operations.

Pre-start up Safety Review (PSSR)

- Timing - Prior to Implementation of 'Significant' Change
- Simple description - A safety review prior to start up by those responsible for operations and maintenance

Change in accordance with specifications	Safety, operating, maintenance, and emergency procedures in place and adequate	Hazard Analysis recommendations resolved
Training completed	PSI current	Safe work practices in place

Mechanical Integrity

- Timing - Continuous
- Simple description - Written procedures to design, fabricate, install, test and inspect, monitor, and maintain 'mechanical intent' of critical process equipment

Pressure vessels, piping systems, pumps	Relief and vent systems	Emergency shutdown systems
Controls, alarms, and interlocks	Deficiencies outside limits are corrected	Suitability with process application

- Tip - Mechanical integrity extends beyond preventive maintenance and testing. It is a means of ensuring the on-going performance of critical equipment.

Summary

- Process safety isn't one department
  – it's everyone's responsibility
- Process safety isn't a one-time job or a project
  – it's everyday, all day
- Process safety isn't paperwork
  – it's a way of thinking and acting
- <u>Process safety is a culture; it's a way of doing business.</u>

14

References

- API RP 75, "Recommended Practice for Development of a Safety and Environmental Management Program for Offshore Operations and Facilities," 3rd Edition, May 2004, Reaffirmed May 2008
- Bridges et. al., "How to Efficiently Perform the Hazard Evaluation Required for Non-Routine Modes of Operation (Startup, Shutdown, Online Maintenance)," AIChE, GCPS, March 2011
- OSHA 1910.119 "Process Safety Management (PSM) of Highly Hazardous Chemicals"
- OSHA 3132, "Process Safety Management," U.S. Department of Labor Occupational Safety and Health Administration, 1994 (reprinted)

Delayed Coking Basics

by Matthew Johnson

with the permission of

FOSTER WHEELER

Introduction

- Matthew Johnson
- B.S. In Chemical Engineering
- Graduated May 2009 from Texas A&M
- Process Engineer @ Foster Wheeler for 2.5 years.
- Spent 1.5 years supporting senior engineers developing about a dozen Delayed Coker Unit (DCU) project proposals.

2 FOSTER WHEELER

Introduction

1. What is a Delayed Coker?
2. Why does a Refinery Need a Delayed Coker?
3. What is the Benefit (aka 'Show Me the Money!')?
4. Process Variables

3 FOSTER WHEELER

What is a Delayed Coker?

<u>A Delayed Coker Unit is a semi-batch process which</u>

- Heats heavy hydrocarbons to a high temperature (~850°F).
- Allows those hydrocarbons to thermally crack.
- Recovers and separates the cracked material.

4 FOSTER WHEELER

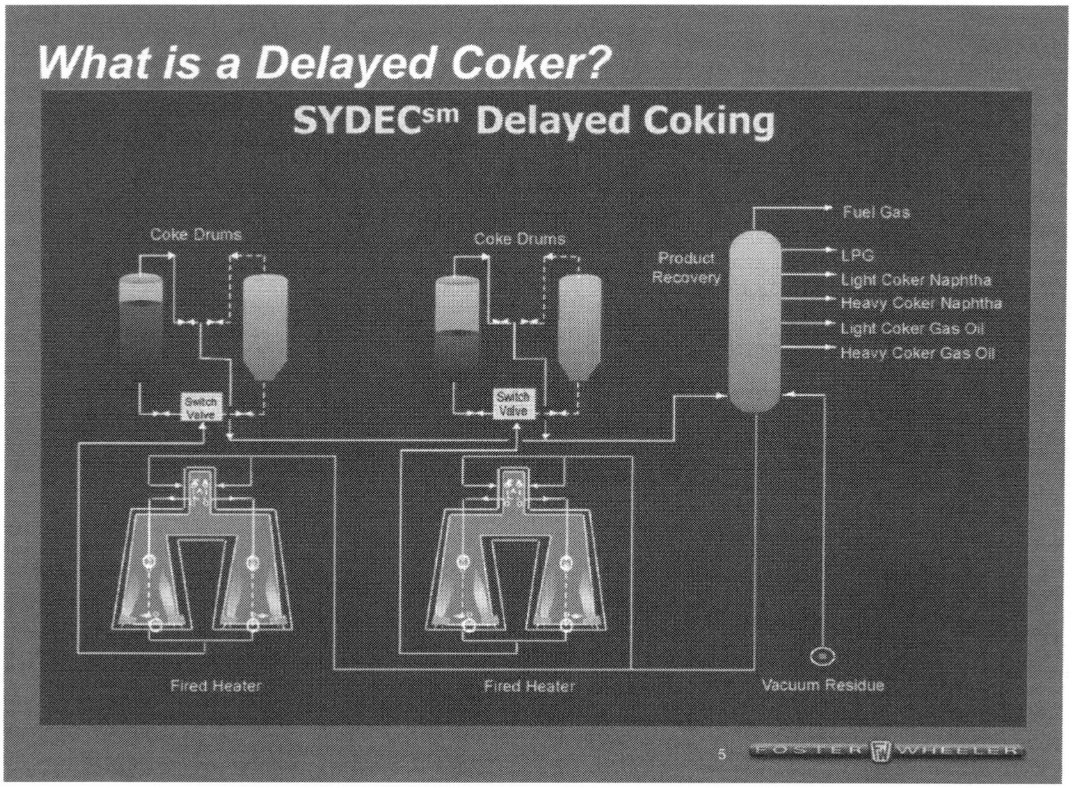
What is a Delayed Coker?
SYDEC℠ Delayed Coking
Coke Drums
Coke Drums
Product Recovery
Fuel Gas
LPG
Light Coker Naphtha
Heavy Coker Naphtha
Light Coker Gas Oil
Heavy Coker Gas Oil
Switch Valve
Switch Valve
Fired Heater
Fired Heater
Vacuum Residue
5
FOSTER WHEELER

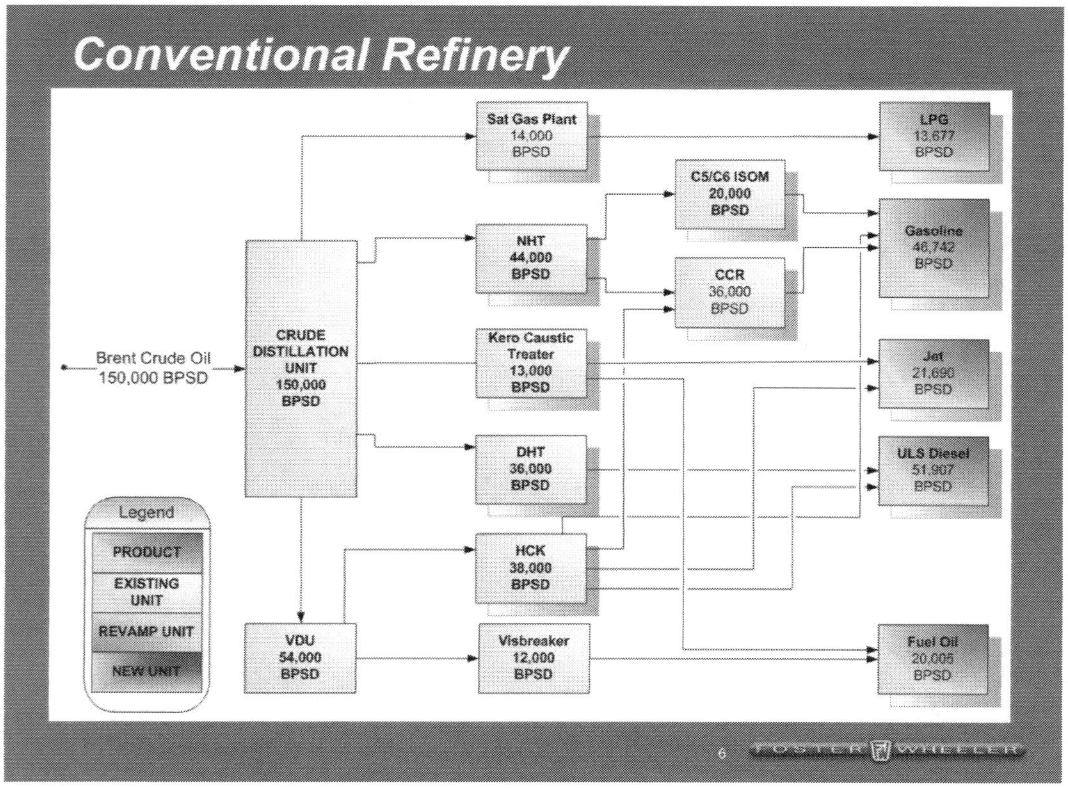
Conventional Refinery
Sat Gas Plant 14,000 BPSD
LPG 13,677 BPSD
C5/C6 ISOM 20,000 BPSD
NHT 44,000 BPSD
Gasoline 46,742 BPSD
CCR 36,000 BPSD
Brent Crude Oil 150,000 BPSD
CRUDE DISTILLATION UNIT 150,000 BPSD
Kero Caustic Treater 13,000 BPSD
Jet 21,690 BPSD
DHT 36,000 BPSD
ULS Diesel 51,907 BPSD
Legend
PRODUCT
EXISTING UNIT
REVAMP UNIT
NEW UNIT
HCK 38,000 BPSD
VDU 54,000 BPSD
Visbreaker 12,000 BPSD
Fuel Oil 20,005 BPSD
6
FOSTER WHEELER

Why Does a Refinery Need a Delayed Coker?

- Historically, the bottoms product from the vacuum distillation unit has been blended with lighter components to yield a heavy fuel oil product which is often high in sulfur (HSFO) .

7 FOSTER WHEELER

Why Does a Refinery Need a Delayed Coker?

- Decreasing demand and lower margins result in fuel oil becoming a waste product.
- Coking offers a way to reduce fuel oil production and improves the margin through increased yield of C5 thru Diesel liquids.

8 FOSTER WHEELER

Why Does a Refinery Need a Delayed Coker?

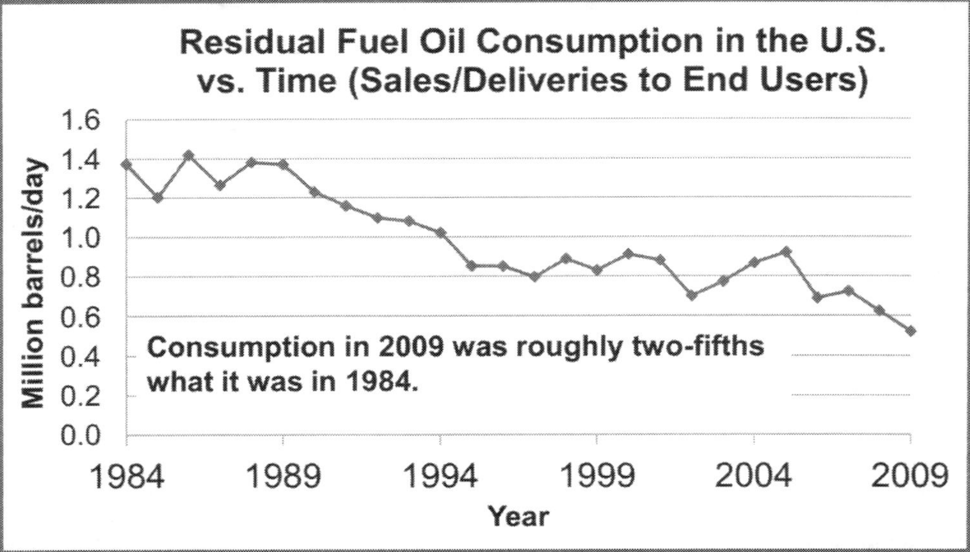

Source: Energy Information Administration (EIA) – http://eia.gov

9 FOSTER WHEELER

Why Does a Refinery Need a Delayed Coker?

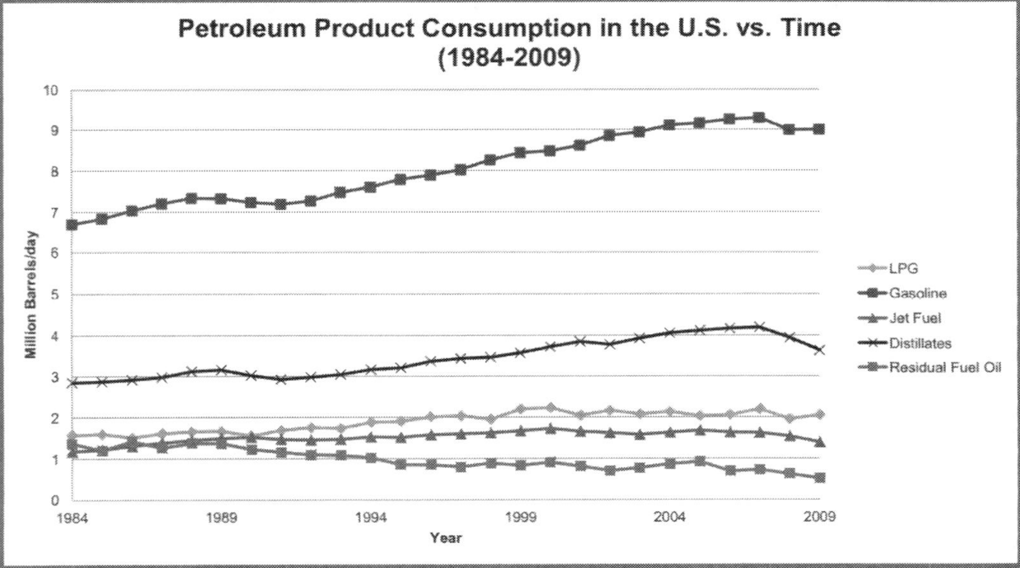

Source: Energy Information Administration (EIA) – http://eia.gov

10 FOSTER WHEELER

Where Has All The Fuel Oil Gone?

Petroleum Consumption in the U.S. (Million Barrels/Day)	1984	2009	% Change
LPG	1.57	2.05	30.5%
Gasoline	6.69	9.00	34.4%
Jet Fuel	1.18	1.39	18.5%
Distillates	2.84	3.63	27.6%
Residual Fuel Oil	1.37	0.51	-62.7%
Total	13.65	16.58	21.4%

Source: Energy Information Administration (EIA) – http://eia.gov

11 FOSTER WHEELER

Why Does a Refinery Need a Delayed Coker?

- A Coker allows the refiner to exploit the "heavy sour" discount on raw crude.
  - Heavy sour crudes are priced at a discount to benchmark crudes such as WTI or Brent.
  - A refiner trying to exploit this discount without a Delayed Coker will generate *huge* amounts of high sulfur fuel oil, which in turn will require a larger amount of diluent.

12 FOSTER WHEELER

Why Does a Refinery Need a Delayed Coker?

- A Coker allows the refinery to improve its "crack spread."
  - The crack spread is the difference between the value of the raw crude and the value of the products generated.
  - A refiner's ability to minimize the cost of the incoming crude and maximize the yield of value product determines its margin.
- Refineries *live* and *die* by their refining margin.

13 FOSTER WHEELER

Conventional Refinery

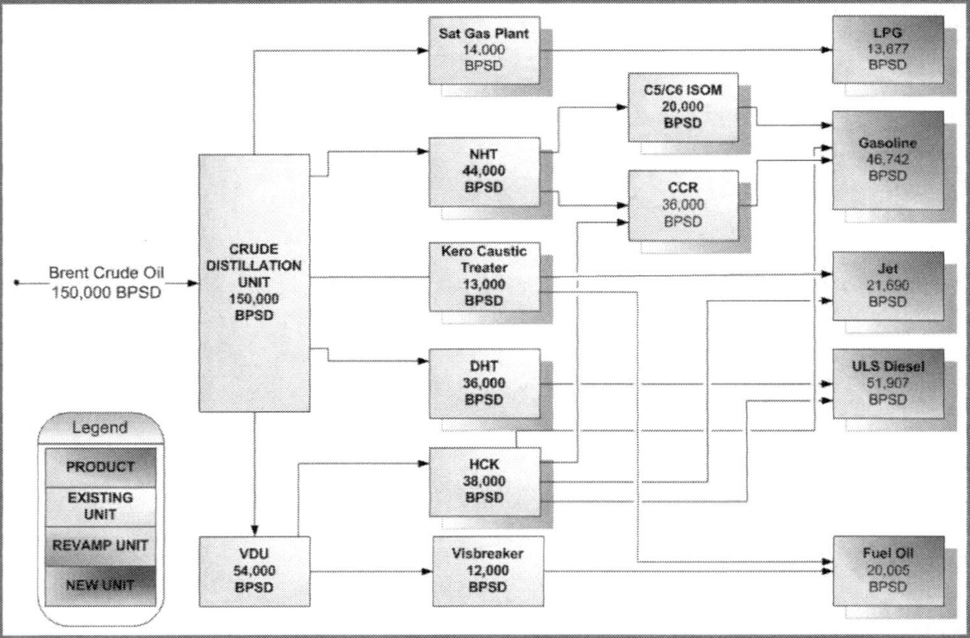

14 FOSTER WHEELER

Full Conversion Refinery

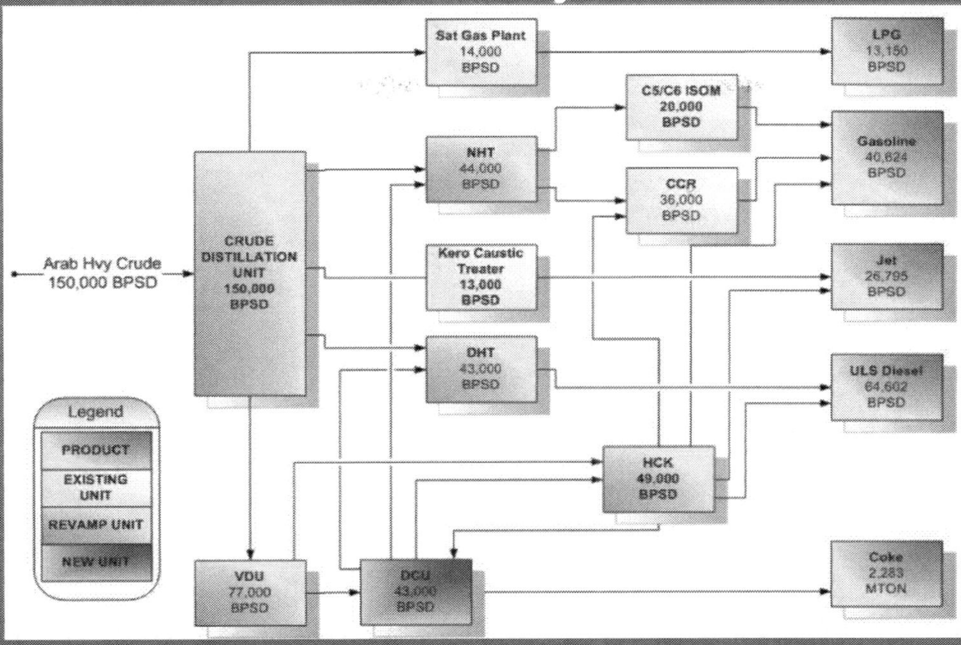

15 FOSTER WHEELER

Conventional Refinery vs. Revamped Refinery + DCU

Refinery Product Yield Comparison	Conventional Refinery (BPSD)	Revamped Refinery + DCU (BPSD)	% Change
Crude	Brent	Arab Heavy	
Crude Charge	150,000	150,000	0
LPG	13,677	13,150	-3.9
Gasoline	46,742	40,624	-13.1
Jet	21,690	26,795	+23.5
ULS Diesel	51,907	64,602	+24.5
Fuel Oil	20,005	-	Exit Market
Coke	-	2,283 MT/day	Export

Source: Wedlake and Elliot, "Paying Residues"

16 FOSTER WHEELER

Conventional Refinery vs. Revamped Refinery + DCU

Project Comparison	Conventional Refinery (BPSD)	Revamped Refinery + DCU (BPSD)	% Change
Refinery Feed Cost (billion USD/yr)	5.885	5.432	-7.70
Refinery Product Revenue (billion USD/yr)	6.286	6.159	-2.02
Gross Refinery Margin (USD/barrel)	7.65	13.85	+81.05

Source: Wedlake and Elliot, "Paying Residues"

17 FOSTER WHEELER

The Delayed Coking Process

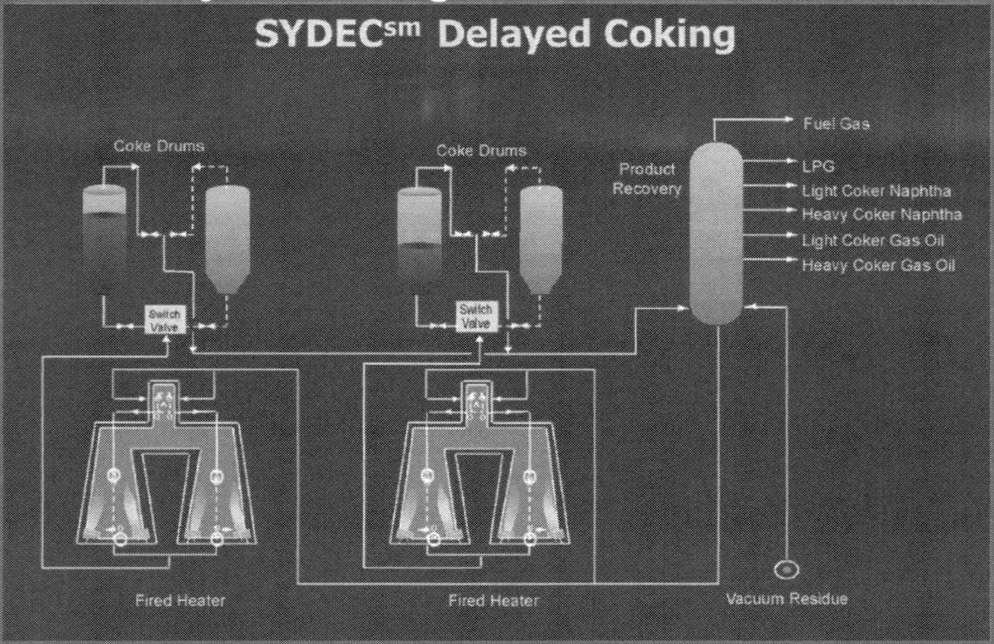

18 FOSTER WHEELER

Affect of Coke Drum Pressure on Yield

- Pressure
  - Increasing pressure in the coke drums suppresses vaporization.
  - Less vaporization means the products are subjected to cracking longer.
  - Therefore higher pressure results in a lower C5+ yield and more coke.
- In practice coke drum pressure can only be minimized so far (10-15 psig).

19 FOSTER WHEELER

Affect of Coke Drum Temperature on Yield

- Temperature
  - Increasing temperature in the coke drums promotes vaporization.
  - More vaporization means the products are recovered as they crack.
  - Therefore higher temperature results in a higher C5+ yield and less coke.
- In practice coke drum temperature can only be maximized so far due to issues with coke hardness.

20 FOSTER WHEELER

Affect of Recycle on Product Yields

- Recycle
  - A recycle stream is often necessary to help meet product specifications.
  - An increase in recycle results in larger equipment as well as more coke and C4- production.
  - Therefore recycle should be minimized to the extent allowable.
- In practice recycle can only be minimized so far because too many impurities in the products will poison catalysts in downstream processing units.

21 FOSTER WHEELER

Maximizing Delayed Coker Yields

- To maximize C5+ yield:
  - Minimize Pressure
  - Maximize Temperature
  - Minimize Recycle
- Why wouldn't I maximize my C5+ yield?
  - You have readily available low sulfur, low impurity crude, and you think you can make money in the anode coke market.

22 FOSTER WHEELER

In Summary

- A Delayed Coker
  - Reduces the amount of fuel oil generated.
  - Reduces the amount of diluent required to blend the fuel oil to the specification.
  - Increases the yield of liquid products which enter the motor fuels blending pool.
    - Coker Naphtha
    - LCGO
    - HCGO

23 FOSTER WHEELER

Works Cited

- U.S. Energy Information Administration, *International Energy Statistics*, http://www.eia.gov/cfapps/ipdbproject, Accessed Sept. 13, 2011.
- Wedlake, David A. and John D. Elliot. "Paying Residues," Hydrocarbon Engineering http://www.fwc.com/publications/tech_papers/files/SYDECJuneHE2011.pdf, June 30, 2011.
- Elliot, John D. "Optimize Coker Operations," Hydrocarbon Processing, Sept. 2003.

24 FOSTER WHEELER

Questions?

This Page Intentionally Left Blank.

Process Simulator Basics...that sometimes you need to learn on your own

Chemstations™

www.chemstations.com

Engineering Advanced

- Chemstations developed the CHEMCAD suite of software to broaden an engineer's capabilities and increase productivity by fitting into the chemical engineering work flow.
- CHEMCAD supercharges an engineer's efficiency when facing the toughest chemical process models or addressing day-to-day challenges.

Chemstations™

Process Simulator Basics…that sometimes you need to learn on your own

- There is a gap between college and professional use of a simulator
- Many engineers are not learning / forgetting important details for simulation
- Some 'best practices' and 'gotchas' will be highlighted

Chemstations™

Why is there a gap? Simulator use timeline

Music	OS	Simulator user
David Bowie	mainframe	PhD research group
Run DMC	DOS	Lead Process Engineer
Pearl Jam	Windows 9x	Project engineer
Nickelback	Win2k	Engineering degree
Adele	Win7	Intern

Chemstations™

Why is this important?

- Simulators get played like a video game
  - Converged doesn't mean correct
  - If you don't know what you're doing, you'll get wrong answer
- Experienced engineers who thought they knew what they are doing have:
  - Built columns that won't work
  - Undersized safety equipment
  - Oversized (or undersized!) reboilers
  - Lots of other mistakes

Chemstations™

Some best practices for simulator use

Chemstations™

One step at a time!

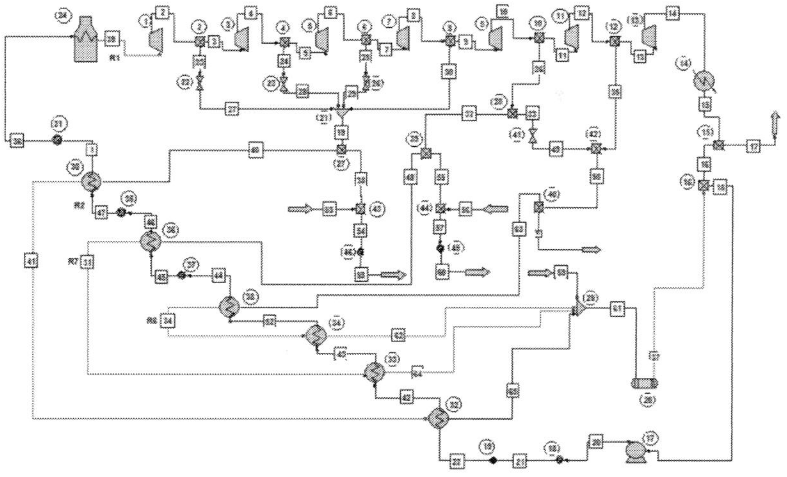

Chemstations™

One step at a time!

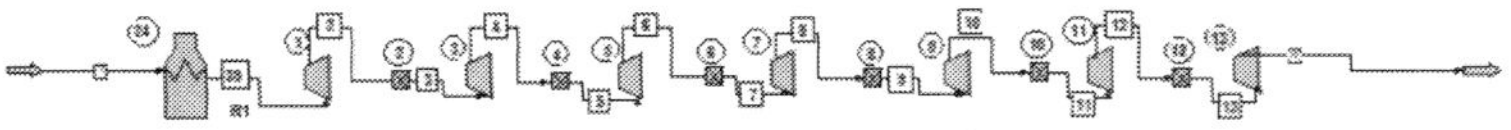

Chemstations™

recycles

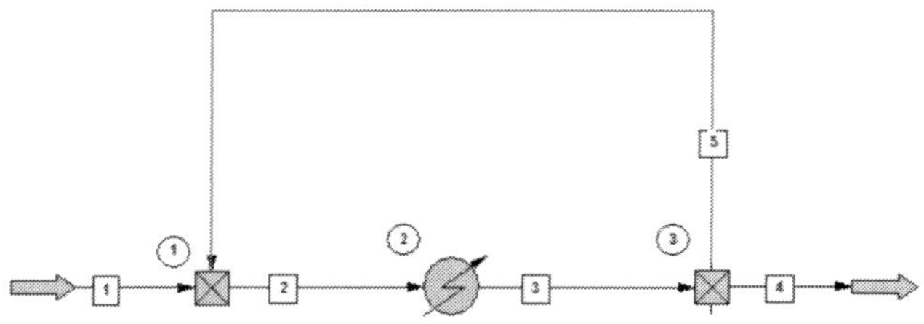

Chemstations

recycles

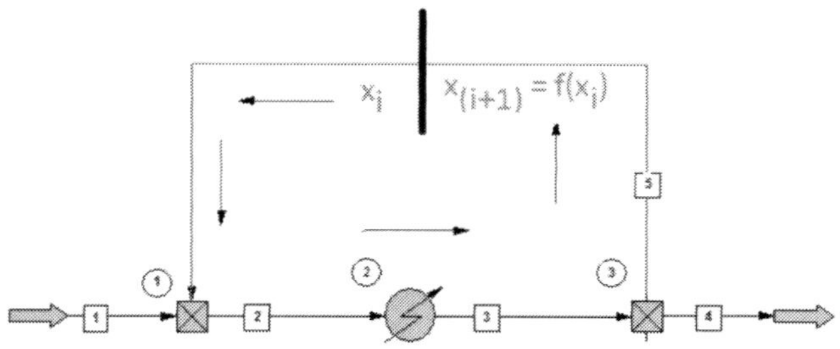

Chemstations

recycles

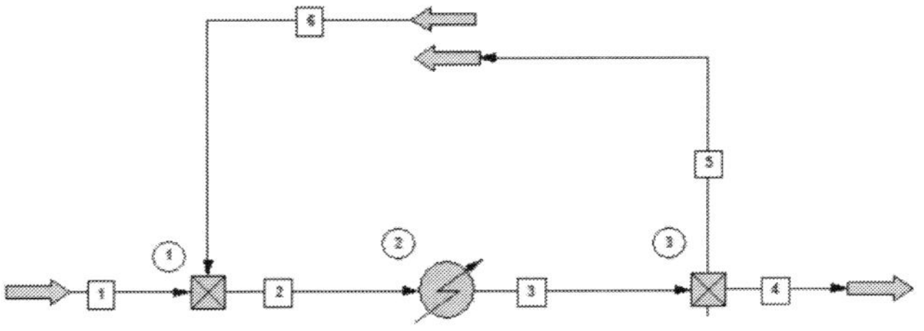

Chemstations

recycles

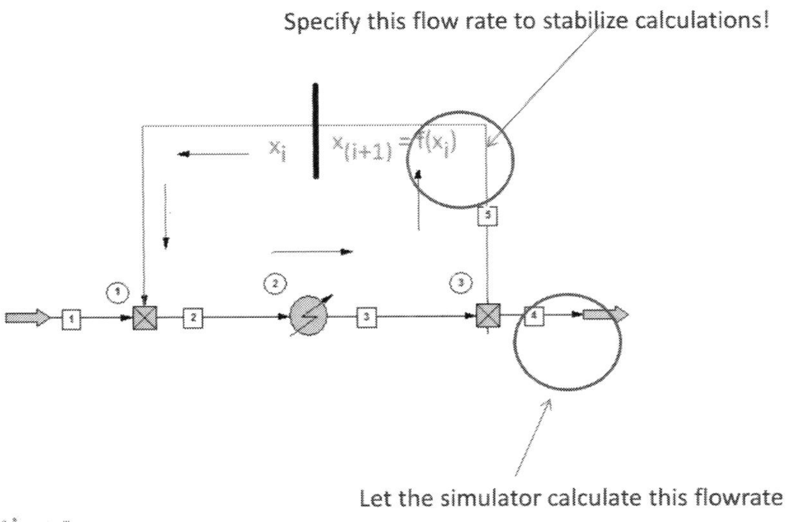

Chemstations

Validate your thermodynamic model

- Inspect TPxy diagrams
- Run some flash calculations, inspect results
- Have a qualitative idea of what the phase separation will look like

- (Step 0: select a thermodynamic vapor liquid equilibrium model)

Chemstations

Learn how to read Txy, Pxy, Txx, xy diagrams

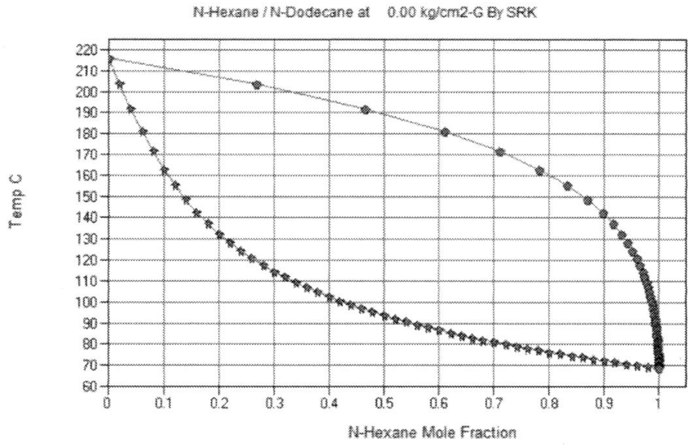

Chemstations

Learn how to choose thermodynamics

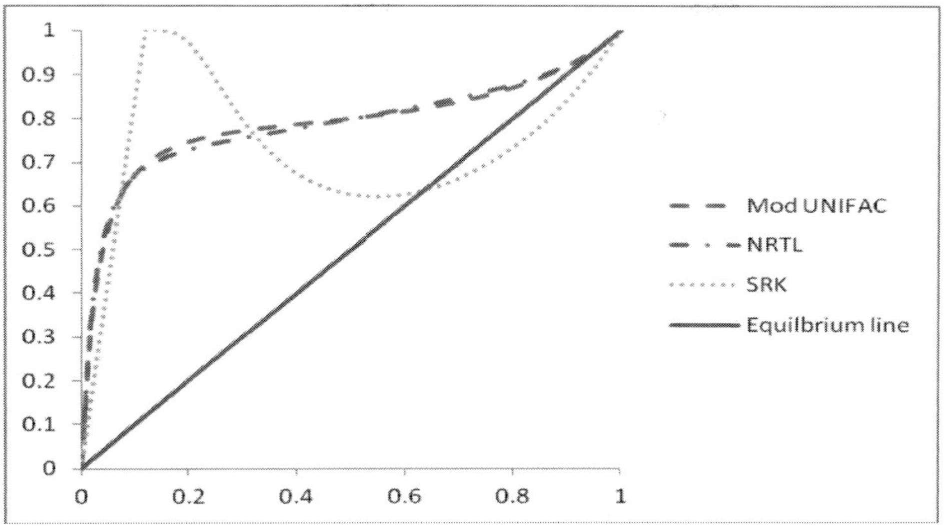

Chemstations

Some things we tend to forget...

Chemstations

Tboil

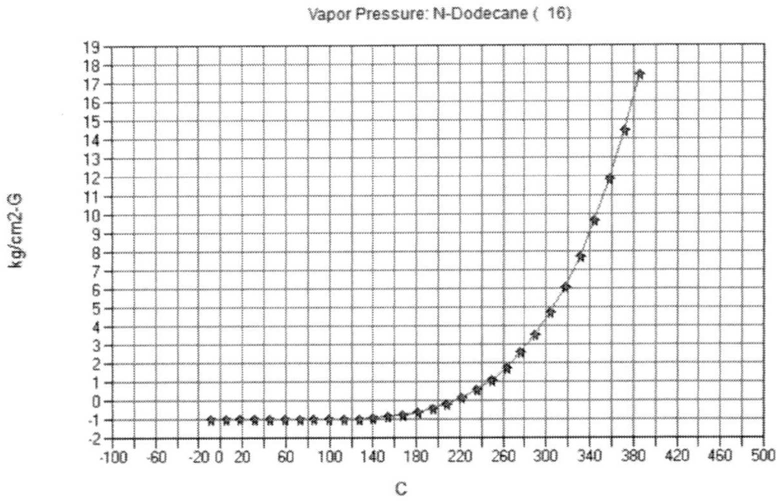

Chemstations

Tdew, Tbubble

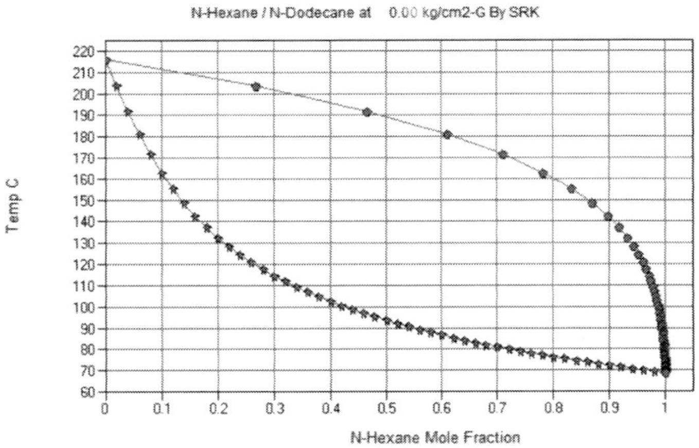

Chemstations™

Negative stream enthalpy?

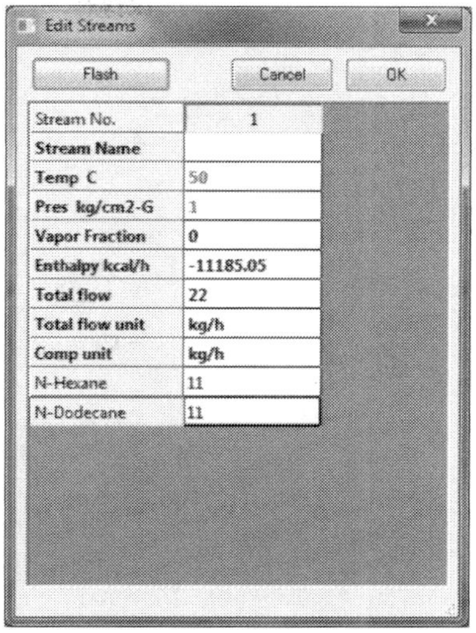

Chemstations™

Negative stream enthalpy?

- Absolute Enthalpy…what is that?
  - H=….??
- Relative Enthalpy…what's the reference?
  - (H1-H0)
- Different tools…different reference…
  - (H1-H0) doesn't match between tools…
- Forget reference, look at the duty
  - (H2-H0) – (H1-H0)

Chemstations™

Beware extrapolation beyond T range

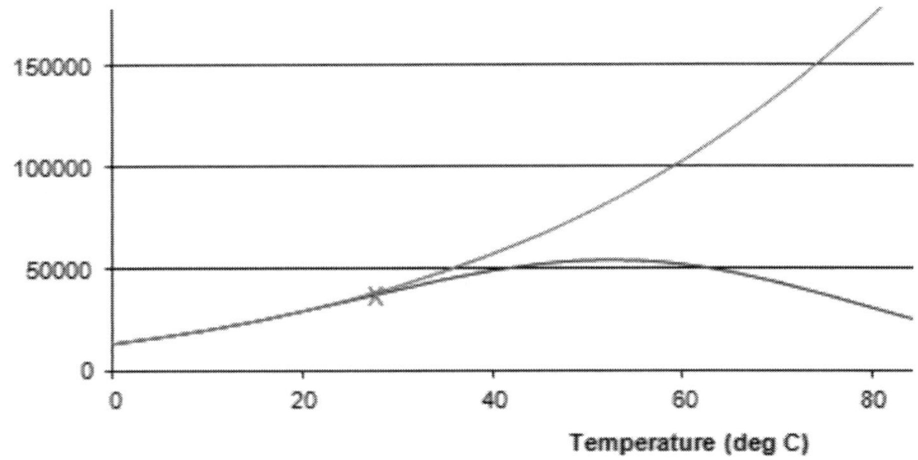

Chemstations

Numbers aren't always reliable

- Matching plant data?
  - How accurate are the measuring devices?
  - Were all measurements taken at same steady state?
  - Are you ignoring trace chemicals?
- T=73.1 is often same as T=74
- Sometimes T=79 is the same as T=74

Chemstations

It’s only a model!

- Learn what happens you pull the levers
- Understand the concepts of math being used by a simulator
  - Review thermodynamic flash calculations
  - Review unit operations
  - Review thermodynamic energy balances

Chemstations™

Learn more!

- Use a simulator to revisit homework problems!
- Find a mentor! Find examples!
- Learn Tpxy plots…applied thermodynamics…
- Read a book!
  - Edwards, _Chemical Engineering in Practice_, kindle edition.
  - Seider, Lewin, _Product and Process Design Principles_
  - Kaes, _Refinery Process Modeling_
- Read a magazine article!

Chemstations™

Bryan Pourchot – Sales & Service Engineer

October 07, 2011

Confidential Information to be used only with permission

HALDOR TOPSOE

Agenda

- Introduction to HTI
  - Markets
  - Products & Technology
  - Environmental Products
- SCR 101
  - SCR Process
  - Why remove NOx
  - SCR Uses
  - Catalyst Arrangements
  - Catalyst Basics
  - Why DNX?
- Catalyst Deactivation
  - What Reduces Catalyst Activity?

HALDOR TOPSOE

Haldor Topsoe, Inc.

- Established in 1965
- Offices located in
  - Fort Mill, SC
  - Houston, TX
  - Los Angeles, Ca
- Employees: ~250
- Manufacturing facility in Houston, TX – for petroleum, sulfuric acid, MTB and DNX catalyst

HALDOR TOPSOE

Industrial Process Market Segments

Methanol

Syngas

Hydrogen

Formaldehyde

Refinery

Ammonia

Sulfuric Acid

- ➢ Catalyst Supply
- ➢ Process Licensing & Engineering

HALDOR TOPSOE

Environmental Process Market Segments

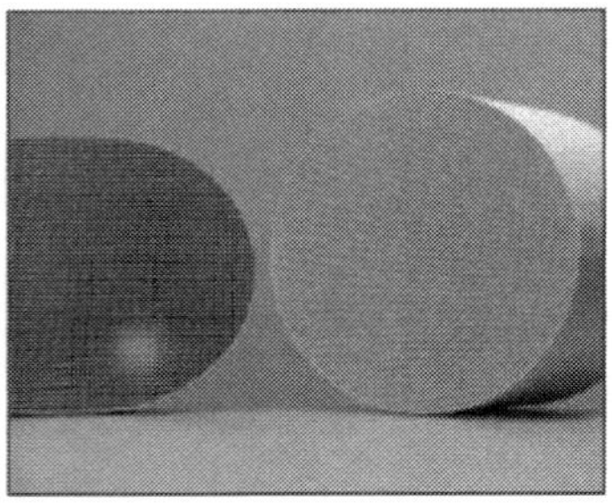

- Selective Catalytic Reduction Catalysts and Technology
- Automotive Emissions Control
- Diesel Emission Control
- Ceramic NOx Control Filters
- Sulfur Dioxide Emission Control
- Combined Sulfur Dioxide & NOx Emission Controls
- VOC Emission Control
- CO Emission Control

HALDOR TOPSOE

SCR Process Basics

Flue Gas: NOx, SOx, CO_2, O_2

- SCR = <u>S</u>elective <u>C</u>atalytic <u>R</u>eduction
- Propose is to reduce NOx from flue gas.
- A reducing agent, most commonly ammonia (NH_3), is injected into the flue gas.
  - The NH_3 is mixed into the gas stream, sometimes using mixing plates.
  - The mixed gas then passes through the catalyst layers where the NH_3 reacts with NOx on the catalyst surface and in the pores to form N_2 and H_20 vapor.

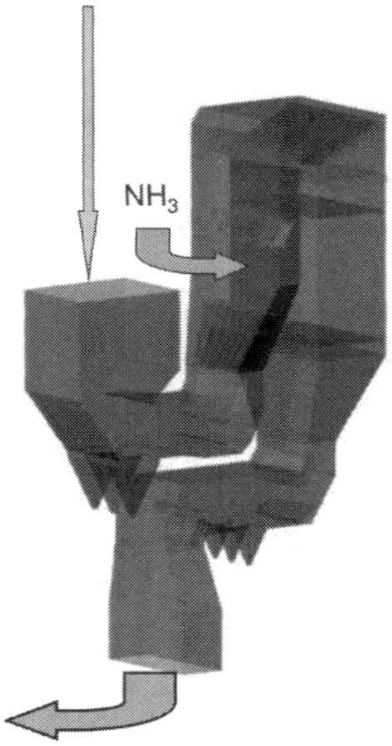

Clean Gas : N_2, H_2O, O_2, SO_2, (SO_3)

HALDOR TOPSOE

Why Remove NOx?

- Formation of atmospheric nitric acid (Acid Rain)
  - Potential damage to building and ecosystem
  - Potential cause / aggravation of health issues
- Formation of tropospheric ozone and other harmful compounds
  - Potential cause / aggravation of health issues
- Causes photochemical smog
- Destruction to stratospheric ozone
  - Increased UV light to Earth's surface

HALDOR TOPSOE

Types of Processes with SCR

- Fired boilers and heaters
- Combustion turbines
- Acid plants
- Waste incinerators
- Large commercial transportation vehicles
- Many other processes
- Multiple fuel types

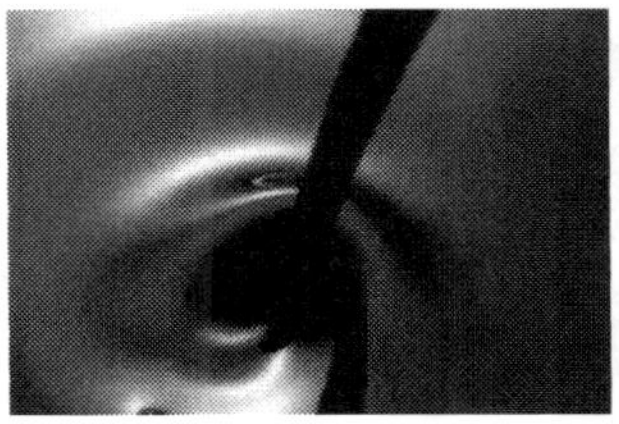

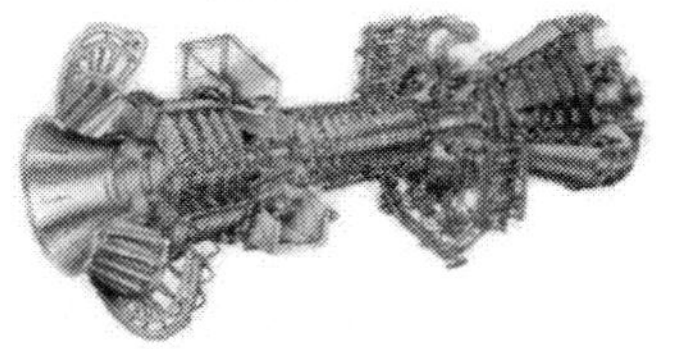

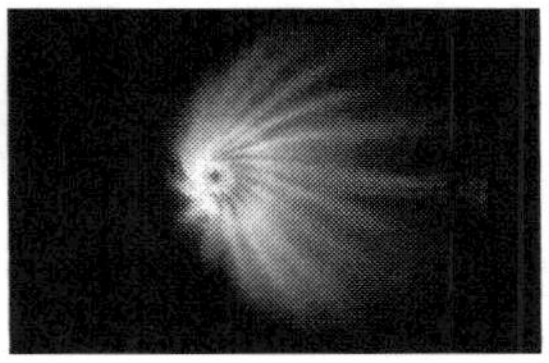

HALDOR TOPSOE

Catalyst Arrangements

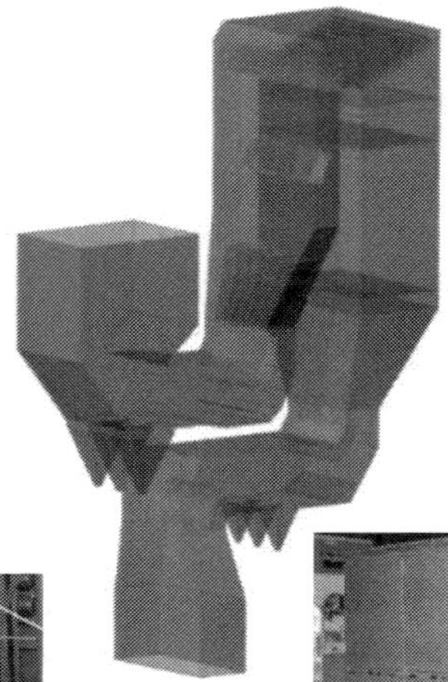

Vertical

Horizontal

HALDOR TOPSØE

SCR DeNOx process

$4\ NO + 4\ NH_3 + O_2 \rightarrow 4\ N_2 + 6\ H_2O$

$NO + NO_2 + 2\ NH_3 \rightarrow 2\ N_2 + 3\ H_2O$

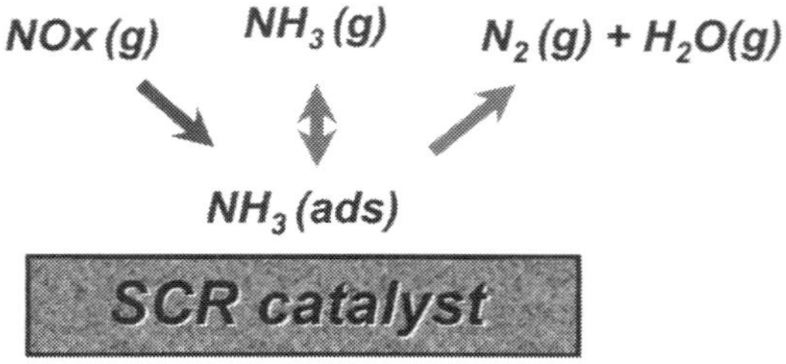

Vanadium pentoxide, V_2O_5
Tungsten trioxide, WO_3
Titanium dioxide carrier, TiO_2

HALDOR TOPSØE

SCR DeNOx Reaction Mechanism

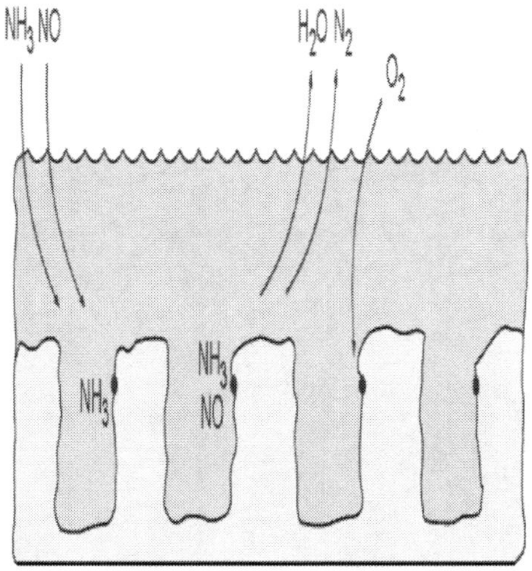

✓Diffusion controlled
- ➢rate increases with smaller pitch, higher velocity, larger pores

✓Kinetic controlled
- ➢Increases with the number of active sites (pore volume / vanadium)

HALDOR TOPSOE

Why Topsoe Catalyst?

DeNOx reaction is diffusion limited
more highways = higher diffusion rate = higher activity

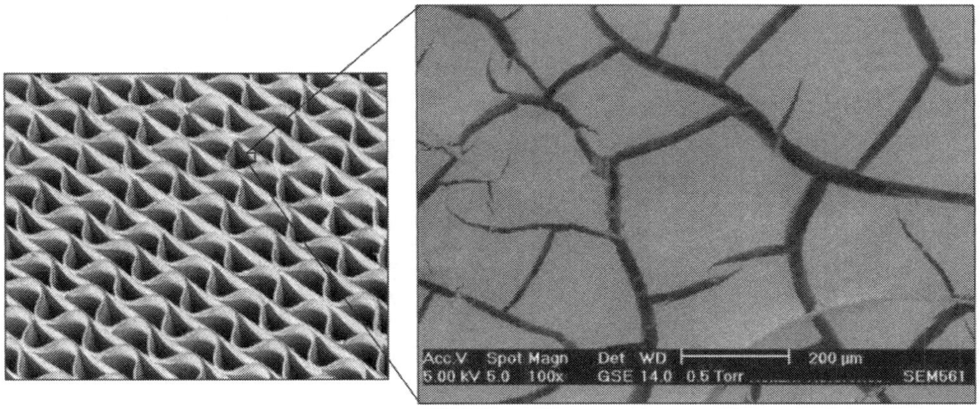

HALDOR TOPSOE

Why Topsoe Catalyst?

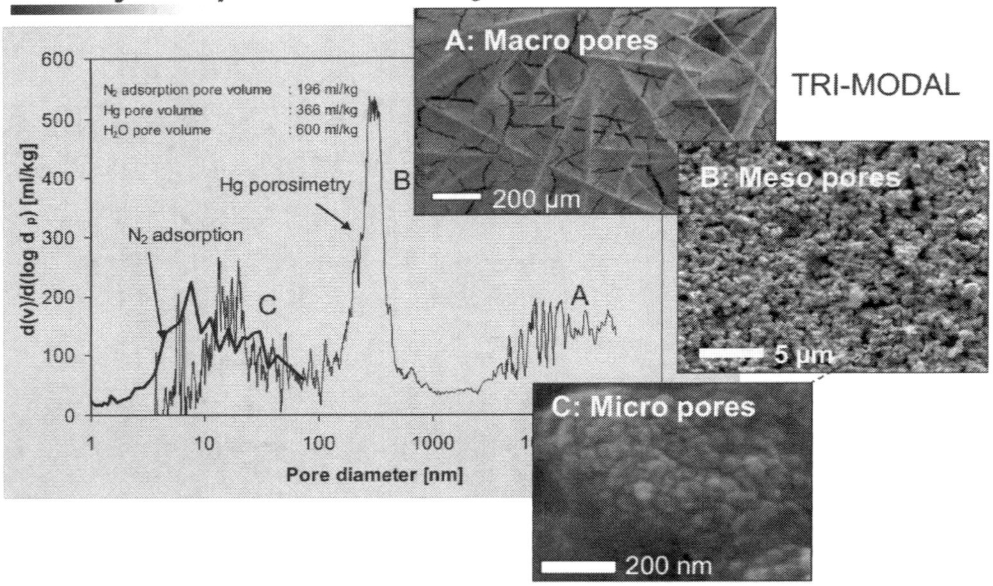

TRI-MODAL

HALDOR TOPSOE

Why Topsoe Catalyst?

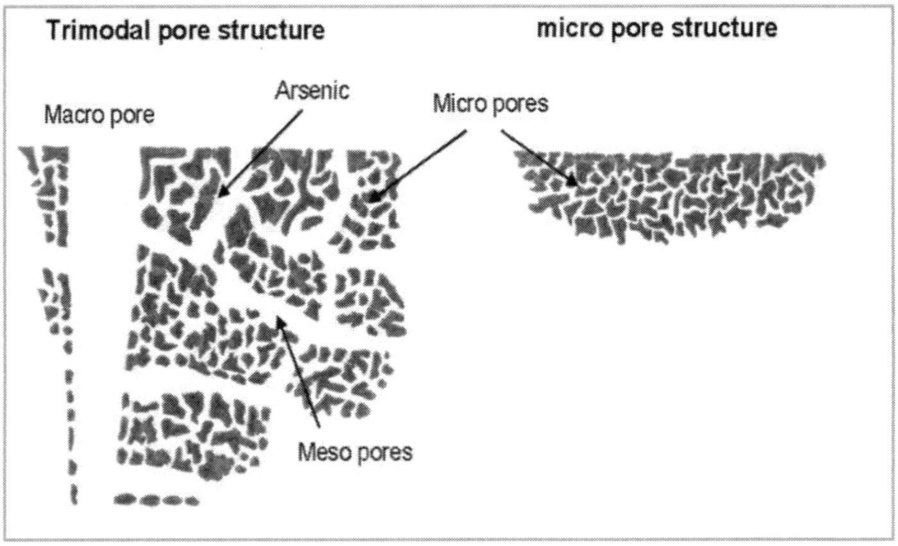

HALDOR TOPSOE

What Reduces Catalyst Activity?

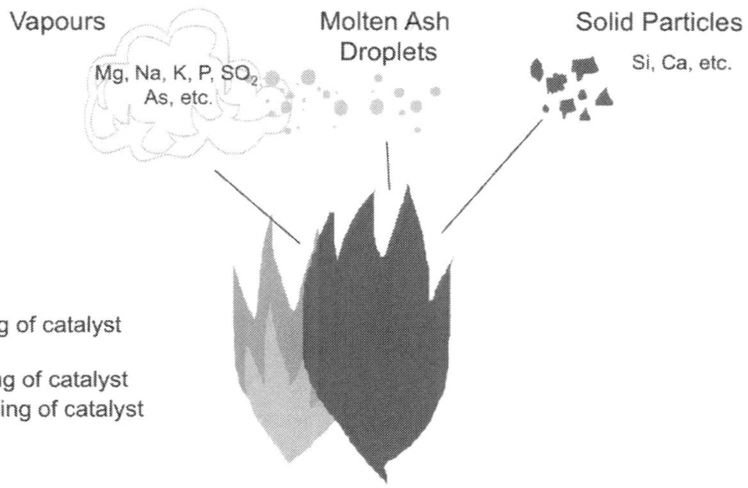

- Fouling or masking of catalyst surface
- Chemical poisoning of catalyst
- Erosion and plugging of catalyst channels

HALDOR TOPSØE

Catalyst Deactivation

Physical - ash and other solid particles

Masking: blocking catalyst surface

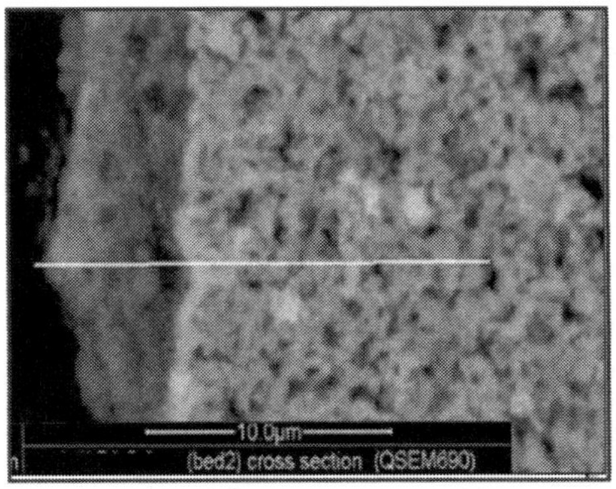

Fouling: physical pore blockage

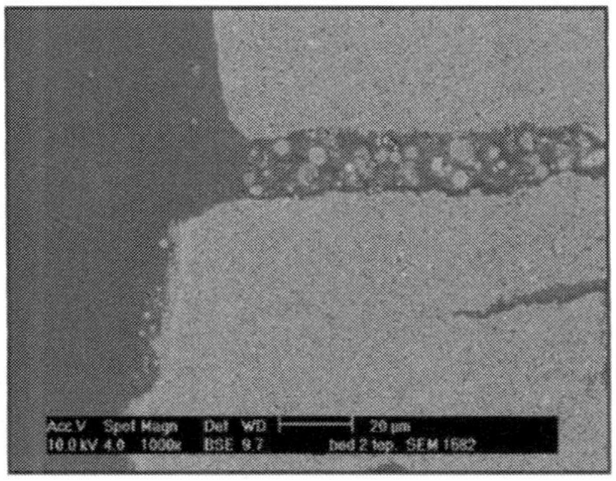

Result: prevents gas diffusion & reaction

HALDOR TOPSØE

Catalyst Deactivation

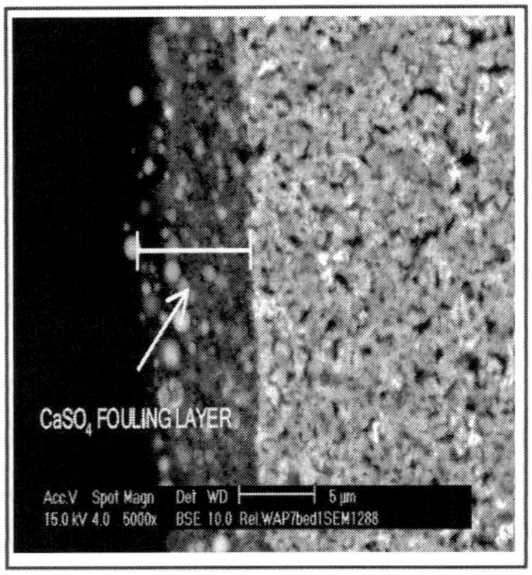

Chemical poisons that physically mask, foul & plug the catalyst surface, pores.

- Alkaline Earth Metals, typically in the form of sulfates, phosphates or arsenates.
  - Common: Calcium, Magnesium
  - Other: Barium, Strontium, Beryllium
- Ammonia-sulfur compounds
- Silicon Compounds

Result: prevents gas diffusion & reaction

HALDOR TOPSOE

Catalyst Deactivation

Poisons that ion exchange with active sites.

- Alkaline Metals
  - Common: Sodium, Potassium
  - Other: Cesium, Lithium, Rubidium

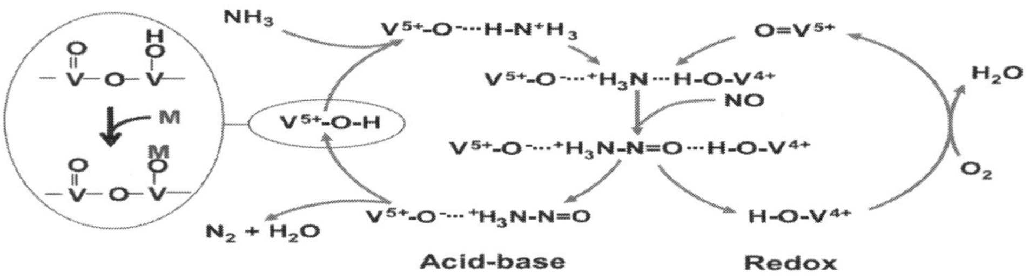

Result: reduces the number of active sites for which NOx can react

HALDOR TOPSOE

Catalyst Deactivation

Halogens, at high levels, react with and volatilize active metal sites on the catalyst.

Result: reduction of active sites

Noble Metals, promote oxidation of ammonia to NOx

Common: Platinum

Other: Palladium, Rhodium

Result: the generation of NOx

Other Elements, cover active sites, deposit on or near catalyst surface, or promote oxidation (ammonia to NOx or SO_2 to SO_3).

Common: Phosphorus, Arsenic

Other: Chrome, Nickel, Iron, Lead, Vanadium, Antimony, Tin, Zinc

Result: reduction of active surface area and sites

HALDOR TOPSØE

DNX Leading the Market

- High NOx removal activity across a large temperature range
  - NOx removal rates up to 98%
  - NH_3 slip of ≤ 1 ppm
  - 300 F to 1050 F
- Low SO_2 oxidation over the life of the catalyst
  - As low as 0.1% oxidation rate
- High mechanical strength
  - Mechanical stability for five or more years

HALDOR TOPSØE

Questions?

Thank you

Contact Information:
Bryan Pourchot
Sales & Service Engineer
(281) 228-5133
bjp@topsoe.com

HALDOR TOPSOE

AUTHOR INDEX

Adarme, Raul 275
Akatay, M. Cem 89
Anderko, Andre 418
Bresland, John..144
Chapman, Walter G...378
Chen, Shirley 239
Chosnek, Jack 161
Chu, Arthur..286
Clark, C. W. 226
Cox, Kenneth R. 397
Delgass, W. Nicholas..89
Dennis, Quin..32
Dolby, Clint 296
Duck, Brian 49
Edwards, Victor H. 161
Ellis, Ralph 161
English, Alan 161
Erazo, Jaime..239
Fischer, Kai 448
Fisher, Doug 21
Griffith, Aron 254
Gripka, Patrick J. 90
Gujar, Arjun 432
Hamdan, Intan Munirah..251
Hawkins, Robert B...296
Hendriks, Eric..448
Hendrix, Bailey..239
Hill, David..484
Hill, Gerald..99
Horovka, Susan D. 126
Ishiyama, Edward M. 349
Johnson, Matthew..470
Koester, Ward..275
Kosinski, J. J. 418
Kraus, Lawrence S...90
Krusche, Maria 73
Kuhne, Eliane..448
Kumana, Jimmy D. 349
Lacasse, Shane..174
Lencka, M. M. 418
Lienhard, John..1
Mannan, M. Sam..143
Marwaha, Bharat..73
McCool, Ben 238
Medhekar, Subodh..167
Miller, Jeffrey T. 89
Nieuwoudt, Izak 188
Olsen, Tim 327
Pazmino, Jorge..89
Peters, Dwight 113
Pless, Lowell 174
Pourchot, Bryan..496
Pugh, Simon J...349
Ray, Myra..161
Reklaitis, Gintaras V. 251
Ribeiro, Fabio H. 89
Riley, Bob..347
Sacco Jr., Albert..237
Sands, Ruth..196
Saraf, Sanjeev 167
Seetharam, Ram 366
Sharma, Pradeep..238
Shekhar, Mayank 89
Singh, Bhari Bhujan..175
Spencer, Calvin F. 432
Springer, R. D...418
Stach, Eric A...89
Summers, Angela E...462
Summers, Daniel R...214
Townsend, Bill 175
Tran, Thienan D. 90
Turini, Kevin 286
Turner, James..286, 314
Vaidyanathan, Shankar 286
Venkatasubramanian, Venkat..251
Wang, Jun 89
Wang, P...418
Wendel, Kirk 239
White, Julie D. 462
Wiehe, Irv 305
Williams, W. Damion 89
Willman, Todd J...432
Wilson, Stacy 14
Wolschlag, Lisa..348
Young, Michael H...126
Young, Wendy 253